I0896723

Redefining Business in Volatile and Ambiguous Times

G. S. Prakasha
https://orcid.org/0000-0002-1287-7606
Christ University, India

Maria Lapina
https://orcid.org/0000-0001-8117-9142
North-Caucasus Federal University, Russia

Francisco David Guillén Gámez
https://orcid.org/0000-0001-6470-526X
University of Málaga, Spain

Juan J. Leiva-Olivencia
https://orcid.org/0000-0002-0082-1154
University of Málaga, Spain

IGI Global
Scientific Publishing
Publishing Tomorrow's Research Today

Published in the United States of America by
IGI Global Scientific Publishing
701 East Chocolate Avenue
Hershey, PA, 17033, USA
Tel: 717-533-8845 | Fax: 717-533-7115
Website: https://www.igi-global.com E-mail: cust@igi-global.com

Library of Congress Cataloging-in-Publication Data

Names: Prakasha, G. S. editor | Lapina, Maria, 1981- editor | Guillén Gámez, Francisco David, 1985- editor | Leiva-Olivencia, Juan J., 1979- editor

Title: Redefining Business in Volatile and Ambiguous Times / edited by G. S. Prakasha, Maria Lapina, Francisco David Guillén Gámez, and Juan J. Leiva-Olivencia.

Description: Hershey, PA : IGI Global Scientific Publishing, [2026] | Includes bibliographical references. | Summary: "This book examines how businesses and professionals can adapt and thrive amid volatility and uncertainty. Through insights on AI, digital transformation, and ethical innovation, it explores strategies for resilience, decision-making, and sustainable growth in today's rapidly evolving global landscape"-- Provided by publisher.

Identifiers: LCCN 2025047446 (print) | LCCN 2025047447 (ebook) | ISBN 9798337379173 hardcover | ISBN 9798337379180 paperback | ISBN 9798337379197 ebook

Subjects: LCSH: Business

Classification: LCC HF1008 .R4 2026 (print) | LCC HF1008 (ebook)

LC record available at https://lccn.loc.gov/2025047446

LC ebook record available at https://lccn.loc.gov/2025047447

British Cataloguing in Publication Data
A Cataloguing in Publication record for this book is available from the British Library.

Table of Contents

Detailed Table of Contents

This article addresses the issue of intellectual well-being in healthcare. A predictive analysis of the advantages and challenges of using artificial intelligence technologies in the professional activities of doctors is provided. The article highlights significant problems related to ethical and legal issues, data quality, lack of algorithmic transparency and the risk of diminishing physicians' professional skills. The forecasting methodology proposed in the article for optimizing diagnosis, treatment, and disease prevention in the professional activities of doctors is based on a comparative analysis of a large volume of digital statistical data. A promising direction for further scientific inquiry, according to the authors, is the consideration of digital healthcare issues using artificial intelligence in combination with other effective and globally recognized tools and technologies. The article is of interest to specialists in healthcare, bioethics, digital medical technologies, as well as to university educators.

The global education landscape has been reshaped by technological advancements, societal disruptions, and the COVID-19 pandemic. The VUCA and BANI paradigms

Markowitz W. Sharpe for the compilation of the investment portfolio; analyzed and selected the tools for developing the application that forms the investment portfolio. During the study the Python programming language was chosen as the tools for developing the application that forms the investment portfolio. Python is a powerful tool for optimizing investment portfolios using the W. Sharpe method, allowing investors and analysts to achieve a balance between risk and return using advanced methods of quantum analysis. The key functions of the developed application are: aggregation of financial data, calculation of expected portfolio return, calculation of variance and standard deviation, portfolio optimization, visualization of results, ease of use and an intuitive interface .

Many energy systems are most widely used as elements of power supply systems for critical infrastructure (such as data centers, server infrastructure, etc.) and in distributed energy systems.But they have various disadvantages, the key of which is the inevitable degradation of battery cells, which can lead to premature failure of the entire system and increase operating costs. This problem can be solved through the use of battery management systems (BMS), allowing to optimize the process of operation, thereby extending the overall service life, one of the most important elements of which is a system for monitoring the parameters of the battery cells. In this regard, we considered some modern approaches aimed at organizing the process of monitoring of operating energy storage systems, which are an integral part of distributed energy systems, the functioning of which must be maintained in real time without failures and outages.

MOOCs have transformed global education by offering unprecedented access to knowledge. Yet, engagement and completion rates remain critical challenges. This study presents a conceptual model that examines the interplay between individual learner characteristics and MOOC design features towards course adoption and use, with special emphasis on the role of gamification in sustaining motivation and retention. By considering elements such as learning styles and gamification preferences of its users, MOOC designers can enhance the learning experience and encourage course completion. Beyond the conceptual model, we explore AI as a complementary technology feature, assuming that AI systems can leverage learner styles and MOOC characteristics, in an adaptive fashion, to further enhance engagement and retention. Overall this chapter highlights how integrating Gamified AI-driven innovations, tailored to individual and course characteristics, can strengthen MOOCs as a scalable and sustainable model of education, ensuring their long-term effectiveness in an evolving learning landscape.

The Agreement on Trade-Related Aspects of Intellectual Property Rights (TRIPS) plays a crucial role in shaping the global intellectual property system that significantly impacts farmers' rights. In agriculture, TRIPS grants intellectual property protection in the form of patents and plant variety protection. Such protection conflicts with the farmer's rights to save, sow and exchange the seeds. This raises human rights concerns related to farmers such as access to genetic resources and their cultural rights, food security, livelihood that are enshrined under various international human rights instruments such as International Covenant on Economic, Social and Cultural Rights (ICESCR) and Universal Declaration on Human Rights (UDHR). This intersection between TRIPS and the human rights of farmers requires a more inclusive approach that balances human rights with innovation. In light of the above perspective, this chapter analyzes farmers' human rights integration under the domain of intellectual property rights protected under TRIPS.

The smart classroom technologies with skill development in a VUCA-BANI context is what this study focuses on. The research uses a sample of 550 students of UG and above in India. It assesses the impacts of IoT, AR/VR, Gamification, and AI on cognitive skills through the mediating effects of student participation and moderating

Influencing consumer decision-making processes in the digital age is vital as E-commerce continues dominating the market. This research work explores the application of nudging to e-commerce or any digital environment involving consumer choices, investigating its efficacy in mitigating cognitive biases that affect online purchasing decisions. We employed a 2 x 2 within-group experimental design to examine how different nudges influence e-commerce choices. Data collected from 88 participants reveal that status-quo nudges can influence decision-making more than social proof nudges or salience nudges in online shopping. These results significantly impact digital marketing strategies, suggesting that carefully designed nudges can guide consumer choices by overcoming ingrained prejudices. This research also provides practical insights into consumer behaviour through subtle interventions in e-commerce settings. While nudging has been studied in offline contexts, applying it to e-commerce and digital consumer choices represents a growing area of study.

Smart pedagogies are transforming conventional teaching methods by incorporating digital technology and innovative approaches to improve the learning process of students. With the advent of the digital age, teachers now utilize sophisticated technologies to create more interesting, personalized, and effective teaching methods. This research discusses the application of Smart Pedagogies, emphasizing the utilization of digital technologies in developing creativity, collaboration, and critical thinking among students. Based on Stacked Artificial Neural Networks (ANNs), we examine the effect of teachers' digital competence on pedagogical practice and student performance. The results show that digitally well-integrated tools enhance learning outcomes considerably, supporting the significance of ongoing professional development in digital competence for teachers. The study underscores the necessity of imaginative forms of teaching which appeal to the requirements of the digital age to equip the students with preparation for their futures, defined by technological forces

Preface

The world of business and enterprise today is confronted with unprecedented levels of volatility, uncertainty, complexity, and ambiguity. Coupled with the brittle, anxious, non-linear, and incomprehensible (BANI) dynamics of modern systems, organizations are compelled to rethink traditional strategies and embrace innovative, technology-driven approaches to remain competitive and sustainable. *Intelligent Enterprise: Redefining Business in Volatile and Ambiguous Times* seeks to capture this transformation by bringing together diverse scholarly and practice-oriented perspectives on how intelligence, digitalization, and innovation are shaping enterprises of the future.

The twelve chapters in this volume traverse multiple dimensions of enterprise intelligence. Beginning with the role of artificial intelligence in the professional activities of physicians and the application of decision-support algorithms in personnel management, the book highlights the critical integration of AI in both human-centered and organizational processes. Learner engagement, resilience building, and e-mentoring are examined as vital tools in adapting to rapidly shifting landscapes, while digital twins and MOOCs underscore the importance of scalable and adaptive learning models in fragile and uncertain contexts.

CHAPTER OVERVIEW

Chapter 1: Artificial Intelligence in the Professional Activity of Physicians

This article addresses the issue of intellectual well-being in healthcare. A predictive analysis of the advantages and challenges of using artificial intelligence technologies in the professional activities of doctors is provided. The article highlights significant problems related to ethical and legal issues, data quality, lack of algorithmic transparency and the risk of diminishing physicians' professional skills.

The forecasting methodology proposed in the article for optimizing diagnosis, treatment, and disease prevention in the professional activities of doctors is based on a comparative analysis of a large volume of digital statistical data. A promising direction for further scientific inquiry, according to the authors, is the consideration of digital healthcare issues using artificial intelligence in combination with other effective and globally recognized tools and technologies. The article is of interest to specialists in healthcare, bioethics, digital medical technologies, as well as to university educators.

Chapter 2: Empowering Learners Through Engagement and E-Mentoring — Building Resilience in a Changing Landscape: Engagement & E-Mentoring for Resilient Learning

The global education landscape has been reshaped by technological advancements, societal disruptions, and the COVID-19 pandemic. The VUCA and BANI paradigms highlight emerging challenges and opportunities in education. This chapter explores how student engagement and e-mentoring foster resilience, adaptability, and inclusivity in this dynamic context. Student engagement is redefined to address digital fatigue, information overload, and psychological stress in hybrid and digital environments. Emerging technologies like AI, gamification, and virtual reality enhance cognitive and emotional engagement while promoting equitable access. E-mentoring provides personalized guidance, emotional support, and professional development through digital platforms, leveraging AI-powered matching and asynchronous tools. The integration of engagement and e-mentoring offers a holistic framework for student retention, learning outcomes, and well-being. Practical recommendations and case studies illustrate their impact in building inclusive, adaptable, and future-ready education systems.

Chapter 3: Digital twins as a method of learning in a fragile world

New technologies in the VUCA and BANI worlds are aimed at the variability of systems and their complexity. The main tasks of such technologies are to be flexible and adapt to a changing situation in the world, but at the same time provide a person with all information at an intuitive level, which in turn can positively affect his mental state. This chapter discusses the use of digital twins as a teaching method according to training simulators in nuclear energy, since it is one of such complex processes, and the integration of technologies such as virtual reality and digital twins is a priority due to the conditions of the BANI world.

Chapter 4: Development of an application for compiling and optimizing an investment portfolio

The article presents the results of the analysis of the theoretical foundations of the formation and optimization of the investment portfolio, considers the models of G. Markowitz W. Sharpe for the compilation of the investment portfolio; analyzed and selected the tools for developing the application that forms the investment portfolio. During the study the Python programming language was chosen as the tools for developing the application that forms the investment portfolio. Python is a powerful tool for optimizing investment portfolios using the W. Sharpe method, allowing investors and analysts to achieve a balance between risk and return using advanced methods of quantum analysis. The key functions of the developed application are: aggregation of financial data, calculation of expected portfolio return, calculation of variance and standard deviation, portfolio optimization, visualization of results, ease of use and an intuitive interface .

Chapter 5: Review of Monitoring Methods in Energy Storage Systems

Many energy systems are most widely used as elements of power supply systems for critical infrastructure (such as data centers, server infrastructure, etc.) and in distributed energy systems.But they have various disadvantages, the key of which is the inevitable degradation of battery cells, which can lead to premature failure of the entire system and increase operating costs. This problem can be solved through the use of battery management systems (BMS), allowing to optimize the process of operation, thereby extending the overall service life, one of the most important elements of which is a system for monitoring the parameters of the battery cells. In this regard, we considered some modern approaches aimed at organizing the process of monitoring of operating energy storage systems, which are an integral part of distributed energy systems, the functioning of which must be maintained in real time without failures and outages.

Chapter 6: Leveraging Gamification, Course and Individual Characteristics to Boost MOOC Adoption

MOOCs have transformed global education by offering unprecedented access to knowledge. Yet, engagement and completion rates remain critical challenges. This study presents a conceptual model that examines the interplay between individual learner characteristics and MOOC design features towards course adoption and use, with special emphasis on the role of gamification in sustaining motivation and retention. By considering elements such as learning styles and gamification preferences of its users, MOOC designers can enhance the learning experience and encourage course completion. Beyond the conceptual model, we explore AI as a complementary technology feature, assuming that AI systems can leverage learner styles and MOOC characteristics, in an adaptive fashion, to further enhance engagement and retention. Overall this chapter highlights how integrating Gamified AI-driven innovations, tailored to individual and course characteristics, can strengthen MOOCs as a scalable and sustainable model of education, ensuring their long-term effectiveness in an evolving learning landscape.

Chapter 7: Trade vis-à-vis Human Rights of Farmers under TRIPS

The Agreement on Trade-Related Aspects of Intellectual Property Rights (TRIPS) plays a crucial role in shaping the global intellectual property system that significantly impacts farmers' rights. In agriculture, TRIPS grants intellectual property protection in the form of patents and plant variety protection. Such protection conflicts with the farmer's rights to save, sow and exchange the seeds. This raises human rights concerns related to farmers such as access to genetic resources and their cultural rights, food security, livelihood that are enshrined under various international human rights instruments such as International Covenant on Economic, Social and Cultural Rights (ICESCR) and Universal Declaration on Human Rights (UDHR). This intersection between TRIPS and the human rights of farmers requires a more inclusive approach that balances human rights with innovation. In light of the above perspective, this chapter analyzes farmers' human rights integration under the domain of intellectual property rights protected under TRIPS.

Chapter 8: Enhancing Cognitive Skills Through Smart Classroom Technologies in The VUCA-Bani World: How Student Participation Channels the Impact and Learning Motivation Shapes This Relationship

The smart classroom technologies with skill development in a VUCA-BANI context is what this study focuses on. The research uses a sample of 550 students of UG and above in India. It assesses the impacts of IoT, AR/VR, Gamification, and AI on cognitive skills through the mediating effects of student participation and moderating effects of learning motivation. Structural equation modeling indicates that IoT, AR/VR, and GAM have substantial direct and indirect impact on cognitive skill development, whereas the impacts of AI are nearly fully mediated through students' participation. The study suggests that student participation is a vital mediator and learning motivation an important moderator. IPMA suggests that learning motivation and student participation have the strongest impact in developing cognitive skills. This study helps to understand the impact of smart education on the formation of basic cognitive skills and provides guidance on the development and implementation of strategic educational policies within the context of an overly complicated and uncertain world

Chapter 9: Smart and Digital Wellbeing Initiatives in Healthcare: The Way Forward

Smart and digital wellbeing initiatives play a significant role in modern healthcare. This chapter explores current initiatives, their use and effectiveness, as well as their challenges and future directions, with a special focus on mental health. It explores digital wellbeing, the growing need for initiatives, and the tools available in India and the world. These initiatives enhance accessibility, reduce stigma, offer personalized interventions, and improve psychological and workplace wellbeing. However, current technology, especially AI-based chatbots, have limited user interaction and authentic therapeutic support. Additionally, biased datasets can lead to flawed responses and issues with inclusive support. Current initiatives face privacy concerns due to a lack of regulatory bodies, emphasizing the need for robust regulation and oversight. Collaborative development, appropriate marketing and improved security are key to improving existing initiatives. Future research needs to address and explore these concerns to ensure the development of effective and ethical initiatives.

Chapter 10: Breaking the Bias: Assessing Nudge Effectiveness in Overcoming Decision-Making Prejudices

Influencing consumer decision-making processes in the digital age is vital as E-commerce continues dominating the market. This research work explores the application of nudging to e-commerce or any digital environment involving consumer choices, investigating its efficacy in mitigating cognitive biases that affect online purchasing decisions. We employed a 2 x 2 within-group experimental design to examine how different nudges influence e-commerce choices. Data collected from 88 participants reveal that status-quo nudges can influence decision-making more than social proof nudges or salience nudges in online shopping. These results significantly impact digital marketing strategies, suggesting that carefully designed nudges can guide consumer choices by overcoming ingrained prejudices. This research also provides practical insights into consumer behaviour through subtle interventions in e-commerce settings. While nudging has been studied in offline contexts, applying it to e-commerce and digital consumer choices represents a growing area of study.

Chapter 11: Enhancing Learning with Smart Pedagogies A Digital Approach to Teaching

Smart pedagogies are transforming conventional teaching methods by incorporating digital technology and innovative approaches to improve the learning process of students. With the advent of the digital age, teachers now utilize sophisticated technologies to create more interesting, personalized, and effective teaching methods. This research discusses the application of Smart Pedagogies, emphasizing the utilization of digital technologies in developing creativity, collaboration, and critical thinking among students. Based on Stacked Artificial Neural Networks (ANNs), we examine the effect of teachers' digital competence on pedagogical practice and student performance. The results show that digitally well-integrated tools enhance learning outcomes considerably, supporting the significance of ongoing professional development in digital competence for teachers. The study underscores the necessity of imaginative forms of teaching which appeal to the requirements of the digital age to equip the students with preparation for their futures, defined by technological forces

Chapter 12: Leveraging Business Intelligence to Explore AI-driven Transformation in the VUCA and BANI World

Artificial Intelligence (AI)-driven technological advancements often create competitive business models especially in the unpredictable BANI and VUCA environments. This chapter focuses on the role AI in transforming Business Intelligence (BI) while allowing efficient decision-making and strategic planning. The AI can transform a business using predictive analytics for volatility, automation to address complexity, decision support in non-linear scenarios, personalized user experiences, reliance on adaptive systems, improving employee productivity, and promoting sustainability. A detailed plan is implemented including resource allocation and timeline for the tasks. Using a mixed class agile methodology, a case study is developed for managing the pandemic crisis in the future for the BANI world. The BI-driven insights regarding mortality rates, infection spread, and economic impact help policymakers form informed and effective decisions. The results demonstrate that AI-driven BI offers improved efficiency, accuracy, scalability, and cost reduction in the BANI world.

Further, the compilation explores investment portfolio optimization, energy storage monitoring, and gamification strategies that enhance technology adoption—domains that demonstrate how innovation reshapes both technical and human systems. The chapters on smart teacher education and cognitive skills development through smart classrooms extend these insights to the education sector, reflecting the need for next-generation pedagogies in a VUCA-BANI environment. Legal and ethical concerns, such as the rights of farmers under TRIPS, and the growing imperative of digital well-being in healthcare, round off this multidisciplinary collection.

This book aims to serve as a guide for business leaders, educators, researchers, and policymakers seeking to understand and implement intelligent enterprise solutions. It underscores that resilience, adaptability, and ethical responsibility will define success in navigating the challenges and opportunities of an ambiguous world.

Chapter 1
Artificial Intelligence in the Professional Activity of Physicians

Ramazan Medzhidovich Abdulgalimov
https://orcid.org/0000-0002-1899-8433
Dagestan State Medical University, Russia

Arif Kamalutdinovich Kasimov
https://orcid.org/0009-0005-4062-9772
Dagestan State Medical University, Russia

Guriyat Nurakhmedovna Abdulgalimova
https://orcid.org/0009-0001-3864-1044
Dagestan State Medical University, Russia

Valentina Nikolaevna Kormakova
https://orcid.org/0000-0001-7768-359X
Belgorod State National Research University, Russia

Tatyana Alexandrovna Peleshenko
https://orcid.org/0000-0003-2761-7847
North-Caucasus Federal University, Russia

ABSTRACT

This article addresses the issue of intellectual well-being in healthcare. A predictive analysis of the advantages and challenges of using artificial intelligence technologies in the professional activities of doctors is provided. The article highlights significant problems related to ethical and legal issues, data quality, lack of algorithmic transparency and the risk of diminishing physicians' professional skills.

DOI: 10.4018/979-8-3373-7917-3.ch001

The forecasting methodology proposed in the article for optimizing diagnosis, treatment, and disease prevention in the professional activities of doctors is based on a comparative analysis of a large volume of digital statistical data. A promising direction for further scientific inquiry, according to the authors, is the consideration of digital healthcare issues using artificial intelligence in combination with other effective and globally recognized tools and technologies. The article is of interest to specialists in healthcare, bioethics, digital medical technologies, as well as to university educators.

INTRODUCTION

In today's world, technological progress relentlessly impacts all spheres of human life, including medicine and healthcare. Modern medicine is facing challenges: a growing amount of data, the need to improve diagnostic accuracy, and the lack of time for personalized treatment. One of the promising innovative trends and a kind of scientific and practical trend is the use of artificial intelligence in medicine, particularly in the professional activities of doctors. Artificial intelligence in medicine refers to the use of machine learning algorithms and other intelligent systems to analyze medical data, automate routine tasks, support decision-making, and develop new methods of treatment and diagnosis. One of the main tasks of using artificial intelligence is to develop machine learning algorithms that can "learn" from large databases to solve various complex tasks. These algorithms are very diverse, including neural networks, decision trees, Bayesian inference methods, etc. Artificial intelligence can perform other tasks ranging from speech recognition to playing chess and driving a car without a driver. Artificial intelligence is already helping to modernize diagnostic technologies and techniques, optimize treatment, predict disease progression, and even create new treatment methods. Thus, it is becoming an integral part of modern medicine, opening new opportunities for diagnosing, preventing, and treating diseases.

Practical experience shows that, on the one hand, artificial intelligence can significantly improve the quality of medical care, increase diagnostic accuracy, and optimize the operations of healthcare institutions; on the other hand, its use raises many questions and problems for practicing doctors. Nevertheless, the implementation of artificial intelligence in medicine and healthcare currently demonstrates significant advantages capable of transforming traditional approaches to diagnosis, treatment, and prevention.

The purpose of this article is to provide a predictive analysis of the advantages and problems of using artificial intelligence technologies in the professional activities of doctors. The forecasting methodology proposed in the article for optimizing diagnosis, treatment, and disease prevention in the professional activities of doctors, based on a comparative analysis of a large volume of digital statistical data, enables the minimization of risks in medicine.

RESEARCH RESULTS

Based on a theoretical analysis of medical, psychological-pedagogical, and other scientific literature (Abdulgalimov, Abdulgalimova, & Trunova, 2024; Amann & Blasimme, 2020; Krittanawong et al., 2017; Klepikova, Kormakova, Satler, & Konopko, 2025; Kormakova, Klepikova, Prokopenko, & Musaielian, 2018; Kormakova, Klepikova, Musaielian, Baybikova, & Lapina, 2019; Liu et al., 2020), the authors concluded that in the application of artificial intelligence in the professional activities of doctors, the following advantages and problems can be identified..

ADVANTAGES OF USING ARTIFICIAL INTELLIGENCE

Improved Accuracy and Acceleration of Diagnostic and Treatment Processes

One of the main advantages of artificial intelligence is its ability to analyze and process vast amounts of clinical data (Topol, 2019) in real time with high accuracy. Artificial intelligence systems trained on large datasets can significantly increase diagnostic accuracy, comparable to or even exceeding that of doctors, in a very short time. For example, artificial intelligence: can process medical images and laboratory results; analyze X-rays, CT (computed tomography) and MRI (magnetic resonance imaging) scans; detect various human pathologies, such as lung and breast cancer; identify (based on ECG analysis) anomalies such as arrhythmias, etc. It should be noted that all this makes it possible to reduce diagnosis time and treat diseases at early stages, especially in such important fields of medical science as oncology and cardiology, where early diagnosis is critically important for successful treatment of the patient (Krittanawong et al., 2017; Char et al., 2018).

Support for Doctors

AI-based decision support systems can provide doctors with up-to-date information and recommendations. This helps doctors make more informed and accurate decisions, especially in complex clinical cases.

Personalized Medicine

Personalized medicine makes it possible to take into account the individual characteristics of each patient, including genetic data, medical history, and lifestyle. This opens the way for the development of a branch of medicine in which treatment and disease prevention are developed based on the unique characteristics of each person (Topol, 2019). For example, artificial intelligence can help determine the optimal dosage of medications or suggest an individualized therapy plan, which increases treatment effectiveness and reduces the risk of side effects.

Personalized Treatment

Personalized treatment allows analysis of a patient's individual data (genetics, medical history, and other factors) and offers personalized treatment plans. In oncology, based on genetic data, artificial intelligence makes it possible to prescribe more effective therapeutic methods tailored to the unique needs of each patient.

Resource Optimization

The use of artificial intelligence in the professional activities of doctors helps optimize the distribution of medical resources by managing physicians' schedules, directing patients to the appropriate departments, and even predicting resource needs based on data analysis.

Optimization of Doctors' Work

Artificial intelligence can automate routine tasks such as analyzing medical records, filling out documentation, and monitoring patient conditions. This frees up doctors' time, allowing them to focus on more complex and important aspects of their work. Additionally, AI can present doctors with recommendations based on data analysis, which helps in making more reasoned and specific decisions.

Disease Prediction and Prevention

With the help of artificial intelligence, patient data can be analyzed to predict the risks of developing diseases (e.g., diabetes, kidney diseases, cardiovascular diseases) and recommend preventive measures (e.g., health monitoring systems based on data from wearable devices). Early identification of risks allows doctors to take preventive measures to reduce the burden on the healthcare system and improve patients' quality of life (Kormakova et al., 2018).

Acceleration of Scientific Research

In this aspect, artificial intelligence is actively used in medical research to analyze large volumes of data, such as genomic sequences, clinical trial results, and scientific publications (Obermeyer & Emanuel, 2016). This speeds up the development of new drugs and treatment methods, helps identify potential molecules for creating new drugs, and predicts their effectiveness and safety.

Improved Access to Medical Care

In regions with limited access to qualified medical care, artificial intelligence can become an important tool for diagnosis and consultation. Telemedicine platforms equipped with AI allow patients to receive consultations and diagnoses remotely, which is especially important during pandemics for residents of remote areas.

Reduced Workload and Time Costs for Medical Personnel

The introduction of artificial intelligence into the professional activities of doctors contributes to reducing costs of treatment and diagnostics. Automation of routine tasks (e.g., filling out medical charts, analyzing lab data) frees up doctors' time for more complex tasks and processes. Early diagnosis and disease prevention reduce expenses on long-term treatment and hospitalization and help optimize the use of human and material resources in medical institutions, such as equipment and staff.

Improved Quality of Life for Patients

Artificial intelligence not only helps with diagnosis and treatment but also improves patients' quality of life. Smart devices and applications can monitor health status in real time, remind patients to take medication, and provide recommendations for a healthy lifestyle. This is especially important for patients with chronic diseases who require constant care, monitoring, and control.

Based on an in-depth theoretical analysis of scientific literature (Abdulgalimov, Abdulgalimova, & Trunova, 2024; Amann & Blasimme, 2020; Obermeyer & Emanuel, 2016; Topol, 2019; Char et al., 2018) and practical experience, the authors believe that in the future, artificial intelligence will be able to play a key role in the professional activities of doctors, significantly improving the quality of healthcare and medical services. Advanced medical technologies will make it possible to create innovative diagnostic and treatment systems based on the analysis of large volumes of data and the use of machine learning algorithms. Artificial intelligence will help doctors establish more accurate diagnoses, predict treatment outcomes, and optimize the provision of medical care (Amann & Blasimme, 2020; Char et al., 2018). However, despite the existing and above-mentioned advantages, it is important to take into account the ethical and legal aspects of using artificial intelligence in the professional activities of doctors to ensure safety, confidentiality, and its effective use (Amann & Blasimme, 2020; Char et al., 2018).

Healthcare is a specific field of activity, and the use of artificial intelligence in this area, as of today, remains a little-studied aspect both theoretically and in terms of practical application. Accordingly, there are significant problems in the use of artificial intelligence in medicine. As shown by work experience and the results of observations and surveys of practicing physicians, as well as students – future doctors and medical residents – the most significant problems in the use of artificial intelligence in the professional activities of doctors include the following.

MAIN PROBLEMS OF USING ARTIFICIAL INTELLIGENCE IN MEDICINE

Ethical and Legal Issues

In the professional activities of a doctor, the use of artificial intelligence raises many ethical problems. One important issue is patient data confidentiality and responsibility for errors (Amann & Blasimme, 2020; Char et al., 2018). The need to protect personal information becomes critically important, especially when it comes to sensitive medical data. For example, who is responsible if artificial intelligence makes an incorrect diagnosis–the developer, the doctor, or the hospital?

Quality and Necessity of Big Data

The effectiveness of artificial intelligence depends on both the quality and quantity of data. Errors in data can lead to incorrect results. If the training data is not representative, artificial intelligence may produce inaccurate results for certain groups of patients. When it comes to rare diseases, an insufficient database can become a limiting factor for algorithm capabilities. Effective AI models require large datasets (Obermeyer & Emanuel, 2016).

Risk of Algorithm Overfitting

Algorithms that are overfitted to specific data usually show reduced effectiveness in real-world conditions. Additionally, an algorithm trained on data from a single medical institution may perform poorly in others due to differences in diagnostic methods.

Lack of Transparency and Trust From Doctors and Patients

AI algorithms operate as a "black box." Older-generation doctors find it difficult to explain AI decisions, prefer traditional methods, and are skeptical of AI-based recommendations, which hinders their implementation.

Regulatory Barriers

The absence of clear regulatory standards for the use of artificial intelligence in the professional activities of doctors can slow down the adoption of new technologies and create legal risks. New regulatory approaches are needed to ensure the safety and effectiveness of AI use in clinical practice.

Technical Limitations and Algorithmic Bias

Artificial intelligence is not always capable of accounting for all nuances and existing risks that a doctor might intuitively consider. This can reduce the use of medical skills in professional practice and the ability to make decisions without the help of algorithms. "It is important to maintain a balance between the use of technology and the preservation of the human factor in professional practice; complex cases requiring an interdisciplinary approach may be beyond the capabilities of artificial intelligence" (Topol, 2019).

AI algorithms may inherit bias from the data on which they are trained. This can lead to unequal access to healthcare services among different population groups and complicate existing social issues faced by medical specialists.

To summarize the above, a comparative analysis of the problems and advantages of using artificial intelligence in medicine and in the professional activities of doctors is presented (Tables 1 and 2). The numerical data indicated in Tables 1 and 2 are taken from various Internet sources (Abdulgalimov, Abdulgalimova, & Trunova, 2024; Abdulgalimov & Abdulgalimova, 2024; Amann & Blasimme, 2020; Char et al., 2018; Liu et al., 2020; Obermeyer & Emanuel, 2016).

Table 1. Indicators of the benefits and challenges of applying artificial intelligence in medicine

Criterion	Advantages	Examples
Improving diagnostic accuracy	Increases diagnostic accuracy by 10–30% depending on the area of application	AI algorithms for mammogram analysis show 94% accuracy (compared to 88% by a radiologist)
Acceleration of diagnostic and treatment processes	Reduces data analysis time by 50–70%	ECG analysis takes less than 10 seconds; manual analysis may take 5–10 minutes
Personalization of treatment	Increases treatment effectiveness by 20–40% through individual approaches	Personalized treatment plans improve patient survival around20%
Reducing the workload on medical staff	Automates routine tasks. Reduces doctor workload by 30–50%	Chatbots collecting anamnesis save up to 20 minutes per patient
Disease prediction and prevention	Reduces disease development risks by 10–20% through early detection and prevention	Health monitoring systems reduce hospitalizations by 15–20%
Problems		
Data quality	Data errors can reduce AI accuracy by 10–25%	If training data is not representative, accuracy may drop to 70% instead of the expected 90%
Ethical and legal issues	60% of doctors and patients express concerns about data confidentiality	40% of cases involve challenges in assigning responsibility for AI errors
High implementation cost	Implementation of AI solutions requires investment: $1.5–5 million for large hospitals	Development and training costs range from $500,000 to $2 million
Lack of trust from doctors and patients	50% of doctors are skeptical of AI-based diagnoses	Only 30% of patients are ready to trust AI without additional verification
Risk of algorithm overfitting	Algorithm accuracy drops by 10–20% in new conditions	Algorithms trained on one hospital's data may show 75% accuracy in another (instead of 90%)
Technical limitations	Lacks context consideration in 20–30% of complex cases	In 25% of cases, doctors adjust AI recommendations based on their experience

Table 2. Advantages and problems of AI use in the professional activity of doctors

Aspect	Advantages	Problems
Diagnostic accuracy	Accuracy increased by 10–30% (up to 94% for mammograms)	Accuracy drop by 10–25% due to poor-quality data
Processing speed	Analysis time reduced by 50–70% (e.g., 10 seconds for ECG)	Overfitting risk reduces accuracy by 10–20%
Personalization	Treatment efficiency increased by 20–40%	Limited data for personalization

continued on following page

Table 2. Continued

Aspect	Advantages	Problems
Time saving	Doctor workload reduced by 30–50% (up to 20 min per patient)	Lack of trust: 50% of doctors and 70% of patients
Forecasting	Disease risk reduced by 10–20%	High implementation costs
Ethics & law	60% concerned about privacy	40% difficulty with assigning responsibility

In this way, as shown by the data in Tables 1 and 2, artificial intelligence represents a promising technology whose implementation in medical practice presents both certain challenges to the scientific and professional community and significant prospects for the development of this area of scientific knowledge and its practical application. Services based on artificial intelligence technologies have many advantages. Firstly, independence - artificial intelligence acts as an expert who makes an objective judgment about the need for treatment. Secondly, the work of neural networks allows you to avoid mistakes caused by the "human factor", specialists often miss important details and signs of illness due to fatigue or inability to concentrate, as well as due to the lack of special professional training. Thirdly, the automatic analysis takes only a few minutes, which saves time for doctors and patients. Among the challenges are issues of safety and ethics in the use of artificial intelligence in the diagnosis and treatment of patients, as well as the need to train medical personnel in working with such technologies.

CONCLUSION

Artificial intelligence is a powerful tool that can significantly improve the professional activity of a doctor. Artificial intelligence does not replace a doctor, it is his partner, which helps to make more informed decisions and improve the quality of medical care. In order to create a future for healthcare, it is necessary to actively introduce artificial intelligence into medicine, where technologies work for the benefit of patients.

The use of artificial intelligence in the professional activities of doctors (diagnosis, treatment, and prevention) represents a promising field capable of transforming healthcare. Although the implementation of artificial intelligence faces certain challenges, such as data privacy concerns and the need to train medical staff, its advantages are clear. However, the prospects for the development of artificial intelligence in medicine include more accurate diagnosis and prediction of possible diseases, personalized treatment, and improved quality of medical care. It is important to consider these factors when integrating artificial intelligence into the professional activities of doctors to achieve the best outcomes. It is important to take these factors into account when introducing artificial intelligence into the professional activities of doctors to achieve the best results. Artificial intelligence makes it possible to remotely monitor the health of patients, create more “intelligent” and autonomous systems, and strengthen the role of healthcare workers' education.

Artificial intelligence is a promising technology that improves the accuracy of medical prescriptions and recommendations, helps reduce time for routine processes and reduces costs. Nevertheless, medical professionals remain the most important link in terms of working with people, since they possess, in addition to professional qualities, the necessary moral and ethical qualities, empathy and understanding of social values. Thus, artificial intelligence in medicine should be understood as a tool to improve the quality of care and reduce the burden on medical professionals. In the near future, artificial intelligence will become an integral part of medicine and healthcare, contributing to the creation of more effective, accessible, and personalized medicine. However, the introduction of artificial intelligence requires very careful consideration of ethical issues and solving technical problems.

REFERENCES

Abdulgalimov, R. M., & Abdulgalimova, G. N. (2024). Digital technologies as a factor in improving the quality of medical training. Bulletin of the Lugansk State Pedagogical University: Collection of Scientific Papers. Series "Pedagogical Sciences. *Education*, (3(120)), 5–9.

Abdulgalimov, R. M., Abdulgalimova, G. N., & Trunova, S. A. (2024). Didactic possibilities of digital technologies and artificial intelligence in the professional training of future medical professionals. *Problems of Modern Pedagogical Education: Collection of Scientific Works*, (84(4)), 4–8.

Amann, J., & Blasimme, A. (2020). Artificial intelligence in healthcare: Predicting ethical, privacy, and bias concerns. *Health Affairs*, *39*(11), 1978–1985. DOI: 10.1377/hlthaff.2020.01345

Char, D. S., Shah, N. H., & Magnus, D. (2018). Implementing machine learning in healthcare—Addressing ethical challenges. *The New England Journal of Medicine*, *378*(11), 981–983. DOI: 10.1056/NEJMp1714229 PMID: 29539284

Klepikova, A., Kormakova, V., Satler, O., & Konopko, E. (2025). Formation of digital literacy of university teachers. In *International Conference on Innovative Approaches to the Application of Digital Technologies in Education and Research (SLET 2022)* (pp. 504–513). Springer, Cham. DOI: 10.1007/978-3-031-78776-8_51

Kormakova, V. N., Klepikova, A. G., Musaielian, E. N., Baybikova, G. V., & Lapina, M. A. (2019). Formation of ICT competencies of postgraduate students of teacher education based on interactive techniques. In International Scientific Conference "Innovative Approaches to the Application of Digital Technologies in Education and Research" (SLET-2019) (Vol. 2494, pp. 11–21). CEUR Workshop Proceedings. https://ceur-ws.org/Vol-2494/invited_paper_2.pdf

Kormakova, V. N., Klepikova, A. G., Prokopenko, Y. A., & Musaielian, E. N. (2018). Quality management principles of scientific and methodological support for students' activity within the e-learning environment. In Proceedings of the International Multidisciplinary Symposium ICT Research in Russian Federation and Europe "Integrating Research Agendas and Devising Joint Challenges" (REMS-2018) (Vol. 2254, pp. 63–75). CEUR Workshop Proceedings. https://ceur-ws.org/Vol-2254/frontmatters.pdf

Krittanawong, C., Zhang, H., Wang, Z., Aydar, M., & Kitai, T. (2017). Artificial intelligence in precision cardiovascular medicine. *Journal of the American College of Cardiology*, *69*(21), 2657–2664. DOI: 10.1016/j.jacc.2017.03.571 PMID: 28545640

Liu, S., Liu, S., Cai, W., Che, H., Pujol, S., Kikinis, R., Feng, D., & Fulham, M. (2020). The role of artificial intelligence in healthcare: A comprehensive review. *Journal of Medical Systems*, *44*(11), 1. Advance online publication. DOI: 10.1007/s10916-020-01662-4

Obermeyer, Z., & Emanuel, E. J. (2016). Predicting the future—Big data, machine learning, and clinical medicine. *The New England Journal of Medicine*, *375*(13), 1216–1219. DOI: 10.1056/NEJMp1606181 PMID: 27682033

Topol, E. J. (2019). *Deep medicine: How artificial intelligence can make healthcare human again*. Basic Books.

Chapter 2

Empowering Learners Through Engagement and E-Mentoring:

Building Resilience in a Changing Landscape

Semiral Öncü
https://orcid.org/0000-0001-8549-094X
Balikesir University, Turkey

Beril Ceylan Nebioğlu
https://orcid.org/0000-0002-2440-0434
Ege University, Turkey

Merve Çolakoğlu
https://orcid.org/0000-0001-5102-0528
Independent Researcher, Turkey

ABSTRACT

The global education landscape has been reshaped by technological advancements, societal disruptions, and the COVID-19 pandemic. The VUCA and BANI paradigms highlight emerging challenges and opportunities in education. This chapter explores how student engagement and e-mentoring foster resilience, adaptability, and inclusivity in this dynamic context. Student engagement is redefined to address digital fatigue, information overload, and psychological stress in hybrid and digital environments. Emerging technologies like AI, gamification, and virtual reality enhance cognitive and emotional engagement while promoting equitable access. E-mentoring provides personalized guidance, emotional support, and professional development through

DOI: 10.4018/979-8-3373-7917-3.ch002

digital platforms, leveraging AI-powered matching and asynchronous tools. The integration of engagement and e-mentoring offers a holistic framework for student retention, learning outcomes, and well-being. Practical recommendations and case studies illustrate their impact in building inclusive, adaptable, and future-ready education systems.

1. INTRODUCTION AND GENERAL CONTEXT

1.1. INTRODUCTION

The global education landscape has undergone seismic shifts due to technological advancements, societal transformations, and the prolonged effects of the COVID-19 pandemic (Bozkurt et al., 2022). These changes have led to a redefinition of the complex dynamics shaping educational systems, bringing the VUCA (Volatility, Uncertainty, Complexity, Ambiguity) paradigm—highlighting unpredictability, complexity, and ambiguity—into focus, alongside the BANI (Brittleness, Anxiety, Nonlinearity, Incomprehensibility) framework, which emphasizes fragility, anxiety, nonlinearity, and incomprehensibility. Both frameworks vividly illustrate the complexities of modern systems, underscoring the necessity of adaptive, innovative, and human-centered approaches in education (LeBlanc, 2018).

To keep up and adapt to societal dynamics, individuals must constantly follow emerging trends and acquire new knowledge and skills. This ever-changing and uncertain environment is referred to as the VUCA world. VUCA is a term used to describe uncontrollable global conditions such as volatility, uncertainty, complexity, and ambiguity. It provides a crucial framework for understanding the difficult-to-manage factors that individuals and institutions face in the business world and society. VUCA enables organizations to recognize, adapt to, and develop strategies in response to environmental changes. Rapidly evolving technology further emphasizes the need for flexibility, adaptability, and innovation (Reeves & Reeves, 2015).

However, some scholars argue that the VUCA era has already passed, and as an alternative, the concept of BANI was introduced in 2016 (Zakharov, 2022). BANI operates on a similar conceptual basis, defining the world through four key dimensions and offering insights into how individuals interact with this environment. Beyond the uncertainty and complexity foreseen by VUCA, the concept of the BANI world emerged in response to the challenges introduced by the COVID-19 pandemic. Based on observations and analyses of these changes, BANI places even greater emphasis on the necessity of flexibility, adaptability, and innovation. Given its core elements—Brittleness, Anxiety, Nonlinearity, and Incomprehensibility—the model encourages organizations to develop resilient and sustainable systems. In the context of crises such as the pandemic or other periods of transformation, it highlights the need for adaptability, innovation, and preparedness for unforeseen circumstances (Bushuyev, Piliuhina, et al., 2023; Melnick, 2022; Nataliia & Olena, 2023).

Viewing VUCA and BANI as entirely separate concepts can be misleading. Instead, BANI can be seen as a reconfiguration of VUCA rather than its outright replacement. This transition became necessary due to increasing complexity in processes and the rise of mass digitalization from the 1980s and 1990s onward. While strategies based on VUCA remain relevant, elements emphasized in BANI—such as flexibility and awareness—are incorporated into these strategies. However, these approaches alone are insufficient, signaling the need for new perspectives and strategies (Zadorina et al., 2023).

In today's context, VUCA continues to serve as a valuable lens for understanding shifting dynamics (Abeysekera, 2023). At the same time, BANI's focus on emotional and psychological dimensions plays a crucial role in helping students develop skills to cope with uncertainty and anxiety. Within this framework, this chapter examines the transformative potential of student engagement and e-mentoring practices in building resilient, inclusive, and future-ready educational systems. The chapter discusses how students' adaptability, flexible thinking, and emotional-psychological resilience can be supported within the challenges posed by the VUCA and BANI worlds. It also presents actionable recommendations that can be integrated into educational practices.

Student engagement and e-mentoring practices, as illustrated in Figure 1, are being restructured within the complex conditions shaped by the VUCA and BANI worlds in contemporary educational environments. The concepts of resilience, adaptability, and inclusivity represent shared objectives and functions that connect these two areas. Student engagement is redefined with a focus on addressing challenges such as fatigue, overload, and stress that impact modern learning processes. E-mentoring emphasizes providing personalized guidance, emotional support, and professional development opportunities. Emerging technologies—including artificial intelligence, gamification, and virtual reality—play a critical role in supporting both engagement and mentoring processes (Inelmen, 2004; Maričić & Lavicza, 2024). The figure as a whole highlights the dynamic, uncertain, and complex nature of today's educational landscape, emphasizing the multidimensional interactions transforming student experiences and mentoring practices, with technology as a pivotal enabler in this evolution.

Figure 1. Student engagement and e-mentoring in VUCA/BANI world

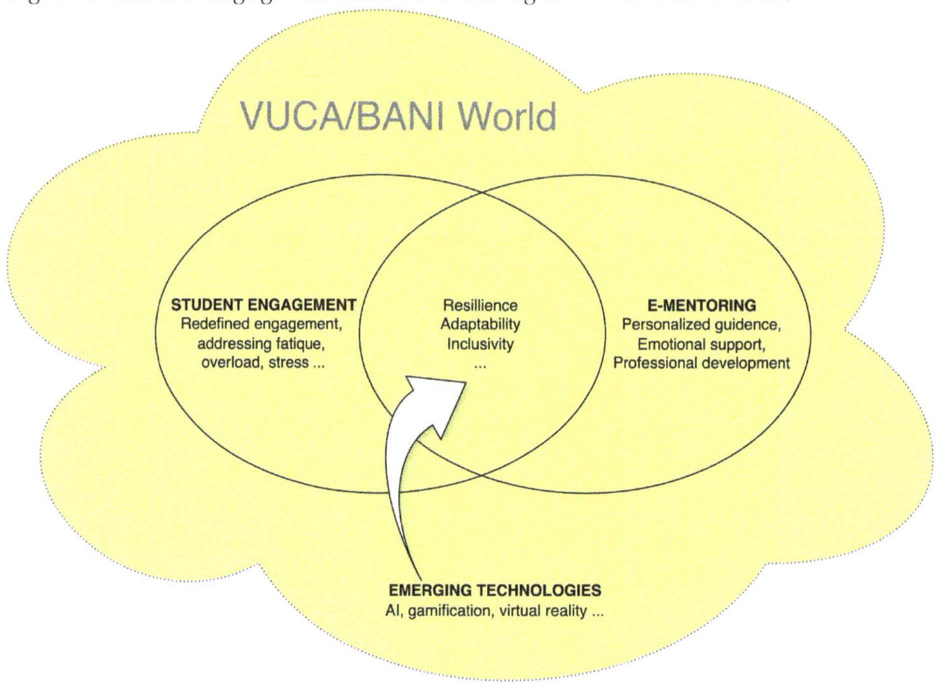

According to data from organizations such as the OECD and UNESCO, the VUCA model has long been a fundamental reference for preparing students for the future (Narot & Kiettikunwong, 2023). However, current trends suggest that BANI is becoming a more dominant paradigm in understanding and preparing for future challenges.

1.2. Context

As is well known, over the past 20–30 years, the VUCA and BANI models have served as tools for organizations to analyze and make sense of a changing environment in the process of developing innovative strategies (Bushuyev et al., 2023). While VUCA helps organizations and educational institutions identify environmental changes, adapt to them, and develop strategic plans, the BANI model emerged particularly in response to the uncertainties, anxieties, and emotional challenges brought about by the COVID-19 pandemic (De Godoy & Filho, 2021). During the pandemic, students had to cope not only with uncertainty and complexity but also with emotional difficulties. This situation has demonstrated that education is not just about transmitting knowledge but also about developing students' abilities to manage stress, anxiety, and uncertainty. Consequently, the BANI approach has highlighted the importance of emotional and psychological support strategies in education while also guiding institutions in developing flexible and sustainable systems (Bushuyev & Verenych, 2018).

Modern educational needs necessitate adopting personalized approaches tailored to individual students rather than one-size-fits-all models. For students to take an active role in the learning process and assume responsibility for their learning, they must develop scenario planning skills, engage in lifelong learning, and build resilience in the face of challenges—key elements for success in today's uncertain and rapidly changing world (Fernandes & Afonso, 2021). In this context, active learning strategies are increasingly being placed at the center of educational models to support both the academic and emotional empowerment of students.

The integration of digitalization into education has brought numerous opportunities alongside significant challenges. The new generation of learners, often referred to as digital natives, must go beyond merely accessing information; they need to navigate through an overload of information, develop flexible thinking skills, and continually renew themselves to tackle the challenges they encounter (Panthalookaran, 2022). In this process, the emotional and psychological flexibility emphasized by the BANI model serves as a critical support mechanism for overcoming the challenges of the digital age. The ability of students to effectively use technological tools to interpret and internalize information has become a cornerstone of success in modern education.

With the impact of digitalization and technological transformation, e-mentoring has become one of the prominent elements of modern education. E-mentoring involves interactive and structured communication networks established between experienced mentors and mentees via digital platforms, offering personalized guidance and continuous support. This approach aims to strengthen students not only in accessing academic knowledge but also in developing the skills to cope with emotional, cultural, and psychological challenges. Arnold et al. (2022) describe a campus-based e-mentoring initiative for multicultural students in which regular, structured online meetings allow mentors and mentees to discuss personal traits, peer relationships, academic progress, and motivational goals. Participants report feeling safe sharing sensitive information with unfamiliar mentors and credit the program with boosting their mental health, self-confidence, digital literacy, autonomy, and receptiveness to guidance. In contrast, Kaufman et al.'s (2022) systematic review finds that, despite these promising outcomes, most e-mentoring research focuses on academic or career domains, with relatively few programs addressing youth health and wellness. Their review underscores that existing studies tend to examine self-efficacy and socio-emotional coping during health crises, but they call for more targeted e-mentoring interventions explicitly designed to enhance young people's overall well-being.

E-mentors go beyond mere knowledge transfer by equipping students with strategies to manage stress, anxiety, and uncertainty, addressing the needs outlined within the BANI framework. Furthermore, by eliminating geographical and time constraints, this digital mentoring system creates flexible learning environments where students can access support at any time. Thus, e-mentoring integrates with active learning strategies, standing out as a key tool in enhancing resilience and adaptability in modern education.

Technological innovations have become transformative elements in education and organizational processes. The Internet of things (IoT), artificial intelligence (AI), machine learning, blockchain, augmented reality (AR), virtual reality (VR), extended reality (XR), mixed reality (MR), the metaverse, and robotics are making educational environments more interactive, immersive, and efficient. For instance, IoT supports the development of smart systems, while blockchain plays a critical role in ensuring data transparency and security. AR, VR, XR, MR, and the metaverse enrich students' experiential learning processes, whereas robotics enhances efficiency by increasing automation and collaboration opportunities. The adoption of these technological advancements by educational institutions and organizations is supported by applications such as cloud computing and AI-powered data analytics, ensuring that processes become more secure, scalable, and flexible. Technology not only solves problems but also creates new opportunities, accelerating digitalization and fostering a more flexible, efficient, and innovative environment (Ciarli et al., 2021).

All these dynamics indicate that global education systems must move beyond focusing solely on academic achievement and instead develop comprehensive approaches that also consider students' emotional and psychological resilience. With the integration of technology in education, constructivist social approaches, active learning strategies, e-mentoring applications, and the opportunities brought by digitalization, it has become inevitable to develop new methods for students and institutions to adapt and engage in a rapidly changing world. This evolving landscape also underscores the necessity of advancing the design and implementation of higher education courses and programs (Zadorina et al., 2023).

2. THE EVOLVING LANDSCAPE OF STUDENT ENGAGEMENT IN THE CONTEXT OF VUCA-BANI

2.1. Basic Definition and Types of Student Engagement

Student engagement refers to students' active involvement in the learning process across cognitive, emotional, and behavioral dimensions (Fredricks et al., 2004; Hasanov et al., 2021). This concept goes beyond mere physical attendance in classes in physical settings; it encompasses students' deep assimilation of learning content, development of competencies such as problem-solving, critical thinking, creativity, and collaboration, as well as the formation of emotional connections. Participation in discussions and problem-solving-focused collaborations significantly contribute to learning outcomes (Meyer, 2014).

While traditional educational models assigned students more passive roles, today's digital and hybrid environments guide them toward active, meaningful interactions. These interactions enhance students' motivation toward course activities, leading to deeper learning experiences. For a deeper learning experience, student engagement is generally examined in three main dimensions (Fredricks et al., 2004):

Behavioral Engagement: Active and regular participation in classroom activities, assignments, group work, and school events.
Cognitive Engagement: The ability to deeply understand course content, think critically, solve problems, and synthesize new information.
Emotional Engagement: The sense of interest, belonging, motivation, and attachment that students feel toward their courses, teachers, and the school community.

Considering these dimensions in an integrated manner not only contributes to students' academic success but also plays a crucial role in their personal and social development.

2.2. Redefining Engagement: Digital Transformation, Global Approaches, and Socio-Emotional Skills in the BANI Era

The rapid proliferation of digital technologies, combined with the uncertainties of global disruptions, has reshaped the way students engage in learning (Crittenden et al., 2019). The integration of innovations into education systems must be based on understanding new technologies and teaching methods. In a world filled with variability and uncertainty, engagement is no longer limited to passive participation in traditional classrooms; it now encompasses active, meaningful interactions across digital, hybrid, and real-world environments (Hodges, 2020). In this context, the BANI approach restructures students' emotional and psychological interaction strategies to foster learning. While VUCA-based strategies remain relevant, elements emphasized by BANI—such as flexibility and awareness—are increasingly crucial. However, these elements alone are no longer sufficient, necessitating a more comprehensive approach and new strategies. Flexibility, in particular, is of great importance, forming the foundation of an expert's competencies in the future BANI world (Zadorina et al., 2023).

The BANI world expects education leaders not only to possess advanced problem-solving and teamwork skills but also to develop emotional intelligence, creativity, and leadership skills that will guide educational institutions through rapid change. Additionally, the BANI world aims to promote learning flexibility and a lifelong learning mindset. Early-stage efforts toward this goal create more learning opportunities by increasing individuals' intrinsic motivation and curiosity (Jiracheewewong, 2023; Panthalookaran, 2022).

Traditional definitions of student engagement should be reconsidered in light of the requirements of the VUCA and BANI worlds. The ability of students to apply their knowledge and skills in real-life contexts is crucial for effective learning. In developed countries, education systems have shifted from educator-centered models to student-centered, collaborative models by enriching curricula with areas such as virtual reality, robotics coding, electronic programming, digital transformation, entrepreneurship, innovation, simulation, mobile programming, the Internet of things, and cybersecurity. To prepare students for VUCA contexts, students must acquire a diverse set of skills. Essential competencies for the VUCA work environment include communication skills, the ability to learn independently and interdisciplinarily, ethics and responsibility, intercultural competence, teamwork in both physical and virtual settings, social intelligence, flexibility, critical thinking, and digital literacy (Fernandes & Afonso, 2021). Therefore, educational institutions must equip students to address the complex and often adverse challenges of the modern world (Luna Scott, 2015). Thriving in today's VUCA world requires adopting innovative teaching methods that support students' active engagement. In the VUCA world, it is essential to create learning environments that promote active participation, encourage collaboration, and provide emotional support. Approaches such as active learning and flipped learning, combined with the effective use of technology, play a significant role in increasing student engagement. Moreover, developing emotional intelligence helps students cope with stress, manage anxiety, and handle uncertainty more effectively. In the VUCA world, emotional and cognitive qualities such as passion and curiosity are essential. These attributes contribute not only to students' acquisition of existing knowledge but also to their ability to develop future skills. Such qualities foster students' active engagement in both their courses and the overall learning process.

Furthermore, in the BANI era, skills such as creativity, collaboration, effective communication, and critical thinking extend beyond cognitive competencies to include emotional intelligence. Social skills help students excel in group work and interpersonal relationships, creating a supportive impact in a world marked by anxiety and instability (Burda et al., 2024). Developing emotional resilience, anxiety management, and coping strategies for uncertainty are essential competencies in the BANI era, as well as cultivating attention, empathy, and awareness.

2.3. Challenges to Student Engagement

In the BANI world, educators face unique challenges in enhancing student engagement within environments characterized by fragility, anxiety, non-linearity, and incomprehensibility. These challenges include both psychological and practical barriers. Information overload, digital fatigue, and the stress caused by unpredictable environments can hinder students' deep learning processes and active participation. To help students navigate these uncertainties, educators must create supportive learning environments while also developing strategies based on shared responsibility and accountability. While the role of e-mentoring in addressing these challenges is explored later in the chapter, it is also valuable to examine additional strategies for enhancing student engagement in the BANI world.

Project-Based Learning (PBL) has emerged as a crucial method for fostering student autonomy and real-world problem-solving skills. By emphasizing active participation, iterative inquiry, and collaborative reflection, PBL enables students to construct knowledge effectively (Chiang & Lee, 2016). This approach not only enhances motivation but also strengthens critical thinking and problem-solving skills, which are essential for coping with the complexities of the BANI world (Bushuyev et al., 2023). Furthermore, PBL aligns with the principles of Responsible Research and Innovation (RRI), which integrate content learning with process learning to foster creativity and critical thinking among students (Ruiz-Mallén et al., 2021).

Collaborative learning, on the other hand, is defined as a set of instructional strategies that promote interdependence and teamwork among students working in small, heterogeneous groups (Le et al., 2018). Beyond improving academic achievement, this method also fosters social interaction and peer support, which are particularly valuable in a world marked by anxiety and instability. Educators play a crucial role in enhancing student engagement by establishing supportive learning environments grounded in shared responsibility and accountability (Ansari & Khan, 2020).

2.4. Technology-Driven Engagement

Digital transformation in education plays a crucial role in enhancing and deepening student engagement. This transformation became necessary due to the increasing complexity of processes and the emergence of new dynamics, such as mass digitalization, after the 1980s and 1990s (Zadorina et al., 2023). The rapid advancements in educational technologies following the COVID-19 pandemic have emphasized the shift toward more flexible and adaptive teaching methods (Sato et al., 2023).

The use of digital tools provides interactive and personalized learning experiences, addressing the diverse needs of students within the BANI context (Lai et al., 2024).

As the educational landscape evolves from the VUCA paradigm to the more nuanced BANI framework, the integration of AI has emerged as a transformative force. AI is no longer a supplementary tool but a foundational element in cultivating meaningful student engagement, fostering emotional resilience, and enhancing adaptability in both learners and educators (GU Rankings, n.d.; Holcombe & Wozniak, 2024; Nguyen et al., 2024). For example, the integration of chatbots and artificial intelligence in educational settings supports students' self-regulated learning by offering instant feedback, thereby increasing engagement (Lai et al., 2024). Through personalized learning pathways, intelligent feedback systems, and adaptive technologies such as virtual agents and immersive simulations, AI enables educational environments to respond more effectively to the diverse and shifting needs of students (Solutions, 2023). These tools serve as important mechanisms for enhancing both cognitive and emotional engagement. This technological integration, thus, not only facilitates more effective learning of course content but also enables students to utilize their creativity, collaborate, and develop critical thinking skills. By centering pedagogy on the learner experience and leveraging technology to anticipate challenges, institutions can empower students not only to acquire knowledge but also to develop the cognitive flexibility and emotional intelligence necessary to navigate an increasingly unpredictable and complex global context.

Furthermore, e-mentoring strategies, which will be discussed in detail in the next section, stand out as one of the key opportunities offered by digital transformation. E-mentoring is an effective method for providing emotional and academic support by offering personalized guidance and communication tailored to students' individual needs. Through this approach, students can benefit more effectively from project-based learning, collaborative learning, and the flexible learning environments provided by digital technologies, helping them navigate the fragility, anxiety, and uncertainty of the BANI world more successfully. At the same time, e-mentoring practices enable educators to implement student-centered approaches more effectively and support their professional development.

2.5. Sustainability of Student Engagement

While student engagement has increasingly expanded beyond mere physical presence in classes as online and digital learning environments become more prevalent, future educational settings must also integrate emotional and psychological dimensions into the learning process. The VUCA and BANI perspectives highlight the importance of environmental factors and individual experiences in developing strategies that ensure students' active participation in the learning process. As discussed so far, strategies such as project-based learning, collaborative learning, and the integration of digital technologies address existing challenges related to uncertainty and anxiety while equipping students with critical skills necessary for success in an increasingly complex and unpredictable world.

Education systems should not only support students' academic achievements but also provide them with the competencies needed to become strong, creative, and effective individuals in both their professional and personal lives. In this context, developing scalable, inclusive, and equitable engagement strategies that cater to diverse student communities is of paramount importance. Digital support systems, such as e-mentoring, offer a holistic solution to the challenges posed by BANI and play a vital role in ensuring the sustainability of engagement. The following section will explore the effects of e-mentoring in reducing uncertainty and anxiety in the BANI era in greater detail.

3. E-MENTORING STRATEGIES AND PERSONALIZED FEEDBACK

3.1. Definition and Types of E-Mentoring

Mentoring is a well-established process in which experienced individuals provide guidance, support, and knowledge to those with less experience, facilitating their personal and professional development (Rowland, 2012). E-mentoring is a structured, systematic, and sustainable framework that allows individuals to receive lifelong support from a diverse pool of experts that does not require physical presence (Ateş Çobanoğlu & Ceylan, 2020). In mentoring practices, the mentor-mentee relationship fosters mutual development through a dynamic process based on satisfaction. Traditional mentoring is typically conducted through face-to-face interactions and one-on-one meetings, enabling direct relationships between mentors and mentees (Insala, 2019). Mentoring focuses on psychosocial development and career advancement, addressing components such as professional expertise, stepping outside one's comfort zone, setting career goals, engaging in diverse tasks, expressing emotions, overcoming the fear of making mistakes, reflection, and empathy.

E-mentoring, on the other hand, is characterized by the use of online tools, computer-mediated communication, and remote technologies (Willbanks, 2014). Rapid technological advancements have led to a redefinition of mentoring practices, giving rise to e-mentoring, which utilizes digital platforms to facilitate mentor-mentee interactions (Malkovich, n.d.). This shift became particularly significant during the COVID-19 pandemic, which necessitated a transition to online mentoring methods in various educational and professional settings (Wright et al., 2024).

As a digital transformation of the traditional mentoring model, e-mentoring effectively eliminates geographical and temporal barriers, enabling mentors and mentees to connect through various online platforms (Kuzu et al., 2012). E-mentoring facilitates equitable communication at a lower cost and is also known as online mentoring, with the potential to connect mentees worldwide with mentors possessing diverse knowledge, skills, and experiences (Wallis et al., 2015). Fundamentally, e-mentoring involves sharing knowledge and advice while also providing support and companionship through online communication tools (e.g., email, instant messaging), fostering a mutually beneficial relationship (Cravens, 2003). E-mentoring can take various forms, including one-on-one interactions and group-based mentoring, utilizing both synchronous (real-time) and asynchronous (flexible timing) communication methods (C. L. Andersen & West, 2021; Kuzu et al., 2012; Rowland, 2012). This categorization highlights the versatility of e-mentoring in accommodating diverse learning needs and preferences. For example, online platforms have emerged as popular tools that demonstrate how modern mentoring practices can be adapted to different contexts and technologies (Ensher et al., 2003).

Furthermore, the diversity of e-mentoring approaches underscores its potential to provide tailored guidance services that address individual characteristics and learning requirements. The flexibility inherent in e-mentoring allows for the development of customized mentoring programs that address the specific challenges faced by mentees, particularly in rapidly changing environments (Ceylan, 2019; McLoughlin et al., 2007). As organizations increasingly recognize the value of digital mentoring, they are exploring innovative frameworks that integrate e-mentoring into broader professional development strategies, thereby enhancing the overall effectiveness of mentoring initiatives (Rowland, 2012). This evolution reflects a growing acknowledgment of the importance of mentoring in fostering professional growth and adaptability in an increasingly digital world (HRbrain.ai, 2024; Kuzu et al., 2012).

By leveraging the extensive capabilities of digital platforms, e-mentoring is gaining prominence as a transformative guidance approach that eliminates geographical, cultural, and temporal boundaries between mentors and mentees. Given the uncertainty, complexity, and emotional challenges introduced by VUCA and BANI environments, e-mentoring extends beyond traditional academic advising, offering a holistic response to students' emotional, psychological, and professional needs. Mentoring programs should be designed with flexible planning, diverse communication tools, and rapid adaptability, allowing them to respond swiftly to changing conditions, new information, and evolving student needs in a VUCA world (Merrick, 2016).

In the context of BANI, e-mentoring can foster adaptation and problem-solving by providing personalized support and feedback, helping mentees navigate complexity and incomprehensibility with confidence. Mentors can act as reflective partners, creating spaces for mentees to think critically and develop resilience (Merrick, 2025). This section will explore the multifaceted contributions of e-mentoring, including its role in enhancing students' adaptive and problem-solving skills, establishing trust-based mentor-mentee relationships, leveraging digital technologies for effective and accessible guidance, and promoting equal opportunities for all students.

3.2. Building Resilience

A forward-thinking approach is required to cope with the BANI world. Addressing concerns, ensuring clarity, and adapting to complex situations are crucial when facing challenges. Transparency and resilience should be prioritized throughout this process. Collaboration, innovation, and continuous learning are key to thriving in the BANI world. Additionally, recognizing and leveraging unfair advantages, as well as maintaining an openness to change and uncertainty, are of great importance. Flexibility is critical in the BANI world, and organizations, communities, and individuals must possess the ability to handle sudden changes and unexpected challenges. BANI helps identify potential threats and facilitates the development of strategies to proactively prepare for these risks.

Many studies emphasize the importance of resilience in the face of uncertainty, complexity, and unpredictability in a VUCA environment. In particular, in a BANI world, the most essential skill students need is antifragility (Gui & Yanchinda, 2025). E-mentoring, by providing personalized support and timely feedback to mentees, enables them to develop adaptability and problem-solving skills. Parallel to student engagement, e-mentoring programs offer structured yet flexible frameworks, allowing mentors and mentees to enhance their interpersonal communication, social skills, and self-efficacy while navigating uncertainty and complexity (Kaçar & Baltacı, 2023). Engagement fosters active participation, deep learning, and motivation, while e-mentoring provides personalized guidance that supports students' emotional well-being, career development, and personal growth (Neely et al., 2017). Thus, mentoring programs help mentees transform challenges into opportunities for growth, enhance their resilience, and foster strategic thinking skills in the BANI world (Merrick, 2025). Rockinson-Szapkiw and Wendt (2021) argue that in e-mentoring experiences, self-efficacy, academic success, and interests are closely linked to a sense of belonging, a driving source of engagement, which helps individuals adapt to new circumstances more quickly and boost their confidence through self-reflection.

Students must develop skills such as flexibility, awareness, adaptability, and transparency to cope with the chaotic BANI world. The process of acquiring these skills should be continuously restructured to consider students' needs and equip them to handle evolving challenges. Additionally, this process should be initiated in a way that strengthens students' psychological, physical, academic resilience, and adaptability (Jiracheewewong, 2023). Dieffenbacher (2023) views resilience as a necessary tool for dealing with fragility. Resilience allows both organizations and individuals to withstand pressure and quickly return to their original state (Dieffenbacher, 2023).

3.3. Ensuring Psychological Safety in Mentoring Relationships

The emotional and psychological challenges students face in their learning processes necessitate the establishment of trust, openness, and emotional security in mentor-mentee relationships. E-mentoring aims to build a strong foundation of trust between mentors and mentees through digital platforms. In this context, mentors not only transfer knowledge and skills but also play a guiding role in shaping mentees' learning and life attitudes. Traditional education methods may be insufficient in addressing unpredictable and multidimensional problems; therefore, the BANI perspective provides mentors with the opportunity to develop strategies that support their mentees' emotional well-being. Socio-emotional and cultural support fosters career development by helping students establish personal and relational similarities (Haeger & Fresquez, 2016). When personalized feedback processes are tailored to students' emotional states and individual needs, they provide actionable support (Hope, 2024). Through this approach, e-mentoring fosters openness, mutual trust, and an environment of emotional security between mentors and mentees, enabling students to better cope with negative emotional states such as isolation, anxiety, and uncertainty.

3.4. The Role of Technology in Scaling E-Mentoring

Although technology in mentoring applications is sometimes perceived as a barrier to communication, it actually is rather effective because it increases communication diversity. E-mentoring offers the flexibility to exist across all digital environments. AI-powered matching systems (e.g., Thunai a mentor-mentee matching platform and and the integrated GURU an AI assistant for task and resources management (Charanya et al., 2025)), can automate logistical tasks such as scheduling calendar appointments, enable secure video communication augmented with dynamic, privacy-enhancing emojis, and leverage intelligent assistants to recommend tailored resources and manage ongoing tasks. These systems continuously analyze user feedback and sentiment to refine mentor–mentee matches, boosting engagement, satisfaction, and goal alignment. Video conferencing, asynchronous communication tools, and digital feedback mechanisms not only accelerate but also further personalize interactions between mentors and mentees. Unlike traditional mentoring methods, these technological tools offer greater scalability — in a way accessibility — and efficiency; mentees can record and review the feedback they receive at any time. However, integrating e-mentoring into a BANI environment is not straightforward. Such an endeavor requires considering various strategies to address challenges related to technology access, communication barriers, anxiety, trust issues, nonlinear progress, cultural differences, and training needs (Jan & Mahboob, 2022; Morellet, 2025).

The digital tools used in e-mentoring foster flexible planning and rapid adaptation skills while supporting leadership development that shapes individuals' career goals. Moreover, strategies focusing on skill development, feedback, and communication-based mentoring help mentees unlock their potential, enhance performance, and boost their confidence. This process aids mentees in acquiring higher-order thinking, metacognitive skills, and self-awareness while providing guidance to overcome isolation and anxiety.

3.5. Accessibility and Equity

E-mentoring significantly eliminates geographical and socioeconomic limitations thanks to the universal accessibility of digital platforms. This approach brings together individuals from different cultures, ages, and experiences, enabling the implementation of more inclusive and equitable strategies in mentor-mentee interactions. It fosters diverse perspectives that contribute to decision-making processes and the development of innovative solutions (Radlick et al., 2020; Rowland, 2012). E-mentoring programs provide flexibility, personalization, and scalability, offering feedback tailored to individual needs. As mentioned earlier, mentees can record and revisit digital feedback at any time, actively engaging in their learning processes. This approach ensures that students receive equitable access to guidance services while supporting both their academic and personal development. These strategies not only strengthen students' agility, resilience, and adaptability in unpredictable and complex environments like VUCA and BANI but also enhance their core leadership and strategic thinking competencies (Cook, 2021).

In summary, e-mentoring emerges as a transformative guidance approach that provides comprehensive solutions for students' personal, academic, and professional development in response to the uncertainty, complexity, anxiety, and ambiguity introduced by VUCA and BANI worlds. While resilience is built through personalized support and feedback processes, psychological safety in mentor-mentee relationships is ensured, the scalability of programs is enhanced through technology, and accessibility and equity principles guide the development of solutions tailored to each student's individual needs. This holistic approach not only strengthens the social and emotional fabric of modern education systems but also equips students with leadership and strategic thinking skills needed to navigate an unpredictable world.

This section introduced e-mentoring and explored its strategies and areas of application in detail. The next main section focuses on how student engagement and e-mentoring activities integrate, complementing each other to create a more inclusive, effective, and sustainable educational experience.

4. INTEGRATION OF STUDENT ENGAGEMENT AND E-MENTORING

Although maneuvers to enhance student engagement and e-mentoring are powerful strategies on their own, their true potential lies in their integration. By combining these approaches, educators and institutions can create a coherent framework that addresses both the cognitive and emotional dimensions of learning. In the BANI era, where emotional resilience and adaptability are just as important as cognitive skills, this integration becomes even more critical.

4.1. Synergies Between Student Engagement and E-Mentoring

Mentors play a key role in maintaining student engagement by providing context, encouragement, and real-world connections (Omar et al., 2012). This synergy ensures that emotional and psychological support complements academic engagement (Dieffenbacher, 2023; Liu et al., 2024), which is particularly important within the BANI framework. Research indicates that students' attitudes toward online learning environments directly influence their e-mentoring experiences (Omar et al., 2012; Williams & Kim, 2011). Within the BANI context, students' emotional states further shape these attitudes. Positive attitudes toward online learning can lead to increased mentor interactions and higher engagement levels (Omar et al., 2012). This interaction supports students' emotional and social adjustment, as those who take responsibility for their learning benefit from the flexibility of online education and communicate with their mentors more frequently (Ensher, 2013). In the BANI world, flexibility, emotional balance, and psychological resilience are crucial (Jiracheewewong, 2023).

Mentoring can be seen as a long-term partnership in which students set goals that may evolve over time. Both the goals and the process belong to the student, while the mentor supports them by helping them become more aware of their experiences, gaining insights and knowledge (Gamage et al., 2021). Within the BANI framework, this support has been restructured to help students cope with insecurity and uncertainty, allowing them to develop their skills and knowledge (Merrick, 2025). By encouraging students to step outside their comfort zones, mentoring facilitates the development of emotional resilience and adaptability, which are essential for success in today's unpredictable world (Ravinutala, 2023). This approach addresses the need for emotional flexibility and adaptation in the BANI world.

As students engage more with their mentors, their level of participation increases, allowing them to benefit from shared valuable insights and experiences (Gamage et al., 2021; Kern et al., 2019). This adaptability enables students to navigate the uncertainties of the BANI world with greater confidence, creating a more responsive mentoring experience. Furthermore, ethical guidance and teacher/mentor empathy not only strengthen student engagement and academic success but also enhance students' emotional intelligence and social skills (Fatima, 2024; Zhang, 2022).

4.2. Designing Comprehensive Educational Models

Integrating engagement and mentoring into curricula, extracurricular activities, and institutional policies is essential for promoting a holistic educational experience. The innovative development of educational systems is becoming increasingly critical in the BANI environment, where variability, uncertainty, unpredictability, and disruptions play a defining role (Bushuyev, Bushuyeva, et al., 2023; Guardian, 2023; Jedaman et al., 2024). BANI principles require these models to prioritize not only cognitive learning but also emotional intelligence and adaptability. Learning is fundamentally a social process, and the integration of network technologies into education has become increasingly important in facilitating this process (Bouton et al., 2021; Bushuyev et al., 2023). From a BANI perspective, these technologies should address emotional and psychological dimensions, thereby enhancing the overall learning experience.

Research has shown that continuous learning opportunities enable online education, making virtual mentoring models like e-mentoring increasingly important (Omar et al., 2012). E-mentoring provides students with a platform to interact with mentors in a way that meets their emotional and academic needs (C. L. Andersen & West, 2021; Nguyen et al., 2024), reinforcing the importance of emotional support in educational environments.

Nguyen et al. (2024) emphasize the importance of integrating artificial intelligence and machine learning into educational settings to create enhanced learning environments, allowing for the development of curricula tailored to individual needs and characteristics. Similarly, Botea-Muntean and Constantinescu (2024) argue that educational institutions can reshape learning and improve efficiency by adopting AI, the Internet of things, blockchain, augmented reality, and robotics. However, the uncertainty and chaos introduced by these technological transformations render traditional strategies insufficient. Educational institutions, students, teachers, and society at large must flexibly adapt to these evolving demands. Innovations in education systems should be based on a deep understanding of new technologies and teaching methods.

To succeed in the BANI framework, universities must anticipate future workforce needs, integrate scientific advancements, and embrace interdisciplinary research approaches. Furthermore, horizontal strategies and personalized educational models will be crucial for institutions to succeed in the BANI world, as they enable tailored learning experiences that address students' diverse needs.

The BANI framework is now being recognized as a core paradigm not only in the business world but also in education. By adopting innovative approaches such as personalized education, institutions can build a sustainable and resilient future in higher education. The evolution of educational systems in the BANI environment requires progressive thinking, flexibility, and openness to new ideas and approaches. This ensures that educational institutions can adapt to society's changing needs and provide students with high-quality education to equip them for future challenges (Bushuyev et al., 2023). Within this evolving landscape, student engagement must be continuously reinforced through dynamic and technology-supported mentoring models, with e-mentoring playing a key role in sustaining meaningful learning experiences.

4.3. Case Studies and Best Practices

The integration of student engagement and e-mentoring strategies has been shown to improve learning outcomes, retention rates, and overall student well-being across various educational contexts. Research suggests that structured online mentoring experiences, in particular, encourage long-term student engagement. For example, students who receive continuous mentoring support demonstrate higher retention rates and greater academic progress, as evidenced by various studies (Gamage et al., 2021; Haeger & Fresquez, 2016). Case studies conducted within the BANI framework similarly indicate that addressing students' emotional complexities can lead to more resilient educational practices.

Students benefit from personalized guidance in developing idea-generation behaviors and self-confidence. When students receive individualized feedback and emotional support, the effectiveness of mentor-student interactions improves. Real-world learning experiences that require both skill development and emotional engagement enhance students' understanding of theoretical concepts learned in the classroom (Nguyen et al., 2024; Officer, n.d.). Additionally, solving real-world problems enables students to apply their own knowledge, leading to a deeper understanding and greater motivation for learning (Dare et al., 2021). Xerri et al. (2018) further support this idea, showing that involving students in live research projects not only enriches their educational experiences but also provides valuable resources for ongoing studies.

However, Anderson and West (2020) highlight several challenges in e-mentoring: administrators often struggle with identifying, selecting, and training mentors and lack robust program-evaluation procedures; mentees experience difficulties forming meaningful connections, face mentor unavailability, and generally lack awareness of e-mentoring opportunities; and mentors report problems with long-term commitment and limited program flexibility. Recognizing and addressing these barriers is a critical first step toward designing more effective online mentoring initiatives.

Several e-mentoring projects illustrate both strengths and areas for improvement: the E-mentorship Project (2022) is a one-to-one program pairing novice preschool teachers with experienced teachers and academics, focusing on e-learning modules in curriculum design, classroom management, and diversity to demonstrate how targeted content can support early-career educators; the Equals Project (2016) is aimed at improving the digital, STEM, ICT, and e-mentoring abilities of 18–25-year-old female students and supports leadership development and career building; Gafni-Lachter et al. (2021) present a model for e-mentoring graduate students that emphasizes wellness and academic success, finding that synchronized online communication delivers instrumental and psychosocial support, improves relationship quality, reduces stress, enhances self-efficacy, and fosters career growth; and Jang et al. (2023) report that in programming learning contexts, mentoring covers informational, psychosocial, and technical support, with female students engaging in a broader range of e-mentoring activities than their male peers.

Institutions that prioritize structured e-mentoring programs report higher student satisfaction levels and improved learning outcomes. Case studies demonstrate that students who develop strong relationships with their mentors benefit from enhanced problem-solving skills, better emotional regulation, and increased career readiness (Carmel & Paul, 2015; Gamage et al., 2021; Haeger & Fresquez, 2016; Xu et al., 2021). The presence of educators and the timeliness of feedback play a crucial role in student engagement. Timely and personalized communication is particularly effective in preventing students from feeling isolated in online learning environments (Meyer, 2014; Rowan & Jin, 2022; Sheridan & Kelly, 2010).

In the VUCA and BANI environments, educational institutions face significant challenges that require rapid adaptation and innovative solutions. These conditions demand that institutions continuously adjust to change and uncertainty while developing new solutions. The rise of digital technologies presents promising opportunities to address these challenges, highlighting the need for solutions that also consider the emotional and psychological dimensions of student engagement (D'Angelo, 2018; Maia et al., 2019). By analyzing successful case studies, institutions can gain valuable insights into best practices for e-mentoring and engagement strategies (T.

S. Andersen & Wellen, 2023; Arnold et al., 2022; Lindsay & Cagliostro, 2020; McAleer & Bangert, 2011; Wallis et al., 2015). This allows them to implement methods that support both academic and emotional development, ultimately enhancing student success.

In conclusion, integrating e-mentoring and student engagement strategies is crucial for fostering resilient educational practices — i.e., student resilience. By leveraging personalized support, real-world applications, and effective communication — while remaining mindful of the administrative, mentee-, and mentor-related challenges identified by Anderson and West (2020) — institutions can enhance student experiences and outcomes in an increasingly complex educational landscape.

5. CONTRIBUTION, CONCLUSION, AND FUTURE RECOMMENDATIONS

5.1. Discussion

This section aims to make a significant contribution to the broader discourse on educational strategies by presenting a detailed examination of student engagement and e-mentoring as interconnected approaches to addressing the challenges present in both VUCA and BANI environments. The VUCA framework has traditionally been used as a tool to understand rapidly changing educational settings. However, the emergence of BANI provides a deeper perspective on the emotional and psychological complexities shaping contemporary education. The integration of BANI principles in this section seeks to offer theoretical and practical tools for educators, researchers, and policymakers to reshape education in an era shaped by external uncertainties and internal emotional turbulence. This approach highlights not only the necessity of adaptation and resilience but also the importance of addressing modern students' emotional needs, thereby contributing to the development of more inclusive and responsive education systems (Hope, 2024).

The BANI framework offers a new perspective on our increasingly complex world by directly addressing challenges such as fragile systems, anxiety-inducing environments, non-linear dynamics, and incomprehensible complexity. Unlike the VUCA model, BANI takes a holistic approach that encompasses not only cognitive but also emotional dimensions, making it more aligned with the real-life experiences of today's students. In this way, BANI plays a crucial role in enhancing the emotional resilience of students and educational institutions. Challenges that were previously defined solely through the VUCA framework now take on a deeper meaning through BANI's approach, which accounts for the emotional and psychological burdens of modern educational environments (Burda et al., 2024; Bushuyev et al., 2023; Zadorina et al., 2023).

This new paradigm is essential for coping with increasing emotional challenges and unpredictable dynamics. Continuous feedback and evaluation processes — gathering insights from both mentors and mentees through surveys and one-on-one feedback sessions — can contribute to improving e-mentoring practices within both VUCA and BANI frameworks. It is critical that these evaluations, with a particular focus on emotional aspects, ensure that program improvements are both comprehensive and responsive.

Educational institutions should not only address the inherent challenges of VUCA environments but also adopt sustainability goals that account for the more fluid and emotionally intense conditions defined by BANI. The rising uncertainty and unpredictability in today's world necessitate the development of approaches that prioritize emotional and psychological resilience over traditional strategies. In this context, integrating student engagement and e-mentoring should be prioritized to meet future workforce needs and ensure adaptation in education (Carmel & Paul, 2015; Gamage et al., 2021) by focusing on key aspects such as uncertainty and anxiety management (Hope, 2024). A model is required that seamlessly blends scientific inquiry with educational practice, supports interdisciplinary research, and leverages innovative approaches to address system fragility.

E-mentoring offers a unique opportunity to continuously monitor and guide student development, creating a more flexible and resilient educational structure. In the context of BANI, where uncertainty and emotional challenges are ever-present, there is a clear need for mentoring approaches that go beyond traditional methods. Guided by BANI principles, mentoring evolves into a more holistic and sensitive process. E-mentoring, which provides flexible and continuous support, not only enhances classroom engagement but also strengthens students' emotional resilience (Mullen, 2021).

Ultimately, these support strategies play a crucial role in enabling educational institutions to adapt to new challenges and produce innovative solutions. Academic leaders are increasingly recognizing the importance of BANI principles in preparing students for a rapidly changing and complex world. This understanding requires educational systems to develop strategies that incorporate not only academic but also emotional and social dimensions into their core practices (Bushuyev et al., 2023; Gui & Yanchinda, 2025). Learning methods should be reshaped to place uncertainty and anxiety — highlighted by BANI — at the center of educational redesign, fostering critical thinking and emotional intelligence. Furthermore, continuous professional development for educators as e-mentors is crucial to adapting to these evolving demands. Institutions should design curricula that emphasize flexibility, collaboration, and real-world problem-solving, which are key indicators of a forward-looking educational vision. Given the increasing emotional and psychological challenges of our time, a BANI-focused approach is not just beneficial but inevitable.

In line with the BANI framework, student engagement is redefined to go beyond mere physical participation, encompassing active involvement in thinking, problem-solving, and collaborative efforts in the digital world. This comprehensive form of engagement plays a critical role in fostering critical thinking and emotional intelligence while mitigating the effects of uncertainty and anxiety. By doing so, it strengthens students' internal resilience and equips them more effectively for future challenges.

Addressing the demands of both VUCA and BANI requires expanding a wide range of skills — including flexibility, empathy, data literacy, media literacy, and critical thinking — to encompass emotional and psychological dimensions in tackling fragility (Bushuyev et al., 2023). E-mentoring is a powerful tool to accomplish this. AI-driven systems further enhance this new form of engagement by offering interactive, adaptive, and personalized experiences that meet learners where they are — emotionally, cognitively, and socially — thereby reinforcing the BANI-aligned goal of holistic education (Bushuyev, Bushuyeva, et al., 2023).

5.2. Conclusion

As the world continues to face challenges such as volatility, uncertainty, complexity, and ambiguity, educational systems must evolve to meet the diverse needs of students (Brylina et al., 2021; Yaccob et al., 2023). In this dynamic landscape, the BANI framework's emphasis on emotional turbulence and non-linearity necessitates a holistic adaptation of educational strategies. When student engagement and e-mentoring are thoughtfully and strategically integrated, they offer robust solutions to overcoming the multifaceted challenges of our time. The BANI perspective underscores the critical importance of addressing not only cognitive development but also emotional and psychological dimensions. These integrated strategies encourage students to take ownership of their learning journeys, build resilience, adapt to rapid changes, and establish meaningful connections in an increasingly fragmented world. Within the BANI framework, students are equipped to handle both external uncertainties and internal emotional challenges, ensuring they are well-prepared for the future.

The VUCA and BANI frameworks provide critical lenses through which the intricacies of contemporary environments can be examined and addressed. Within this context, e-mentoring emerges as a strategically relevant and adaptive mechanism, fostering essential competencies such as resilience, adaptability, and innovative thinking. These attributes are imperative for individuals and organizations aiming to navigate and flourish amidst volatility, uncertainty, fragility, and accelerated change. E-mentoring offers a psychologically safe space for students to articulate their concerns and seek guidance, thereby addressing the emotional challenges emphasized by the BANI framework. Moreover, it can play a pivotal role in nurturing key competencies — such as resilience, adaptability, empathy, and critical thinking — that are essential for navigating both VUCA and BANI environments. By meeting students where they are and adapting to their unique needs and circumstances, digital mentoring provides personalized support that enhances their capacity to thrive in increasingly complex and unstable contexts.

The shift from VUCA to BANI has deepened the challenges of student engagement, underscoring the need for emotional support and adaptability. In this context, e-mentoring offers a flexible and effective means of addressing both academic and emotional needs, making it a valuable component of future-ready educational strategies (Vatansever & Ceylan, 2021). To foster meaningful student engagement within VUCA and BANI conditions, education must go beyond cognitive development and intentionally integrate emotional support, empathy, and personalized learning approaches (Zadorina et al., 2023). This holistic focus strengthens students' ability to remain connected and thrive despite external instability.

This section has highlighted the vital role of integrating student engagement with e-mentoring as a means of creating educational systems that not only respond to today's demands but also anticipate the unpredictable challenges of the future. Informed by insights from BANI, these strategies become even more resilient and address the full spectrum of modern challenges. By focusing on the symbiotic relationship between engagement and mentoring, this educational vision aims to be inclusive, adaptable, and deeply human-centered. In the BANI era, such a vision requires constructing an educational system that is deeply attuned to students' emotional and psychological landscapes — one that is resilient, responsive, and directly prepared for the future.

While the discussion in this section has provided a deep analysis supported by concrete examples and theoretical insights, the conclusion presents comprehensive recommendations for shaping future educational policies and strategies. Together, they lay the foundation for resilient and emotionally intelligent educational systems that are ready to face both today's uncertainties and tomorrow's challenges.

5.3. Recommendations

Future research could aim to provide experimental or qualitative data demonstrating improvements in students' emotional well-being, adaptability, or academic performance through e-mentoring. To enhance these processes, e-mentoring could be further supported by generative AI tools. Practitioners should plan the mentoring process by clarifying responsibilities, offering constructive feedback, setting detailed expectations, and fostering emotional clarity. Mentoring practices could then be evaluated using reflective, structured guidelines. Longitudinal studies or design-based research could be conducted in online and face-to-face classroom settings to examine the interaction between student engagement and the BANI context. Mentees and mentors' relations on the issue of engagement and BANI will contribute to designing models or instruments. Institutions, policymakers, and academics should emphasize VUCA and BANI implications in education and explore their practical applications. Furthermore, sustainable educational systems grounded in the BANI context should be integrated across disciplines and at a global scale.

REFERENCES

Abeysekera, N. (2023). *Institutional and action research for VUCA and BANI World.* ResearchGate. DOI: 10.13140/RG.2.2.34878.66888

Andersen, C. L., & West, R. E. (2021). "We overwhelm them with hope": How online mentors can support online learners. *Online Learning : the Official Journal of the Online Learning Consortium, 25*(4), 388–415. DOI: 10.24059/olj.v25i4.2440

Andersen, T. S., & Wellen, H. (2023). Being a mentor in the digital era: An exploratory study of the benefits undergraduate student mentors derived from providing virtual mentoring to youth. *Journal of Community Psychology, 51*(7), 2635–2651. DOI: 10.1002/jcop.23051

Ansari, J. A. N., & Khan, N. A. (2020). Exploring the role of social media in collaborative learning the new domain of learning. *Smart Learning Environments, 7*(1), 9. DOI: 10.1186/s40561-020-00118-7

Arnold, K. D., Deutschlander, D. T., & Lewis, J. S. (2022). Someone on my side: The experiences of high-achieving, underserved high school students in an e-mentoring college access intervention. *Mentoring & Tutoring, 30*(4), 381–408. DOI: 10.1080/13611267.2022.2091193

Ateş Çobanoğlu, A., & Ceylan, B. (2020). E-mentorluk, mesleki gelişim ve yaşam boyu öğrenme. [E-mentoring, professional development and lifelong learning] In Öngöz, S. (Ed.), *Eğitim Bilimleri Alanında Elektronik Mentorluk* [Electronic Mentoring in Educational Sciences]. (pp. 23–45). Pegem Akademi Publishing. DOI: 10.14527/9786257052801.02

Botea-Muntean, D.-R., & Constantinescu, R. (2024). VUCA and BANI Worlds-challenges for nowadays business models. Could circular economy and various digital instruments like internet of things help a busıness model to respond to such challenges? An examinatıon of solutions and future strategies through marketing lens. *Journal of Information Systems & Operations Management, 18*(1), 38–54.

Bouton, E., Tal, S. B., & Asterhan, C. S. C. (2021). Students, social network technology and learning in higher education: Visions of collaborative knowledge construction vs. the reality of knowledge sharing. *The Internet and Higher Education, 49*, 100787. DOI: 10.1016/j.iheduc.2020.100787

Bozkurt, A., Karakaya, K., Turk, M., Karakaya, Ö., & Castellanos-Reyes, D. (2022). The impact of COVID-19 on education: A meta-narrative review. *TechTrends, 66*(5), 883–896. DOI: 10.1007/s11528-022-00759-0

Brylina, I. V., Okonskaya, N. K., Ermakov, M. A., & Brylin, A. V. (2021). Education of the future in the conditions of VUCA World. *European Proceedings of Social and Behavioural Sciences, Economic and Social Trends for Sustainability of Modern Society (ICEST-II 2021)*, 1372–1380. DOI: 10.15405/epsbs.2021.09.02.154

Burda, Y., Samus, T., Bazhmina, E., Bondarenko, O., & Myrnyi, D. (2024). Shaping the competencies of the future: The importance of developıng soft skills in higher education within the landscape of information technologies. *AD ALTA: Journal of Interdisciplinary Research, 14*(1), 111–116. DOI: 10.33543/j.140140.115121

Bushuyev, S., Bushuyeva, N., Murzabekova, S., & Khussainova, M. (2023). Innovative development of educational systems in the BANI environment. *Scientific Journal of Astana IT University, 14*(June), 104–115. DOI: 10.37943/14YNSZ2227

Bushuyev, S., Piliuhina, K., & Chetin, E. (2023). Transformation of values of the high technology projects from a VUCA to a BANI environment model. *Innovative Technologies and Scientific Solutions for Industries, 2*(24), 191–199. DOI: 10.30837/ITSSI.2023.24.191

Bushuyev, S., & Verenych, O. (2018). The blended mental space: Mobility and flexibility as characteristics of project/program success. *2018 IEEE 13th International Scientific and Technical Conference on Computer Sciences and Information Technologies (CSIT), 2*, 148–151. DOI: 10.1109/STC-CSIT.2018.8526699

Carmel, R. G., & Paul, M. W. (2015). Mentoring and coaching in academia: Reflections on a mentoring/coaching relationship. *Policy Futures in Education, 13*(4), 479–491. DOI: 10.1177/1478210315578562

Ceylan, F. (2019). *Mentorluk kavramı ve mentorluğun bireysel kariyer üstündeki etkisi* [The concept of mentoring and the effect of mentoring on the effect on individual career]. Tekirdağ Namık Kemal Üniversitesi.

Charanya, P., Nidhish, M., Prajesh, U. R., Saktheeswar, K., & Sanjana, S. P. (2025, February). Enhancing career development: An AI-driven mentorship platform for student-mentor connections. In *2025 International Conference on Computational, Communication and Information Technology (ICCCIT)* (pp. 147–152). IEEE.

Chiang, C. L., & Lee, H. (2016). The effect of project-based learning on learning motivation and problem-solving ability of vocational high school students. *International Journal of Information and Education Technology (IJIET), 6*(9), 709–712. DOI: 10.7763/IJIET.2016.V6.779

Ciarli, T., Kenney, M., Massini, S., & Piscitello, L. (2021). Digital technologies, innovation, and skills: Emerging trajectories and challenges. *Research Policy, 50*(7), 104289. DOI: 10.1016/j.respol.2021.104289

Cook, S. (2021). *Educators: Do you know about VUCA? Here is why you need to ASAP!* https://www.linkedin.com/pulse/educators-do-you-know-vuca-here-why-need-asap-spike-cook

Cravens, J. (2003). Online mentoring: Programs and suggested practices as of February 2001. *Journal of Technology in Human Services, 21*(1–2), 85–109. DOI: 10.1300/J017v21n01_05

Crittenden, W. F., Biel, I. K., & Lovely, W. A.III. (2019). Embracing digitalization: Student learning and new technologies. *Journal of Marketing Education, 41*(1), 5–14. DOI: 10.1177/0273475318820895

D'Angelo, C. (2018). The impact of technology: Student engagement and success. In *Technology and the curriculum: Summer 2018*. Power Learning Solutions. https://pressbooks.pub/techandcurriculum/chapter/engagement-and-success/

Dare, E. A., Keratithamkul, K., Hiwatig, B. M., & Li, F. (2021). Beyond content: The role of STEM disciplines, real-world problems, 21st century skills, and STEM careers within science teachers' conceptions of integrated STEM education. *Education Sciences, 11*(11), 11. Advance online publication. DOI: 10.3390/educsci11110737

De Godoy, M. F., & Filho, D. R. (2021). Facing the BANI World. *International Journal of Neurology, 14*(02), e33–e33. DOI: 10.1055/s-0041-1735848

Dieffenbacher, S. F. (2023, September 3). BANI World: What is it and why we need it? *Digital Leadership*. https://digitalleadership.com/blog/bani-world/

E- mentorship Project (2022). What is E-Mentorship Project? https://ementorship.eu/

Ensher, E. A. (2013). E-mentoring as a critical e-learning approach: The impact of social presence on e-mentoring. *Communications and Network, 05*(03), 1–3. DOI: 10.4236/cn.2013.53B1001

Ensher, E. A., Heun, C., & Blanchard, A. (2003). Online mentoring and computer-mediated communication: New directions in research. *Journal of Vocational Behavior, 63*(2), 264–288. DOI: 10.1016/S0001-8791(03)00044-7

Equals Project. (2016). EQUALS her digital skills. https://www.equalsintech.org/her-digital-skills

Fatima, N. (2024). The impact of teacher's empathy on students' self-compassion: An evidence-based intervention study. *Annals of Human and Social Sciences*, *5*(2), 2. Advance online publication. DOI: 10.35484/ahss.2024(5-II)38

Fernandes, J. M., & Afonso, P. (2021). Engineering education in a context of VUCA. *2021 4th International Conference of the Portuguese Society for Engineering Education (CISPEE)*, 1–8. DOI: 10.1109/CISPEE47794.2021.9507229

Fredricks, J. A., Blumenfeld, P. C., & Paris, A. H. (2004). School engagement: Potential of the concept, state of the evidence. *Review of Educational Research*, *74*(1), 59–109. DOI: 10.3102/00346543074001059

Gafni-Lachter, L., Niemeyer, L., Doyle, N., Norcross, J., & Jacobs, K. (2021). Equal peer e-mentoring for online graduate students: A case study and mediation model. *Mentoring & Tutoring*, *29*(5), 545–564. DOI: 10.1080/13611267.2021.1986799

Gamage, K. A. A., Perera, D. A. S., & Wijewardena, M. A. D. N. (2021). Mentoring and coaching as a learning technique in higher education: The impact of learning context on student engagement in online learning. *Education Sciences*, *11*(10), 574. DOI: 10.3390/educsci11100574

Guardian. D. (2023, June 11). Education may bid 'bye to VUCA, hi to BANI. *Daily Guardian*. https://dailyguardian.com.ph/education-may-bid-bye-to-vuca-hi-to-bani/

Gui, J., & Yanchinda, J. (2025). *Building counselor competencies for supporting students in a VUCA and BANI World: A training and coaching model*. EasyChair Preprint No: 15732

Haeger, H., & Fresquez, C. (2016). Mentoring for inclusion: The impact of mentoring on undergraduate researchers in the sciences. *CBE Life Sciences Education*, *15*(3), ar36. DOI: 10.1187/cbe.16-01-0016

Hasanov, Z., Antoniou, P., Suleymanov, E., & Garayev, V. (2021). The impact of behavioural, cognitive and emotional dimensions of student engagement on student learning: The case of Azerbaijani higher education institutions. *International Journal of Knowledge and Learning*, *14*(1), 10–38. DOI: 10.1504/IJKL.2021.115027

Hodges, L. C. (2020). Student engagement in active learning classes. In Mintzes, J. J., & Walter, E. M. (Eds.), *Active learning in college science* (pp. 27–41). Springer International Publishing., DOI: 10.1007/978-3-030-33600-4_3

Holcombe, A., & Wozniak, S. (2024). *Using AI to fuel engagement and active learning*. ASCD. https://www.ascd.org/el/articles/using-ai-to-fuel-engagement-and-active-learning

Hope, M. (2024). *Mastering leadership in a BANI World: Key strategies*. Https://Asymmetric.Pro/. https://asymmetric.pro/navigating-a-bani-world-strategies-for-future-leadership/

HRbrain.ai. (2024). *How mentors accelerate professional growth*. https://hrbrain.ai/blog/how-mentors-accelerate-professional-growth-and-success/

Inelmen, E. (2004). E-mentoring: A novel approach in the use of technology in education. *Information Technology Based Proceedings of the Fifth International Conference on Higher Education and Training, 2004. ITHET 2004.*, 183–186. DOI: 10.1109/ITHET.2004.1358160

Insala. (2019). *Traditional mentoring vs modern mentoring*. Insala. https://www.people360ai.com/blog/what-are-the-differences-between-traditional-and-modern-mentoring

Jan, S., & Mahboob, U. (2022). Online mentoring: Challenges and strategies. *Pakistan Journal of Medical Sciences*, *38*(8), 2272–2277. DOI: 10.12669/pjms.38.8.5804

Jang, Y., Choi, S., Kim, S., & Kim, H. (2023). The SNS-based e-mentoring and development of computational thinking for undergraduate students in an online course. *Journal of Educational Technology & Society*, *26*(2), 147–164. DOI: 10.30191/ETS.202304_26(2).0011

Jedaman, P., Pitchaya-Auckarakhun, M., Kasorn, K., & Kenaphoom, S. (2024). Sustainability education management scenarios under changes in the BANI era. In Wongmahesak, K., Wekke, I. S., & Suanpang, P. (Eds.), *Sustainable Development, Humanities, and Social Sciences for Society 5.0* (pp. 195–210). IGI Global Scientific Publishing., DOI: 10.4018/979-8-3693-7989-9.ch011

Jiracheewewong, E. (2023). Tools for transforming from VUCA World to BANI World: A case of educational sector in Thailand. *STOU Education Journal, 16*(2), Article 2.

Kaçar, I. G., & Baltacı, H. Ş. (2023). Designing a complementary e-mentoring program for pre-service ELT teachers: Online co-mentoring project. *Mentoring & Tutoring*, *31*(3), 352–380. DOI: 10.1080/13611267.2023.2202473

Kaufman, M. R., Levine, D., Casella, A., & DuBois, D. L. (2022). E-mentoring to address youth health: A systematic review. *Adolescent Research Review*, *7*(1), 63–78. DOI: 10.1007/s40894-021-00172-3

Kern, L., Harrison, J. R., Custer, B. E., & Mehta, P. D. (2019). Factors that enhance the quality of relationships between mentors and mentees during check & connect. *Behavioral Disorders*, *44*(3), 148–161. DOI: 10.1177/0198742918779791

Kuzu, A., Kahraman, M., & Odabaşı, H. F. (2012). Mentörlükte yeni bir yaklaşım: E-mentörlük [E-mentoring: A new approach in mentoring]. *Anadolu University Journal of Social Sciences, 12*(4), 173–183.

Lai, J. W., Qiu, W., Thway, M., Zhang, L., Jamil, N. B., Chit, L. S., Hwee, S. N. S., & Lim, F. S. (2024). *Leveraging process-action epistemic network analysis to illuminate student self-regulated learning with a socratic chatbot. EdArXiv*. https://doi.org/DOI: 10.35542/osf.io/b9vq6

Le, H., Janssen, J., & Wubbels, T. (2018). Collaborative learning practices: Teacher and student perceived obstacles to effective student collaboration. *Cambridge Journal of Education, 48*(1), 103–122. DOI: 10.1080/0305764X.2016.1259389

LeBlanc, P. J. (2018). Higher education in a VUCA World. *Change, 50*(3–4), 23–26. DOI: 10.1080/00091383.2018.1507370

Lindsay, S., & Cagliostro, E. (2020). A web-based intervention for youth with physical disabilities: Comparing the role of mentors in 12- and 4-week formats. *JMIR Pediatrics and Parenting, 3*(1), e15813. DOI: 10.2196/15813

Liu, Y., Ma, S., & Chen, Y. (2024). The impacts of learning motivation, emotional engagement and psychological capital on academic performance in a blended learning university course. *Frontiers in Psychology, 15*, 1357936. DOI: 10.3389/fpsyg.2024.1357936

Luna Scott, C. (2015). *The futures of learning 2: What kind of learning for the 21st century?* [Education Research and Foresight • Working Papers]. United Nations — Educational, Cultural and Scientific Organization. https://repositorio.minedu.gob.pe/handle/20.500.12799/3709

Maia, M. C., Figueiredo, J. C. A. D., & Serey, D. (2019). Online student engagement: A case study in teaching of programming. *Anais Do XXX Simpósio Brasileiro de Informática Na Educação (SBIE 2019), 30*, 51–60. DOI: 10.5753/cbie.sbie.2019.51

Malkovich, A. (n.d.). *Mentorship in the digital age: How technology can enhance the mentor-mentee relationship*. PushFar. Retrieved February 16, 2025, from https://www.pushfar.com/article/mentorship-in-the-digital-age-how-technology-can-enhance-the-mentor-mentee-relationship/

Maričić, M., & Lavicza, Z. (2024). Enhancing student engagement through emerging technology integration in STEAM learning environments. *Education and Information Technologies, 29*(17), 23361–23389. DOI: 10.1007/s10639-024-12710-2

McAleer, D., & Bangert, A. (2011). Professional growth through online mentoring: A study of mathematics mentor teachers. *Journal of Educational Computing Research, 44*(1), 83–115. DOI: 10.2190/EC.44.1.e

McLoughlin, C., Brady, J., Lee, M. J. W., & Russell, R. (2007). *Peer-to-peer: An e-mentoring approach to developing community, mutual engagement and professional identity for pre-service teachers.* http://bibliotecadigital.mineduc.cl//handle/20.500.12365/17803

Melnick, M. (2022). *The BANI world is a new world that has replaced the world of VUCA.* https://hackernoon.com/the-bani-world-is-a-new-world-that-has-replaced-the-world-of-vuca

Merrick, L. (2016, May 18). Mentoring in a VUCA World—'Future proofing' your people. *Coach Mentoring.* https://www.coachmentoring.co.uk/blog/2016/05/mentoring-in-a-vuca-world/

Merrick, L. (2025, January 17). Mentoring in a BANI World—Five free mentor CPD sessions. *Coach Mentoring.* https://www.coachmentoring.co.uk/blog/2025/01/mentoring-in-a-bani-world-five-free-mentor-cpd-sessions/

Meyer, K. A. (2014). Student engagement in online learning: What works and why. *ASHE Higher Education Report, 40*(6), 1–114. DOI: 10.1002/aehe.20018

Morellet, P. (2025, January 23). *Coaching in a BANI World: Building resilience amid uncertainty and complexity.* Speexx. https://www.speexx.com/speexx-blog/coaching-in-a-bani-world/

Mullen, C. A. (2021). Online doctoral mentoring in a pandemic: Help or hindrance to academic progress on dissertations? *International Journal of Mentoring and Coaching in Education, 10*(2), 139–157. DOI: 10.1108/IJMCE-06-2020-0029

Narot, P., & Kiettikunwong, N. (2023). *Interdisciplinary perspectives on special and inclusive education in a volatile, uncertain, complex & ambiguous (VUCA) world.* Emerald Publishing. https://bookstore.emerald.com/interdisciplinary-perspectives-on-special-and-inclusive-education-in-a-volatile-uncertain-complex-ambiguous-vuca-world.html

Nataliia, H., & Olena, M. (2023). The key administrative competencies of managers required for company development in the BANI World. *Economics - Innovative and Economics Research Journal, 11*(1), Article 1. DOI: 10.2478/eoik-2023-0012

Neely, A. R., Cotton, J., & Neely, A. D. (2017). E-mentoring: A model and review of the literature. *AIS Transactions on Human-Computer Interaction, 9*(3), 220–242. DOI: 10.17705/1thci.00096

Nguyen, A., Kremantzis, M., Essien, A., Petrounias, I., & Hosseini, S. (2024). Enhancing student engagement through artificial intelligence (AI): Understanding the basics, opportunities, and challenges. *Journal of University Teaching & Learning Practice, 21*(06), 06. Advance online publication. DOI: 10.53761/caraaq92

Nguyen, L., Tomy, S., & Pardede, E. (2024). Enhancing collaborative learning and e-mentoring in a smart education system in higher education. *Computers, 13*(1), 28. DOI: 10.3390/computers13010028

Officer, M. J. Chief Product. (n.d.). *From theory to practice: Designing real-world learning experiences*. Retrieved February 22, 2025, from https://blog.learnamp.com/from-theory-to-practice-designing-real-world-learning-experiences

Omar, N. D., Hassan, H., & Atan, H. (2012). Student engagement in online learning: Learners attitude toward e-mentoring. *Procedia: Social and Behavioral Sciences, 67*, 464–475. DOI: 10.1016/j.sbspro.2012.11.351

Panthalookaran, V. (2022). Education in a VUCA-driven world: Salient features of an entrepreneurial pedagogy. *Higher Education for the Future, 9*(2), 234–249. DOI: 10.1177/23476311221108808

Radlick, R. L., Mirkovic, J., Przedpelska, S., Halvorsen Brendmo, E., & Gammon, D. (2020). Experiences and needs of multicultural youth and their mentors, and implications for digital mentoring platforms: Qualitative exploratory study. *JMIR Formative Research, 4*(2), e15500. DOI: 10.2196/15500

Rankings, G. U. (n.d.). *Trends in Higher Education*. GU Rankings, London. Retrieved April 21, 2025, from https://www.globaluniversityrankings.com/trends-in-higher-education

Ravinutala, S. (2023, June 14). *How can coaching help in maneuvering through the BANI World?* https://uexcelerate.com/how-can-coaching-help-in-maneuvering-through-the-bani-world/

Reeves, T., & Reeves, P. (2015). Educational Technology Research in a VUCA World. *Educational Technology Archive*. https://www.semanticscholar.org/paper/Educational-Technology-Research-in-a-VUCA-World.-Reeves-Reeves/94de9dca801e3707e5d21e40ec8eff343a31b0c6

Rockinson-Szapkiw, A., & Wendt, J. L. (2021). The benefits and challenges of a blended peer mentoring program for women peer mentors in science, technology, engineering and mathematics (STEM). *International Journal of Mentoring and Coaching in Education, 10*(1), 1–16. DOI: 10.1108/IJMCE-03-2020-0011

Rowan, L. M., & Jin, Y. (2022). Conversations that count in online student engagement – A case study. *ASCILITE Publications*, e22162. DOI: 10.14742/apubs.2022.162

Rowland, K. N. (2012). E-mentoring: An innovative twist to traditional mentoring. *Journal of Technology Management & Innovation*, *7*(1), 228–237. DOI: 10.4067/S0718-27242012000100015

Ruiz-Mallén, I., Heras, M., & Berrens, K. (2021). Responsible research and innovation in science education: Insights from evaluating the impact of using digital media and arts-based methods on RRI values. *Research in Science & Technological Education*, *39*(3), 263–284. DOI: 10.1080/02635143.2020.1763289

Sato, S. N., Moreno, E. C., Rubio-Zarapuz, A., Dalamitros, A. A., Yañez-Sepulveda, R., Tornero-Aguilera, J. F., & Clemente-Suárez, V. J. (2023). Navigating the new normal: Adapting online and distance learning in the post-pandemic era. *Education Sciences*, *14*(1), 19. DOI: 10.3390/educsci14010019

Sheridan, K., & Kelly, M. A. (2010). The indicators of instructor presence that are important to students in online courses. *Journal of Online Learning and Teaching*, *6*(4), 767–779.

Solutions, L. (2023). *Student engagement strategies: Leveraging the power of AI.* https://www.lambdasolutions.net/en/blog/student-engagement-strategies-leveraging-the-power-of-ai

Vatansever, Ç., & Ceylan, F. (2021). Gençlerin kariyer gelişiminde e-mentorluk sürecinin etkisi [Effect of the e-mentoring process on youth's career development]. *Istanbul Management Journal*, *90*(90), 91–114. DOI: 10.26650/imj.2021.90.0005

Wallis, J. A. M., Riddell, J. K., Smith, C., Silvertown, J., & Pepler, D. J. (2015). Investigating patterns of participation and conversation content in an online mentoring program for northern canadian youth. *Mentoring & Tutoring*, *23*(3), 228–247. DOI: 10.1080/13611267.2015.1072395

Willbanks, J. E. (2014). E-mentoring: Examining the feasibility of electronic, online, or distance mentoring. *Journal of Higher Education Theory and Practice*, *14*(5), 24–28.

Williams, S. L., & Kim, J. (Jin-H. (2011). E-mentoring in online course projects: Description of an e-mentoring scheme. *International Journal of Evidence Based Coaching and Mentoring*, *9*(2), 80–95.

Wright, K., Levine, D. K., Salcido, M., Garringer, M., Almendra, T., Bazell, A., & Kaufman, M. R. (2024). From crisis to capacity: Lessons learned from youth e-mentoring during the COVID-19 pandemic. *Computers in Human Behavior Reports, 14*, 100400. DOI: 10.1016/j.chbr.2024.100400

Xerri, M. J., Radford, K., & Shacklock, K. (2018). Student engagement in academic activities: A social support perspective. *Higher Education, 75*(4), 589–605. DOI: 10.1007/s10734-017-0162-9

Xu, S., Liu, P., Yang, Z., Cui, Z., & Yang, F. (2021). How does mentoring affect the creative performance of mentors: The role of personal learning and career stage. *Frontiers in Psychology, 12*, 741139. DOI: 10.3389/fpsyg.2021.741139

Yaccob, N. S., Yunus, M. M., & John, D. S. (2023). Global education movement: English as a second language teachers' perceptions of integrating volatility, uncertainty, complexity, and ambiguity elements in lessons. *Frontiers in Psychology, 14*, 1007970. DOI: 10.3389/fpsyg.2023.1007970

Zadorina, O., Burchak, L., Panas, O., Ardelian, O., & Apalat, H. (2023). Shaping the competencies of the future: The importance of developing soft skills in higher education. *Cadernos de Educação Tecnologia e Sociedade, 16*(2), 361–371. DOI: 10.14571/brajets.v16.n2.361-371

Zakharov, Y. (2022). Comparative analysis of approaches to world concepts: SPOD Society, VUCA Society and BANI Society. *Society and Economy, 64*(64), 149–158. DOI: 10.26565/2524-2547-2022-64-13

Zhang, Z. (2022). Toward the role of teacher empathy in students' engagement in English language classes. *Frontiers in Psychology, 13*, 880935. DOI: 10.3389/fpsyg.2022.880935

Chapter 3
Digital Twins as a Method of Learning in a Fragile World

Rebrov Egor D.
https://orcid.org/0009-0006-1702-2937
Moscow State Technical University of Civil Aviation, Russia

Guleshov Alexey V.
https://orcid.org/0009-0002-4629-5388
Moscow State Technical University of Civil Aviation, Russia

Romancheva Nina I.
https://orcid.org/0000-0003-0522-3855
Moscow State Technical University of Civil Aviation, Russia

ABSTRACT

New technologies in the VUCA and BANI worlds are aimed at the variability of systems and their complexity. The main tasks of such technologies are to be flexible and adapt to a changing situation in the world, but at the same time provide a person with all information at an intuitive level, which in turn can positively affect his mental state. This chapter discusses the use of digital twins as a teaching method according to training simulators in nuclear energy, since it is one of such complex processes, and the integration of technologies such as virtual reality and digital twins is a priority due to the conditions of the BANI world.

DOI: 10.4018/979-8-3373-7917-3.ch003

INTRODUCTION: PROBLEMS AND RESEARCH OBJECTIVES

The modern world is characterized by a high degree of uncertainty and rapid change, which is reflected in the concepts of VUCA (volatility, uncertainty, complexity and ambiguity) and BANI (fragility, anxiety, nonlinearity and incomprehensibility). In such conditions, traditional methods of forecasting and training become insufficient and lose their effectiveness. New approaches are needed that will not only adapt to the changing environment, but also safely test various scenarios.

This is particularly relevant in areas with high safety and reliability requirements, such as nuclear power, aviation, and medicine, where mistakes can have disastrous consequences. The world of VUCA requires organizations to be flexible in decision-making and able to respond quickly to challenges, while the world of BANI emphasizes the importance of dealing with fragility of systems, anxiety and complexity of processes.

In such conditions, educational processes should be focused not only on the assimilation of knowledge, but also on the development of skills for dealing with uncertainty, analyzing complex systems and adapting to sudden changes. These are the essential skills required to thrive in a VUCA world (Salun & Zaslavska, 2024). Modern technologies, including digital twins and virtual reality, are becoming key tools for minimizing uncertainty and increasing the resilience of specialists to complex and unstable situations.

The purpose of this paper is to show the role of digital twins as a learning method capable of providing safe modeling of real—world processes and decision-making training in conditions of instability. The research is aimed at developing approaches to the effective use of digital doubles for training specialists working in the VUCA and BANI worlds, with an emphasis on improving adaptability and crisis preparedness.

The chapter focuses on practical recommendations for industry trainers who train specialists in critical areas. As an illustration, an example of using a digital twin based on a simulator for an operator of a control system unit at a nuclear power plant (NPP) is given, demonstrating the possibilities of safe modeling and testing actions in complex emergency scenarios.

THE WORLDS OF VUCA AND BANI: CHALLENGES AND UNCERTAINTY IN EDUCATION

Modern reality is characterized by increasing instability, uncertainty and complexity, which is reflected in the concepts of VUCA and BANI. The term VUCA, which originated in the military sphere at the end of the 20th century, refers to the volatility, uncertainty, complexity and ambiguity of modern conditions. Volatility is characterized by rapid and unpredictable changes that require organizations and individuals to be highly flexible and prompt in making decisions. Uncertainty manifests itself in a lack of information about current events and their future development, which makes forecasting and planning difficult. Complexity: globalization, disappearing cultural differences in the developed countries and extension of IT-technologies utilization in all areas of life made the world interconnected, multi-angled, multi-tooled and complicated which makes it almost impossible to understand in full scope (ability to build clear structures) (Bushuyev, Piliuhina & Chetin, 2023). Finally, ambiguity is expressed in the absence of clear interpretations and unambiguous answers to emerging questions, which further complicates the decision-making process.

The development of the BANI concept was a reaction to new challenges that are not fully covered by the VUCA model. The world of BANI highlights the fragility (brittle) of systems, their susceptibility to destruction even with small impacts, anxiety as an emotional reaction to instability and uncertainty, nonlinear processes in which small changes can lead to large—scale and unexpected consequences, as well as incomprehensible - the inability to fully understand and explain many events and processes. Thus, BANI highlights the vulnerability of modern systems and the need to develop new approaches to management and training that can take these features into account.

These changes significantly affect the sphere of education and professional training. Traditional teaching methods, focused on the assimilation of fixed knowledge and standard algorithms of actions, are insufficient to train specialists who are able to function effectively in a rapidly changing and uncertain environment. In such conditions, the development of flexibility of thinking, critical analysis, resistance to stress and the ability to work with complex and ambiguous situations become key. Educational programs should be transformed to ensure the formation of these competencies and to prepare specialists for the continuous process of adaptation and decision-making in conditions of instability.

Modern digital technologies play an important role in this transformation. In particular, digital doubles and virtual reality provide unique opportunities for modeling complex processes and creating a safe environment for decision-making training in crisis situations. These tools contribute not only to reducing learning risks, but also to developing the skills necessary for successful work in the VUCA and BANI worlds, which makes them indispensable components of the modern educational process.

MODERN EDUCATION AND PEDAGOGICAL THEORIES

In the context of the rapid changes characteristic of the VUKA and BANI worlds, traditional educational models focused on the assimilation of structured information and template solutions are insufficient to form flexible and adaptive specialists. As a result, a person ceases to develop the ability to analyze, identify causation, place accents (Dureev & Dureeva, 2023). Modern challenges require a revision of learning approaches, an emphasis on the development of critical thinking, the ability to make independent decisions, risk management and adaptation to uncertainty.

One of the key responses to these challenges is the concept of lifelong learning, which involves constant updating of knowledge and skills throughout life. This approach is reflected in the practice of micro-education, personalized educational trajectories and digital learning formats that allow adapting the educational process to the individual needs of the student and the requirements of a rapidly changing market.

As early as 1972, Edgar Faure's report "Learning to Be" emphasized the role of technology in making education accessible and flexible. Today, this idea is more relevant than ever: the technological transformation of education sets pedagogical science the task of understanding new learning formats not only in terms of the effectiveness of knowledge transfer, but also in terms of the formation of stable cognitive and emotional skills.

Despite the declared orientation of educational models towards the development of independence, in real practice, education often boils down to the reproductive assimilation of pre-prepared material. This model reduces the ability to analyze, critically comprehend and synthesize new information, which is especially critical in conditions of instability and high uncertainty.

Next, modern pedagogical theories and approaches are considered that can provide the necessary level of adaptability, stability and independence of students. Special attention is paid to constructivism, cognitive theories, adult education, as well as the role of emotional intelligence and psychological stability in the educational processes of modern times.

CONSTRUCTIVISM AND ADULT EDUCATION

One of the priorities of modern education is the formation of students' flexibility of thinking — the ability to quickly rebuild cognitive circuits, perceive new information and adapt to changing conditions. These qualities are becoming particularly important in the context of accelerating digitalization and automation, which are fundamentally transforming the requirements for professional skills. In these circumstances, the development of individual potential as a key resource for sustainable personal and professional growth becomes particularly relevant.

The constructivist approach in adult education offers an effective methodological framework for the implementation of this task. It is based on the idea that knowledge is not transmitted in a ready-made form, but is formed by the student himself through active interaction with information, the environment and other participants in the educational process. This contributes not only to deeper learning of the material, but also to the formation of the ability to critically comprehend reality, analyze complex relationships and make non-standard decisions.

Practice-oriented techniques, such as project-based learning and case studies, make it possible to simulate real professional situations in which the student independently develops a solution by integrating theoretical knowledge and practical experience. In project activities, attention is focused on solving urgent tasks, which requires not only cognitive efforts from participants, but also initiative, creativity, the ability to work in a team and adapt to unpredictable changes in conditions.

The case method contributes to the formation of critical thinking and analytical abilities, as it requires the analysis of situations that do not have an unambiguous solution. This approach develops students' flexibility in reasoning and the ability to justify alternative behavioral strategies — important elements of individual potential in the digital age.

Modern interactive formats, especially immersive technologies (virtual, augmented and mixed reality), enhance the effect of engagement and immersion in the educational process. These technologies provide the opportunity to model complex professional scenarios, creating conditions for the formation of skills of adaptation and self-determination, which are so necessary in conditions of instability and constant transformation of the labor market.

Thus, the constructivist model of adult education not only meets the challenges of flexible thinking and adaptability, but also creates the basis for the development of individual potential. In conditions when routine tasks are transferred to automated systems, it is the ability to self-study, critical thinking, knowledge integration and innovative thinking that becomes the determining factor of a person's professional viability and sustainability in the digital world.

COGNITIVE APPROACHES AND CRITICAL THINKING

Cognitive learning theories focus on the processes of perception, processing, and storage of information in the learner's mind. Unlike behavioral models, which focus on external stimuli and reactions, cognitive approaches consider learning as an internal process of active comprehension and organization of knowledge. This includes the development of cognitive skills such as attention, memory, perception, analysis and synthesis of information.

Of particular importance in the context of modern education is the development of critical thinking — the ability to objectively evaluate information, identify logical errors, form reasoned judgments and make informed decisions. Critical thinking allows students not only to assimilate facts, but also to analyze them, question established beliefs and develop their own views based on evidence.

In the world of VUCA and BANI, where information is often contradictory and incomplete, and situations are complex and ambiguous, the ability to think critically becomes a key skill. This is especially important for professionals working in high-risk fields such as nuclear power, medicine, and aviation, where mistakes can have serious consequences. Cognitive approaches in teaching contribute to the formation of students' ability to analyze systems, adapt to new conditions and make decisions in conditions of uncertainty.

Thus, the integration of cognitive methods and the development of critical thinking in educational programs ensures an improvement in the quality of training specialists who are ready to act effectively in the complex and rapidly changing conditions of the modern world.

PSYCHOLOGICAL RESILIENCE AND EMOTIONAL LEARNING

Psychological stability, as a person's ability to maintain internal stability, motivation, and functionality in the face of uncertainty, stress, and crisis changes, is becoming an integral part of the modern educational paradigm. Against the background of global transformations caused by digitalization, automation and social instability, learning should go beyond knowledge transfer, focusing on the formation of sustainable behavioral strategies, emotional self-regulation and analytical thinking.

Emotional learning, aimed at developing a conscious attitude to one's feelings and an emotional response to external stimuli, forms the student's ability to effectively interact in a team, constructively resolve conflicts and maintain productivity in conditions of increased pressure. These skills are especially important in professional activities involving a high degree of responsibility, where mistakes can have critical consequences.

The modern educational environment focuses not only on the formation of emotional stability, but also on the development of critical thinking, a skill necessary for effective navigation in an information—saturated world. The information revolution has led to the fact that access to data is no longer a limiting factor. Today, it is important not just to have information, but to be able to interpret it, identify sources of manipulation, recognize unreliable or distorted data, and build reasoned judgments. Critical thinking, as the cognitive basis of resilience, allows a person to resist emotional pressure, information distortions and act rationally in difficult circumstances.

In response to these challenges, educational programs increasingly include tasks that require students to conduct independent research, analyze conflicting information, and choose strategies in conditions of uncertainty. This not only helps to develop intellectual flexibility, but also strengthens emotional stability, as the student learns ways of making decisions based on evidence, rather than on fear or anxiety reactions.

It should be noted that the development of critical thinking and psychological stability is closely related to the unfolding of individual potential. A personality-oriented approach that recognizes the uniqueness of each student makes it possible to build flexible educational routes that promote self-confidence, the ability to adapt and act in a rapidly changing environment. In such conditions, the emphasis is on developing the ability to introspect, the formation of internal motivation and a conscious approach to learning.

Additional opportunities are provided by the use of immersive technologies — virtual and augmented reality, as well as simulators that allow students to experience simulated critical situations. Such environments form behavioral and emotional readiness for real challenges by integrating cognitive, emotional, and behavioral responses into a single learning context. The learning process has become more interactive and accessible, allowing students to learn in a more flexible way (Erlangga, Perdamen, Ternasih & Umar, 2024).

Thus, in the context of digital transformation and instability of the world, the key objectives of education are the development of psychological stability, emotional intelligence and critical thinking as interrelated components of a holistic personality. Their formation determines not only academic and professional success, but also the individual's ability to maintain subjectivity and autonomy in the conditions of turbulence inherent in the era of VUCA and BANI.

THE ROLE OF MODERN TECHNOLOGIES IN THE TRANSFORMATION OF EDUCATIONAL PROCESSES

One of the key trends in the transformation of education has been the introduction of digital twins and virtual reality, which make it possible to create immersive educational environments. These technologies open up new opportunities for modeling complex processes, developing professional skills and increasing the level of involvement of students. Digital twins were developed for use in the engineering industry, but they also quickly found application in other fields, including education. Education virtualization is an actively developing process and the use of digital twins is one of the stages of its development (Hohlov & Korobova, 2024). This technology allows you to recreate real objects and systems in digital format, which makes learning practice-oriented and adaptive. The virtual environment, in turn, provides a full immersive effect, allowing you to learn disciplines not only theoretically, but also in practice, interacting with virtual objects in an interactive environment.

The use of these technologies is especially important in areas that require a deep understanding of processes and work with high-precision equipment, creating the conditions for adaptive and personalized learning.

In this work, the digital twin will be considered as a digital model of material objects using virtual reality to fully simulate work with an object that is a computer-generated representation of tangible (object, system, service, product etc.) or intangible (world, environment etc.) assisted by the use of digital technologies that intends to provide fundamental improvement and operational revolution to an entity (a business organizations, an industry or society) by exchanging the information (Awasthi, Padwekar & Misra, 2025). The virtual environment is of particular value in educational institutions that train specialists, where access to existing equipment may be limited.

The use of VR in learning also contributes to more flexible adaptation to rapidly changing world conditions. Virtual reality allows you to instantly respond to change and simulate new scenarios. This makes training more responsive and resilient to external challenges.

In a world where traditional teaching methods are becoming insufficient to solve problems associated with increased complexity of processes and technological instability, VR simulations enable students not only to assimilate knowledge, but also to actively apply it in practice, simulating situations that could arise in real work.

With this approach, the basic concepts of digital twins can be distinguished, which provide interactive and immersive educational environments, allowing students to model and train their skills in real and virtual conditions:

- Interactive digital twin
- Digital twin for personalized learning
- Simulation-based digital twins
- Digital life cycle twin

An interactive digital twin is actively used to create simulators that allow students not only to observe objects and processes, but also to interact with them in real time. This approach allows you to train behavior in extreme situations without human risks and harm to the environment, material objects and the living conditions of people. This concept focuses on the user's interaction with the object or system being modeled at the time of execution. Models based on this concept can be used in training the management of application systems, improving decision-making skills and analyzing their consequences in conditions close to real.

A digital twin for personalized learning tracks a student's individual characteristics, including their learning style, progress, weaknesses and strengths. These models can adapt the learning process depending on the level of knowledge and preferences of each student.

Simulation digital twins allow realistic processes and systems to be simulated, creating hands-on training for learners. Such models allow students to interact with simulations online, creating an atmosphere as close as possible to the real one. Unlike the concept of interactive digital twins, this concept focuses on imitating the processes themselves. Models recreated in this way allow replicating complex or emergency situations for direct learning.

Digital twins covering the full life cycle of an object allow trainees to see how different stages - from design to disposal - affect the functionality of a system or device. This helps students better understand the life cycle dynamics of objects in aspects such as engineering, project management and environmental sustainability.

EDUCATION IN VIRTUAL WORLDS: DIGITAL TWINS AS A RESPONSE TO VUCA AND BANI CHALLENGES

In the context of the functioning of modern organizations, the economy and society as a whole, VUCA and BANI, modern technologies, including digital twins, are becoming an important tool for creating educational models that can effectively cope with the limitations and challenges inherent in traditional methods of education.

Modeling of rapidly changing scenarios, where events and characteristics over time can change unpredictably and at high speed, provides an opportunity to train specialists in making effective decisions in conditions when the external environment undergoes global changes. Virtual reality digital twins include many variables, help students develop the ability to analyze and make decisions in uncertain condition.

The advantages of using digital twins for modeling complex systems, where many interconnected factors affect the result, are especially clear. In such simulations, students can study the interaction of various elements of the system, see the consequences of their decisions and understand how small changes in one of the processes can lead to serious consequences. The learning model under consideration helps you develop skills in working with multi-level systems and find optimal solutions in conditions of high complexity.

A virtual copy of an object is configurable and more manageable than a real physical object. The ability to set specific initial system parameters allows you to simulate situations with multi-valued and uncertain outcomes, where it is not always clear which solution will be effective.

The use of digital twins effectively reduces the anxiety and tension that comes with encountering unforeseen or critical situations. By recreating realistic scenarios in a secure virtual environment, trainees can repeatedly hone their work responsibilities and knowledge without worrying about the consequences of their mistakes.

For these reasons, the integration of digital twins is becoming a priority in a BANI world where instability and unpredictability require flexibility and responsiveness. Nevertheless, this process faces a number of significant challenges that make their successful implementation and use difficult.

One major difficulty is the difficulty of adequately modeling and processing the data. Complex systems that digital twins try to recreate often defy accurate and unambiguous description. In such conditions, digital twins can produce incomplete or even contradictory results, which reduces their usefulness.

In addition, there is a problem of the resistance of digital twins to unpredictable changes. Often, models are developed for specific purposes without the task of its adaptability to unpredictable changes. Under conditions of constant changes, this can lead to failure or loss of relevance of the model.

To solve these problems, consider the difficulty of recreating some processes. So, to some subprocesses, it is enough to use statistical data. This approach allows you to simplify the model, save resources and get fairly accurate results faster when there is no complete information about the process or when detailed modeling is too complicated.

When creating a model, you should take into account the possibility of cardinal changes in the system in advance. Models should be sufficiently flexible and adaptable to modern trends of technological development.

To solve these problems, you should adhere to certain principles of building digital twins. The development of digital twins for modeling complex technical systems requires taking into account many factors that ensure the reliability, adaptability and stability of the created model. The creation of a digital twin is based on the need to reflect the physical object, its dynamics and interaction with the external environment as accurately as possible, which allows not only analyzing the behavior of the system, but also predicting possible scenarios for its functioning.

Among the basic principles for constructing such digital twins are the following:

1. Reliability
2. Reality and detail
3. Modularity

The principle of reliability in the construction of digital twins is the need to accurately reproduce the characteristics of a real object or system solely on the basis of mathematical modeling. Since physical equipment are not used in this case, reliability is ensured through the development and validation of models that are as close as possible to the actual processes.

A key aspect is the construction of simulation models that are able to correctly describe the dynamics of the system in various operating modes, including normal and emergency situations. For this, complex mathematical dependencies are used that describe physical and technological processes. The accuracy of such models is determined by comparing their calculations with known computational data, experimental results, or historical information about the functioning of similar systems.

Additionally, validation of the simulation model itself is an important element. Digital twin validation is the process of confirming that a model accurately reflects the behavior and characteristics of a real object or system (Kuzmin, 2024). One approach is to create reference calculations for individual elements of the model and then compare them with the simulation results. If deviations are detected, the model parameters and calculation algorithms are refined, which makes it possible to increase the degree of its compliance with the real object.

The reliability of digital twins in this approach is ensured through the comprehensive development of mathematical models, their testing on reference scenarios and constant refinement of calculation algorithms, which makes them a reliable tool for analyzing and optimizing complex technical systems. It should be borne in mind that in this case the reliability of digital twins should be determined precisely at the stage of development, and not operation. Since the system does not receive data from physical sensors, it is important to ensure in advance that the mathematical models used correctly reflect the physical object.

At the stage of operation, the digital twin should already be calibrated system and ready for use in modeling and training. If deviations are detected during operation, this indicates either the need to adjust the model or its limited applicability in specific conditions.

The principle of correspondence to the actual system says that the digital twin must recreate the key characteristics of the object, including its physical parameters, workflows and dynamics of change. For this, mathematical models are developed that describe the interaction of system elements, control processes and the influence of external factors.

However, in some cases, full compliance with the actual system with maximum detail is not required. Simulating situations such as disconnection or board damage can not only be unnecessary, but also interfere with effective learning. Instead of focusing on physical accuracy, you should focus on shaping the correct behavior of the system in emergency situations.

When building a digital twin for learning, it is important to find the optimal balance between the accuracy of the model and its objective function in order to focus on those aspects that are most important for an adequate response of the system in a critical situation. At the same time, it is not necessary to fully reproduce all physical and functional details of the object in order to simulate unlikely events that do not significantly affect the final behavior of the system in a critical situation.

For effective modeling and learning using a digital twin, it is important to abstract from stochastic interference that can affect system performance, such as noise, weak contact, disconnection, and other similar factors. These elements, while relevant to the actual system, can unnecessarily complicate the model in the context of a digital twin, distracting from the main goals of decision-making in an environment of instability and developing critical thinking in staff.

The principle of modularity implies that a digital twin must be built as a collection of separate, interdependent components, each of which performs its own function. Modularity allows you to create a flexible structure in which individual parts can be replaced, updated or improved without the need to redesign the entire model. This principle is especially important when scaling a system, adding new features, or integrating with other systems.

Each module in the digital twin is responsible for a specific task, be it collecting data from sensors, processing this data, or performing specific calculations related to the model. Splitting the system into separate modules also facilitates testing and modernization. Each component can be tested independently, accelerating development and increasing reliability.

In the event of a failure in one of the modules, the system can continue to function due to the autonomy of individual parts. If, for example, one component fails, the other modules continue to operate, ensuring the continuity of the process.

In the conditions of VUCA and BANI worlds, the principle of modularity becomes especially important. In such an unstable and rapidly changing environment, systems must be as flexible and adaptable as possible without incurring a significant cost to recycle the entire solution.

In addition, the ability to independently modify individual components helps to respond faster to surprises arising in conditions of high uncertainty and instability. This gives the system stability and allows you to effectively manage the risks that inevitably arise in the conditions of VUCA and BANI worlds.

COMPARATIVE ANALYSIS OF DIGITAL TWINS WITH ALTERNATIVE INTENSIVE LEARNING TECHNOLOGIES AND INTERACTION OPPORTUNITIES

Considering digital twins as an effective learning tool in modern conditions, it is also important to compare them with other key technologies of immersive learning — virtual reality and artificial intelligence. Each of these technologies has unique capabilities and limitations that affect their application in educational processes, especially in rapidly changing and unstable conditions.

Virtual reality

Digital twins and virtual reality (VR) are often viewed as complementary immersive learning technologies, but there are fundamental differences between them that affect their educational application.

A digital twin is a highly accurate digital copy of a real object, system, or process that includes not only a visual representation, but also modeling of dynamic behavior and interactions in real time. Due to this, digital twins allow students not only to "immerse themselves" in a virtual environment, but to interact with real processes and change model parameters, observing their impact on the system. This approach is especially useful in training operators of complex technical systems, where it is necessary to analyze the consequences of various actions and develop reactions to non-standard situations.

Virtual reality, in turn, creates a completely immersive space in which the user can find himself "inside" the simulated environment. This provides unique opportunities for the development of spatial perception, motor skills and emotional engagement, which significantly improves the quality of practical skills and motivation for learning. VR is well suited for practicing procedures, simulating interaction with equipment, and creating scenarios for extreme situations in which it is important not only to understand, but also to feel the specifics of what is happening.

However, VR itself is often limited to visualization and simulation of predefined scenarios, without deep modeling of complex dynamic processes that can be observed in digital counterparts. In turn, digital doubles can integrate VR components, using virtual environments to enhance perception and interaction with the model.

In addition, the combined use of virtual reality and digital doubles opens up unique synergistic opportunities. The digital twin provides an accurate and dynamic model of an object or process, while virtual reality provides an immersive environment for interacting with this model. This combination allows trainees not only to visualize and analyze complex systems in real time, but also to fully immerse themselves in the context of their functioning, which contributes to a deeper understanding and consolidation of skills. This integrated approach is especially valuable for working out complex scenarios and adapting to unexpected situations typical of unstable and rapidly changing worlds.

In rapidly changing and unstable worlds, where adaptability and prompt decision-making are required, the combination of digital doubles and virtual reality allows for both a deep understanding of processes and direct interaction experience in an immersive environment. This combined use of technology significantly increases the effectiveness of training and preparation of specialists to work in conditions of uncertainty.

Artificial intelligence

If virtual reality primarily provides the effect of full immersion and visual interaction with the learning environment, and the digital twin reproduces the dynamic behavior of real objects and systems for practical modeling, then artificial intelligence (AI) acts as an intelligent component that significantly expands the possibilities of adaptive learning.

AI does not work with visualization or a physical model, but with data analysis, forecasting, and decision-making. It is able to process large amounts of information about a student's actions, identify their strengths and weaknesses, predict possible errors, and create personalized learning scenarios. This allows you to make learning not only interactive, but also maximally adapted to the individual needs of the user.

In combination with a digital twin, AI plays the role of a "smart mentor" who not only provides an environment for practicing skills, but also actively manages the learning process — adjusts the complexity of tasks, offers recommendations, and generates new non-standard situations for developing critical thinking and stress tolerance. This integration is especially valuable in the VUCA and BANI worlds, where instability and uncertainty require high flexibility and the ability to adapt quickly to new challenges.

Thus, the digital twin creates a realistic and functional model of the object, and the AI provides intellectual support, making learning dynamic, personalized and more effective in the complex, unstable conditions of the modern world. Together, these technologies form a powerful platform that surpasses the capabilities of a single virtual reality, providing not only visual immersion, but also deep intellectual analysis and adaptation of the learning process.

VIRTUAL TRAINING SIMULATOR AS AN ADAPTIVE LEARNING TOOL

Earlier it was mentioned the use of such a concept of digital doubles as interactive, for interacting with objects and processes online. In particular, this practice is used in simulators. A specialized tool designed for training and training, which simulates actual processes, situations or systems, allows the student to develop the necessary skills in a safe and controlled environment. Simulators are used to train specialists in various fields, aviation and medicine to industry and military affairs.

Modern simulators can be both physical and virtual. Unlike traditional training methods, simulators make it possible to interact with simulated processes at the current moment, which speeds up the learning process and contributes to better assimilation of the material.

Thanks to the integration of modern technologies, virtual simulators offer a much higher degree of interaction with simulated processes. These technologies allow not only to simulate real conditions, but also to reproduce complex, multivariable scenarios, including those that are difficult or impossible to recreate on a physical copy.

Digital twins as simulators are not just static models, but dynamically changing replicas that accurately reflect all the processes that occur on a real object, including reactions to external influences and changes in parameters. This allows users to interact with the system at the current time, observing the reaction of the object to its actions.

In combination with virtual simulators, digital twins allow you to create the most accurate and adaptable model that can interact with the student directly in the process, providing a high level of realism and immersion.

Virtual simulators also provide trainees with experience in stressful situations, modeling critical moments when important decisions need to be made with limited information or in time-poor conditions. These workouts develop resilience to stress and help learn to stay calm during emergency moments, which is becoming an important skill in today's world. In addition, such simulators are trained to work with uncertainty, where information may be incomplete or inconsistent, and to make decisions based on incomplete data.

One of the striking examples of the use of virtual simulators using digital twins is simulators for training operators of control units at nuclear power plants (MCR), which are responsible for maintaining stable and safe reactor operation, monitoring process parameters, timely response to deviations and preventing emergencies. The simulators under consideration make it possible to create valid conditions for working with high-tech potentially dangerous systems, where accuracy and speed of making decisions are critical.

Virtual simulators for MCR, using digital twins, provide not only accurate reconstruction of technical characteristics and parameters of actual systems, but also imitation of various emergency situations, such as equipment failure or exposure to external factors. This allows operators not only to debug their skills in typical situations to perfection, but also to prepare for unlikely situations at a nuclear power plant, improving their ability to react quickly and correctly. Such simulators serve as an important tool for training personnel in an area where errors can have serious consequences, ensuring safety and efficiency during the operation of critical facilities.

The use of digital twins in MCR simulators allows you to integrate data in real time, ensuring the accuracy of models and improving interaction with students. This opens up new horizons for deeper, adaptive and safer learning, enabling operators to recreate application scenarios in complex and unstable situations, which is an integral part of training specialists in such high-risk areas.

TRAINING OF NPP OPERATORS: ROLE OF VIRTUAL SIMULATORS IN ADVANCED TRAINING

Main control room simulators (MCR) play a key role in ensuring the safety and efficiency of nuclear power plants. They allow you to simulate a wide range of scenarios, including transient processes, sudden equipment failures and external impacts, thereby increasing the level of personnel readiness for possible emergency events. Nuclear reactor control requires high accuracy, efficiency and deep understanding of all technological processes occurring in the system. The slightest mistakes of the operator can lead to critical consequences; therefore, personnel training should take place in conditions as close as possible to real.

Such simulators have a multilayer structure aimed at accurately reproducing the processes occurring in the operating conditions of a nuclear power plant. One of the key components is the server module, which acts as a central node that receives information from client modules and transmits it to the UCR system. The server module processes data and controls the processes taking place in the simulator. It manages communications between different parts of the system, processes information and sends commands to each client module depending on the current situation. This includes the integration of various technological processes, such as heat exchange, reactor control and the functioning of auxiliary systems. Models based on digital twins reproduce the behavior of physical equipment, allowing training of personnel both in normal working conditions and in emergency situations, which allows you to practice actions in actual scenarios.

Simulator client modules are physical or virtual simulations of equipment connected to the server module. These modules simulate the operation of various elements of a nuclear plant, such as a reactor, cooling systems or other auxiliary systems. The client modules display all the necessary parameters that the operator can adjust by interacting with the equipment. This allows the operator to work with systems as close to valid as possible and test their actions in various scenarios, including non-standard or emergency situations. Visualization of processes on these modules helps to better understand the dynamics of parameter changes and train in real time.

Each client module reproduces the behavior of its part of the equipment, including reactions to external influences, as well as possible malfunctions or emergencies, creating a virtual model of the actual system.

The division into several client modules, each of which is responsible for a separate system or object, in combination with the server module, creates a flexible and adaptive simulator architecture. This approach not only allows you to easily update and scale the system, but also guarantees its stability and efficiency when conditions change or new elements are added.

Additionally, MCR simulators are equipped with a level of scenarios and analytics, which allows you to flexibly configure various scenarios depending on the training goals. This provides the opportunity to train personnel in both standard and emergency situations, simulating events that operators may encounter in actual work. An important part of the system is monitoring the operator's actions and analyzing his response to changes, which allows you to quickly correct training, identify errors and increase its efficiency.

Traditional simulators of the main control room at nuclear power plants are developed on the basis of physical models and hardware complexes that reproduce the control panels of the power unit. They provide training for personnel in conditions as close as possible to operation on site, but their creation requires a significant investment of time and resources. In addition, upgrading such simulators with reactor design changes or software upgrades is a complex and costly task.

The use of digital twins as virtual simulators solves the problem of the versatility of training systems for fourth-generation reactors. Unlike physical simulators of main control room, which require expensive modernization or complete replacement of hardware components when switching to a new type of reactor, virtual simulators provide flexibility due to software modularity. This means that instead of creating a new physical complex, it is enough to replace program modules, adjust the mathematical model and adapt the input parameters to the features of a particular reactor.

Thanks to this architecture, digital twins can cover a wide range of nuclear reactor concepts, modeling various operating principles, coolants and neutron characteristics. This not only reduces the cost of developing new simulators, but also speeds up the process of preparing operators to control different types of reactors. Moreover, the possibility of flexible adaptation of the mathematical model allows taking into account not only the basic operating characteristics, but also the behavior of the system in non-standard situations, which is critical for the development of emergency scenarios.

This makes digital twins not only a convenient training tool, but also a platform for continuous improvement of reactor operating methods in conditions of uncertainty and variability of modern energy systems.

Digital twins provide modeling of not only standard, but also emergency situations with a high degree of detail. Unlike traditional simulators, where certain algorithms of actions are laid in advance, digital models can take into account changing parameters and train personnel to make decisions flexibly in conditions of uncertainty. The virtual environment allows personnel to safely work out scenarios of equipment failures, work in non-standard situations and actions in case of possible operator errors. Unlike traditional hardware simulators, software models can be quickly updated in accordance with new regulations or changes in the operation of equipment.

In addition, the use of digital twins in combination with virtual reality technologies opens up new opportunities for training specialists. Immersive simulators allow operators to interact with UCR interfaces in a three-dimensional environment, which increases the realism of training and speeds up the memorization of correct action algorithms. This is especially important when preparing personnel for work at stations with different reactor configurations, since digital models are easily adapted to specific technical requirements.

The use of virtual reality in combination with digital twins enhances the immersive effect, helping students not only work out action algorithms, but also psychologically adapt to work in stressful situations. The presence effect achieved in a virtual environment allows you to prepare specialists in advance for actual operating conditions, minimizing the risk of errors due to unexpected events.

At the same time, access to other systems through a graphical control interface will provide convenience and flexibility in monitoring and controlling components that do not require full visualization. Such an interface can display not only the state of the equipment, but also the system operation logs, signal changes, data from sensors, as well as the process of transmitting information between various elements. Managing this process from the GUI allows you to focus on other aspects of the system, such as diagnosing, configuring parameters, or analyzing ongoing events.

It is especially important that this approach makes it possible to train operators without overloading the virtual reality interface with unnecessary details that are not related to tasks. The virtual environment approach for MCR allows you to focus on key aspects of working with the system, such as making decisions in critical situations and adjusting work in real time, without distracting to hardware parts that do not affect operational actions.

One of the main advantages of using VR in this context is the ability to play emergency situations in a safe environment. Operators will be able to train under stress, making decisions without endangering physical equipment and people. This allows not only to increase safety, but also to reduce the likelihood of errors in everyday work, in the event that such emergencies do occur.

The development of a digital twin requires a clear structure that ensures its functionality, adaptability and the ability to integrate with real and virtual systems.

The architecture of the digital twin includes several key levels, each of which is responsible for certain aspects of modeling, analysis and user interaction.

The digital twin completely exists in a virtual environment, and all its processes and interactions are modeled within this world, without influence from physical objects and stochastic events. The basis of his work is mathematical models that reproduce key characteristics and processes of the system, including physical, chemical and technical aspects. These models can be both simple and more complex, using various equations such as differential and stochastic, depending on the required accuracy and complexity of the problem.

It is important that the models take into account not only the static, but also the dynamic characteristics of the system, which allows you to simulate the change in the state of objects and their interaction over time. During the simulation, you can reproduce both normal modes of operation and emergency or non-standard situations, such as equipment malfunctions or system overloads.

Since the digital twin operates in a completely virtual space, it does not depend on external physical factors. Instead, all information and adjustments that are made to the model are based on scenarios defined in a virtual environment. This allows you to train users in system management in various, including critical, situations, without risks.

The structural diagram of the digital twin for the MCR simulator (main control panel) in the context of nuclear energy involves the use of mathematical models to describe the physical processes occurring in various systems. These models can be used to reproduce the characteristics of the reactor, ventilation system and other key objects, which allows training personnel in controlling the reactor and other critical systems in a virtual environment. The use of mathematical models allows not only to simulate physical behavior, but also to adapt the system to test various scenarios, including emergency and non-standard situations.

Systems that can be virtualized, in particular those that include computers or programmable logic controllers (PLCs), can be represented as full-fledged digital twins. Thanks to modern virtualization capabilities on computers, you can simulate the operation of these systems, reproducing not only their behavior, but also interaction with other objects within the entire simulated installation.

It is assumed that the entire system will be simulated on one computer, which combines all components into a single virtual environment. This allows not only efficient use of resources for modeling, but also significantly reduce the cost of physical equipment. Virtualization allows operators to train on digital twins of different systems without the need for material equipment, which increases the flexibility and safety of the training process. This approach corresponds to the creation of simulators of the first type. Simulator of the first type - an exact copy of the control panel (devices, panels, indicators). The dynamic and static characteristics of the equipment fully correspond to those of the prototype object (Magid & Arkhipova, 2017). The main disadvantage of the first type simulator is its high price and rather long manufacturing time, as a rule, from two to three years, but the use of digital technologies neutralizes these disadvantages.

An important part of the digital twin is the archiving of the entire process of interaction between the operator and the simulator. This includes recording all actions and decisions taken by the operator in the process of working with the system. Data archiving allows not only to track the results of the operator's work, but also provides an opportunity for a deeper analysis of his behavior in various situations. All data, including system operation parameters, decisions made and operator actions, shall be stored for further analysis.

This data can be used to assess operator performance, identify errors, or improve operator response in unstable situations. For example, you can analyze how an operator reacts to abnormal or emergency situations, how quickly he makes decisions, and how he controls the system under uncertainty. This approach helps not only to increase the level of training, but also to improve the quality of training, preparing personnel for the actual conditions of working with technological systems.

In the process of interacting with the digital twin, the operator is faced with various emergency situations, problems and non-standard events, which forms the necessary experience and confidence in his actions. Virtual provides a unique opportunity to work out problem solving algorithms and correct behavior in conditions of uncertainty. These workouts activate psychological mechanisms that develop the operator's ability to remain calm, make informed decisions and respond quickly to changes in the situation.

It is important that such workouts with virtual models help not only learn to act within predetermined scenarios, but also contribute to reducing stress levels in real life. When the operator has already encountered possible situations in a virtual environment, his psyche becomes more prepared for similar stress in the modern world. Even with the realization that virtual reality is not physically real, the training process itself contributes to the formation of cognitive resilience, which helps to maintain clarity of thinking and calm in a critical situation. This reduces the level of stress in real life, because a person already has experience in effective solutions.

CONCLUSION

The digital twin is an important component of the virtual process control model, which allows you to reproduce not only physical objects and their interactions, but also complex behavioral and algorithmic processes that can occur in real-life conditions. This technology allows not only to train specialists based on static models, but also to create adaptive systems that emulate the dynamics of the system and its behavior in various situations.

Despite some disadvantages, the advantages of using virtual reality and digital twins in the context of operator training certainly outweigh these limitations, especially in terms of psychological training. The main argument in favor of such technologies is that they help develop critical thinking and correct behavior algorithms in the operator, which is extremely important for a successful response in emergency situations. Even if there are no stress factors in the virtual environment, such as physical sensations, the training itself helps to develop the necessary reactions and resistance to stress.

The main advantage of digital technologies in training operators is their flexibility, accessibility and adaptability. Virtual simulators allow you to safely work out complex and potentially dangerous scenarios without risking life and equipment, as well as reduce personnel training costs. Digital twins enable rapid curriculum updates and integration of new data, making learning more dynamic and relevant.

In addition, modern virtual reality and modeling technologies increase the psychological readiness of operators to work in conditions of uncertainty and stress. in the conditions of instability, uncertainty and changes characteristic of the VUCA and BANI worlds. These worlds require operators to adapt to rapidly changing conditions, make decisions in conditions of limited information and take into account many unpredictable factors. Gradual adaptation to different scenarios modeled in a virtual environment helps develop stress tolerance, critical thinking, and decision-making ability in emergency situations. These skills are key to working at critical infrastructure sites where the cost of error can be extremely high.

The use of these technologies contributes to the development of key competencies, such as the ability to respond quickly to unforeseen events, interact effectively with systems under stress, and increase flexibility in decision-making. In the context of VUCA and BANI, it is important not only to have technical knowledge, but also the ability to perceive complex and ambiguous situations, effectively manage risks and quickly adapt strategies. Virtual simulators and digital twins give operators the opportunity to repeatedly encounter various non-standard situations, developing the ability to act not according to a template, but in conditions of uncertainty.

Thus, the combination of traditional methods of training in the capabilities of digital twins and virtual reality allows you to create a comprehensive training system that provides a high level of professional competence of operators - critical thinking, emotional stability and confidence in decision-making in difficult unpredictable conditions.

REFERENCES

Awasthi, K., Padwekar, K., & Misra, S. C. (2025). Digital Twin: A Unified Definition, Issues, Challenges, and Opportunities. In M. Khosrow-Pour, D.B.A. (Ed.), Encyclopedia of Information Science and Technology, Sixth Edition. Advance online publication. DOI: 10.4018/978-1-6684-7366-5.ch074

Bushuyev, S., Piliuhina, K., & Chetin, E. (2023). TRANSFORMATION OF VALUES OF THE HIGH TECHNOLOGY PROJECTS FROM a VUCA TO a BANI ENVIRONMENT MODEL. *Innovative Technologies and Scientific Solutions for Industries*, *2*(24), 191–199. DOI: 10.30837/ITSSI.2023.24.191

Dureev, S., & Dureeva, N. (2023). TRANSFORMATION OF THE MODERN EDUCATIONAL PROCESS. *Bulletin of Chelyabinsk State University*, *474*(4), 43–49. DOI: 10.47475/1994-2796-2023-474-4-43-49

Erlangga, I. S., Tarigan, F. P. P., Ternasih, L., & Umar, D. (2024). Digital transformation in education: Challenges and opportunities in the modern era. [INJOTEL]. *International Journal of Teaching and Learning*, *2*(12), 3251–3260.

Hohlov, M., & Korobova, M. (2024). The usage of digital twins in the field of virtual education. *Education, Innovation. Research as a Resource for Community Development*, *2*, 36–38. DOI: 10.31483/a-10620

Kuzmin, A. (2024). ANALYSIS OF THE REQUIREMENTS FOR THE VALIDATION AND VERIFICATION OF DIGITAL TWINS. *Science Bulletin*, *4*(3 (72)), 340–343.

Magid, S. I., & Arkhipova, E. N. (2017). Simulators-aided training for personnel of power industry. History. State. Norms. In NPO Energobezopasnost eBooks. DOI: 10.24223/978-5-600-02001-6

Salun, M., & Zaslavska, K. (2024). Strategies for Resilience in a Dynamic World: from VUCA to BANI. *Proceedings of Socratic Lectures*. DOI: 10.55295/PSL.2024.I23

Sergeev, S. (2010). Virtual Training Simulators: Problems of Theory and Methodology of Design. *Biotechnosfera*, *2*, 15–20.

Chapter 4
Development of an Application for Compiling and Optimizing an Investment Portfolio

Liza Muratovna Elkanova
https://orcid.org/0000-0003-3973-1874
North-Caucasus State Academy, Russia

Natalya Vladimirovna Kononova
MIREA - Russian Technological University, Russia

Milana Bayramukova
North-Caucasian State Academy, Russia

Kamilla Gerieva
North-Caucasian State Academy, Russia

Sapiyat Bidzhieva
North-Caucasian State Academy, Russia

ABSTRACT

The article presents the results of the analysis of the theoretical foundations of the formation and optimization of the investment portfolio, considers the models of G. Markowitz W. Sharpe for the compilation of the investment portfolio; analyzed and selected the tools for developing the application that forms the investment portfolio. During the study the Python programming language was chosen as the tools for developing the application that forms the investment portfolio. Python is a power-

DOI: 10.4018/979-8-3373-7917-3.ch004

ful tool for optimizing investment portfolios using the W. Sharpe method, allowing investors and analysts to achieve a balance between risk and return using advanced methods of quantum analysis. The key functions of the developed application are: aggregation of financial data, calculation of expected portfolio return, calculation of variance and standard deviation, portfolio optimization, visualization of results, ease of use and an intuitive interface .

1. INTRODUCTION

The relevance of the problem is due to the fact that the dynamic development of the economy in the modern world, constant fluctuations in exchange rates, inflation lead to the fact that many organizations, as well as individuals, consider investments as a way to create a certain financial safety net. One of the attractive aspects of investments is their focus on long-term goals. However, the formation of an effective investment portfolio requires a certain financial literacy, the ability to navigate the diversity of assets, assess possible profitability and existing claims.

The problem of developing effective income portfolios is considered in the research of D. Blanchett and H. Ratner (2015), risk management through diversification when choosing a portfolio with non-historical information is devoted to the research of C. Calvo et al (2018), the problem of creating a portfolio using graphical characteristics was considered by J. Danko and V. Šoltés, (2018), the method of combining basic portfolios into a single portfolio that imitates a factor closely related to the optimal portfolio was developed by M. Grinblatt and K. Saxena, (2018), the problem of dynamic portfolio allocation under time-varying shock risk is considered in C. Zhou et al (2019). The problem of optimizing investment portfolio management is considered in the study by V. Oleynik and O. Kuzmenko (2019).

It should be noted that the use of mathematical methods to formalize the problem of creating an effective investment portfolio and building a mathematical model allows us to find the optimal balance between profitability and risks, taking into account the individual preferences of investors, allowing us to adjust the criteria for optimizing the investment portfolio. The use of mathematical models creates the basis for creating an effective, balanced investment portfolio, while the development of an application for implementing algorithms for calculating the main indicators of an investment portfolio allows us to automate the processes of building and optimizing investment portfolios based on regression analysis methods.

2. OBJECTIVES OF THE STUDY

Based on the above, the objectives of the study are following:

- study the theoretical foundations of forming and optimizing an investment portfolio;
- consider models for compiling an investment portfolio;
- analyze and select tools for developing an application that forms an investment portfolio;

-develop an application for optimizing an investment portfolio;
-conduct an empirical analysis and validation.

The purpose of the work is to implement and analyze an application for compiling and optimizing.

3. LITERATURE REVIEW

The development of IT technologies, various online services has made the topic of investments accessible and relevant for a wide range of private investors. To improve the efficiency of investments, to be able to control the level of profitability and the degree of risk, many investors form an investment portfolio. The use of IT technologies in this area is the subject of the works of many researchers. X. Zhong and D. Enke (2019) developed a comprehensive procedure for analyzing big data using hybrid machine learning algorithms to predict the daily direction of the SPDR S&P 500 index (ticker: SPY). The work of K. Waliszewski and M. Zieb-Szklarska (2020) is devoted to the analysis of the history, rules of use, functionality and development of Robo-Advice. A critical analysis of the concept of robo-advisory, carried out in their work, showed that "robo-advisers do not pose a threat to traditional financial advisers, but complement them, which makes a hybrid model combining a physical adviser with technological solutions (robo-advisers) most likely" (Waliszewski and Zieb-Szklarska, 2020, p.155). The article by V. Kobets et al (2021) presents a model for calculating the insurance premium for personal insurance for investment portfolios with different levels of risk using a robot adviser. Three types of investors are analyzed: conservative, aggressive and moderately aggressive. The model helps to determine the individual size of the insurance premium for each investor profile, taking into account their attitude to risk. The architecture of a deep convolutional generative adversarial network (DCGAN) for solving the problem of predicting the closing price of stocks was presented in his work by A.Staffini (2022). The results of one-stage and multi-stage forecasting showed that the developed model works better than standard widely used tools, based on which the author concludes "that deep learning (and in particular generative adversarial networks) is a promising area for forecasting financial time series" (Staffini, 2022, p.14). The problem of forecasting the stock market index using LSTM is considered in their work by H.N. Bhandari et al (2022). The results of the experimental study showed that "the single-layer LSTM model provides excellent adaptation and high forecasting accuracy compared to multi-layer LSTM models" (Bhandari et al,2022,p.1).

4. RESEARCH MODEL

There are many theories and methods that can be used in the formation of an investment portfolio, the key ones are::

- G. Markowitz's method (Mangram, 2013);
- W. Sharpe's method (CAPM model) (Sharpe et al, 2018).

Harry Markowitz's model, called the "Portfolio Choice Theory", is a fundamental cornerstone in the field of modern investing and asset management. First published in 1952, this theory proposed a quantitative method for compiling an investment portfolio in such a way as to maximize the expected return at a given level of risk.

As a criterion for the expected income from a series of possible incomes, which we consider to be normally distributed, we take the mathematical expectation, which is calculated using the formula (1):

$$m_i = \sum_{j=1}^{n} R_i P_{ij} \tag{1}$$

where Ri is the possible income on the i-th security, rubles;

Pij is the probability of receiving income;

n is the number of securities.

To calculate the risks, we take the standard deviation, which is a measure of dispersion. The greater the spread of possible values, the greater the risk of not receiving income (2).

$$\sigma_i = \sqrt{\sum_{j=1}^{n} P_{ij} \left(R_{ij} - m_i\right)^2} \tag{2}$$

Thus, the investment portfolio is characterized by the following parameters.

Expected return (3):

$$m_p = \sum_{i=1}^{n} X_i m_i \tag{3}$$

where X_i is the investment amount coming to the i-th asset;
m_i is the expected income from the i-th asset;
m_p is the expected return on the portfolio.
Risk measure (4)

$$\sigma_p = \sqrt{\sum_{i=1}^{n} \sum_{j=1}^{n} X_i X_j \sigma_{ij}} \tag{4}$$

where σ_p is the measure of portfolio risk;
σ_ij is the covariance between the returns of the i-th and j-th assets.
We calculate the covariance using formula (5):

$$\sigma_{ij} = \rho_{ij} \sigma_i \sigma_j \tag{5}$$

where ρ_ij is the correlation coefficient of the returns of the i-th and j-th assets. If we consider the limiting case, when theoretically an infinite number of securities can be included in the investment portfolio, we find that the risk measure indicator of the investment portfolio will approach the average value of covariance (see Figure 1).

Figure 1. Portfolio risk measure

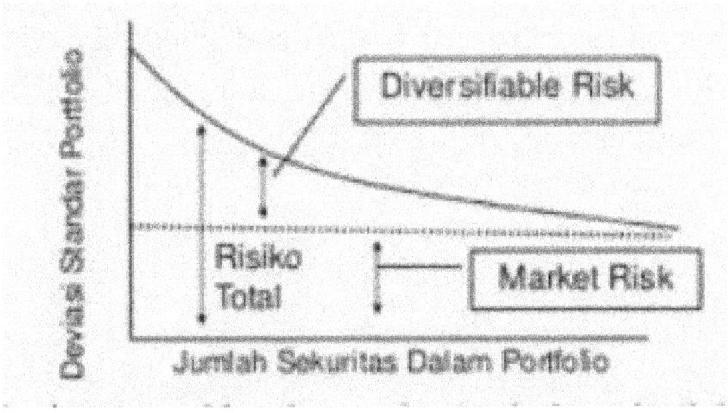

The key concept in the Markowitz model is the "efficient frontier" – a line that passes through a set of potential portfolios and indicates those that have the best ratio of expected return to risk level (see Figure 2).

Figure 2. Efficient Frontier

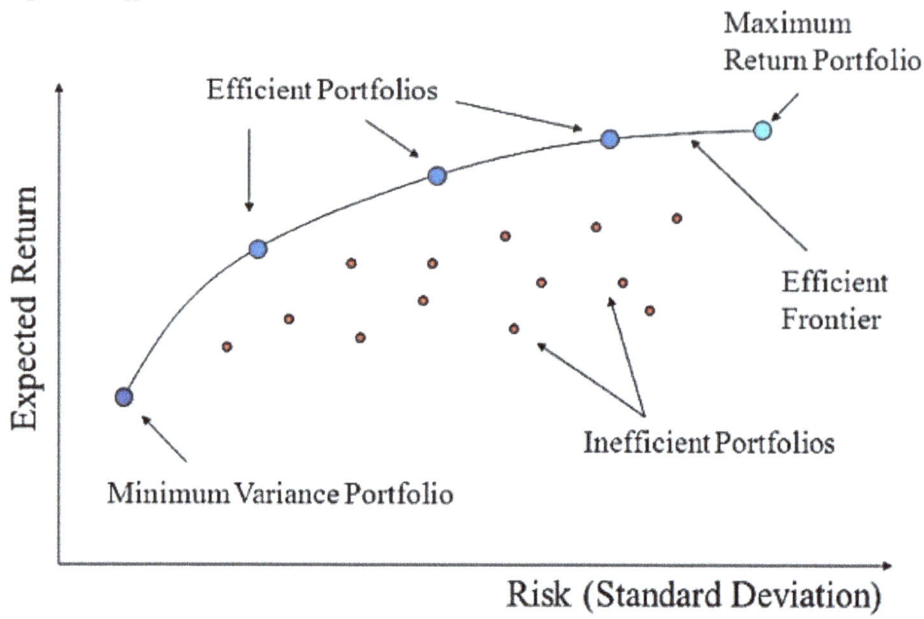

Portfolios on this frontier are considered optimal; they provide the maximum expected return for a given level of risk or the minimum risk for a desired level of return.

With information on the expected returns and standard deviations of possible portfolios of securities, it is possible to construct a map of indifference curves reflecting investors' preferences (see Figure 3).

Figure 3. Investor Indifference Curve Maps

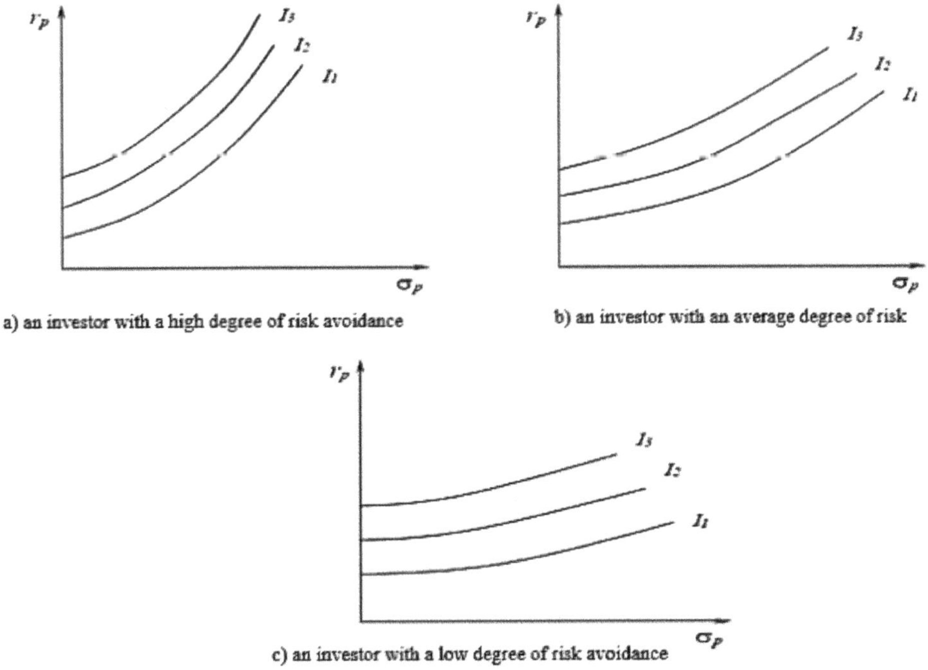

The investor should therefore choose a portfolio that lies on the indifference curve that is above and to the left of all other curves. The efficient set theorem states that the investor should not consider portfolios that do not lie on the upper left boundary of the availability set, which is its logical consequence. Based on this, the optimal portfolio is located at the contact point of one of the indifference curves of the most efficient set. In Figure 4, the optimal portfolio for some investor is denoted by O*.

Figure 4. Choosing the optimal portfolio

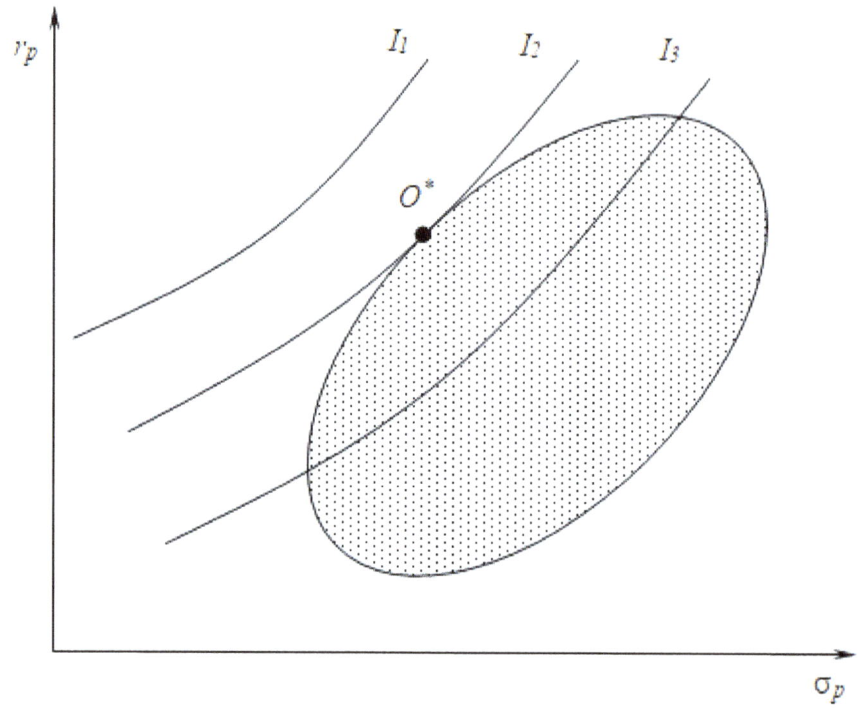

Additionally, it is important to note the concept of the "market portfolio", which contains all assets available on the market in proportion to their market capitalizations. According to the Capital Asset Pricing Model (CAPM), developed later based on the work of Markowitz, this portfolio is considered optimal in terms of the ratio of return to systematic risk (risk of non-diversifiable factors).

The application of the Markowitz model has had a significant impact on various aspects of the financial sector: from approaches to managing personal investments to hedge fund strategies and pension plans. Financial managers use it as a basis for creating balanced investment strategies.

However, it is important to mention some limitations of the Markowitz model. It operates on the assumption of a normal distribution of asset returns, which is not always true. In addition, the model is based on historical data to predict future results, which may not always be accurate due to market volatility.

The basis of the Sharpe method is the ratio of the excess of a portfolio's return over the risk-free rate to its volatility (risk). The risk-free rate is a theoretical indicator of investment profitability that can be obtained without any risk of loss (Sharpe et al, 1995).

First of all, the Sharpe method is one of the basic concepts of the modern theory of portfolio investments. It helps investors not only to choose the best capital allocation strategies, but also to evaluate the work of asset managers.

On the other hand, portfolio volatility measures the degree of variability of its profitability and serves as a measure of investment risk. The higher the volatility, the higher the risk and the higher the potential profitability.

The Sharpe ratio is defined as the difference between the expected portfolio profitability and the risk-free rate, divided by the standard deviation of the portfolio profitability (its volatility) (6):

$$S = \frac{P_p - P_f}{\sigma_p}$$

where S is the Sharpe ratio;

P_p is the expected portfolio return;

P_f is the risk-free rate;

σ_p is the standard deviation of portfolio return.

The W. Sharpe index model uses a close correlation between the change in the prices of individual stocks. It is assumed that the necessary input data can be approximately determined using only one basic factor and the relationships linking it to the change in the prices of individual stocks. As a rule, the value of some index is taken as such a factor. The dependence of the security's return on the index is described by the following formula (7):

$$r_i = \alpha_{ij} + \beta_{ij} + r_I + \varepsilon_{ij} \qquad (7)$$

where r_i is the yield of security i for a given period;

r_I is the yield on market index I for the same period;

α_ij is the bias coefficient;

β_ij is the slope coefficient;

ε_ij is the random error.

The linear regression equation (7) considered allows us to quantitatively assess the relationship between the asset yield and the market yield.

The main role in this equation is played by the coefficient β_ij, by the value of which we can determine the correlation between the yield of a specific security and changes in the overall market yield.

If this coefficient has a positive value β_ij>0, then the asset is more volatile than the market, that is, when the market changes by k%, the asset yield will change by more than k%.

If β_ij<0, then when the market changes by k%, the asset will change by less than k%.

When β_ij=0, the yield of the security changes equally with the market yield.

Let us consider the relationship between the Sharpe beta coefficient and linear regression in risk assessment and investment portfolio optimization. A graphical representation of this relationship is given in Figure 5.

Figure 5. Visualization of the relationship between the Sharpe beta coefficient and linear regression

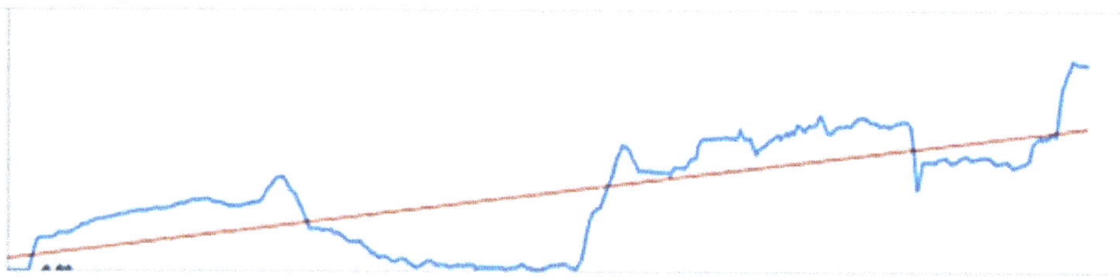

The horizontal axis reflects the market return r_I, and the vertical axis reflects the return of a specific asset r_i, linear regression finds a line that minimizes the distance between the points expressing the values of the beta coefficient and the points of the line.

The greater the slope of the regression line, the greater the Sharpe ratio and the more sensitive the asset return is to changes in market returns.

Let's consider the advantages and disadvantages of the Sharpe ratio to understand how it can be used in the decision-making process.

The main advantages of the Sharpe ratio, which are the basis for using this method in the decision-making process, are: efficiency, ease of calculation, intuitive interpretation. But along with these advantages, there are some disadvantages that need to be taken into account when using this method: risk does not correspond to volatility, short-term focus, does not take into account the shape of the distribution of returns

Despite these shortcomings, the Sharpe ratio remains a powerful tool for analyzing the effectiveness of portfolio investments due to its simplicity and ability to present complex concepts in an easily interpretable form, and with proper use of the Sharpe ratio, you can significantly improve the efficiency of investment management due to a more balanced approach to the profit/risk ratio.

To create a functional and reliable application, you need to carefully consider its architecture. It must be flexible to ensure the integration of new solutions and expansion of functionality without loss of performance. In architectural terms, the application can be divided into the following main components:

Database (DB): Stores information about assets, historical price data, users and their portfolios.

Server part (backend): Provides the logic for processing data from the DB, performing calculations and providing an API for interaction with the client side.

Frontend: The user interface of the application through which users interact with the functionality.

The key functions of such an application are:

Aggregation of financial data. The application must provide up-to-date information on all available assets on the market. This requires integration with various exchanges and financial services to obtain real quotes.

Calculation of the expected portfolio return. One of the key tasks is to assess the potential profit from investments. Statistical analysis of past data using various mathematical models will help with this.

Calculation of dispersion and standard deviation. To understand the level of investment risk, it is necessary to analyze fluctuations in asset returns, which allows you to determine the dispersion and standard deviation of the portfolio.

Portfolio optimization. Based on the Sharpe ratio maximization criterion, the application will help you select a ratio of assets in the portfolio that will bring maximum returns with minimal risks.

Visualization of results. Graphs and charts greatly simplify the process of analyzing large amounts of data and help make the right investment decision faster.

The ease of use. An intuitive interface will allow the user to start working with the investment optimization system without special knowledge.

In today's digital world, where the speed and efficiency of data processing play a key role, Python stands out as one of the most preferred programming languages in the financial sector. As a high-level, dynamic language, Python has a rich ecosystem of libraries and frameworks, making it an ideal tool for developing various financial applications.

Python has many libraries for data analysis, such as NumPy for numerical calculations, pandas for processing tabular data, Matplotlib and Seaborn for data visualization (9). These tools allow financial analysts to effectively work with large amounts of data to predict trends, optimize portfolios and manage risks. The combination of Python with SQL and NoSQL databases makes it easy to manipulate and store huge amounts of financial information. The SQLAlchemy and PyMongo libraries provide powerful interfaces for interacting with databases. Thanks to libraries such as scikit-learn, TensorFlow, and Keras, Python has become a standard choice for machine learning. In finance, this is especially valuable when creating predictive analytics models to assess customer creditworthiness or forecast asset price movements. Tools such as Zipline or Backtrader are used to test investment strategies on historical data. They allow you to simulate trading in stocks or other financial instruments without actually investing money. Python easily integrates with many financial service APIs. This allows you to automate the process of obtaining data from exchange platforms or other independent information providers. The online platform Fin-Skill uses Python in the process of training "financial advisors." The ability to perform complex calculations as part of training programs makes the language an invaluable resource. In addition to the libraries discussed, Python has a special module designed to optimize investment portfolios. PyPortfolioOpt is a Python library that implements portfolio optimization methods, including the classic mean-variance optimization methods and the Black-Litterman distribution, as well as more recent developments in this area. PyPortfolioOpt contains functions that make portfolio optimization an efficient and simple task: computing the efficient frontier, optimal portfolio allocation, risk models, factor models, plotting.

Let's take a look at two more popular Python libraries: CVXPY and Quantopian's pyfolio. CVXPY is a Python library for convex optimization. It provides a framework for defining and solving convex optimization problems, making it highly flexible and capable of solving a wide range of problems beyond just portfolio optimization. However, CVXPY's strength can also be its weakness. Its broad scope makes it a more challenging tool to work with, especially for people without a deep knowledge of convex optimization theory. Additionally, CVXPY does not provide ready-made solutions for portfolio optimization, requiring users to define these problems from scratch. In contrast, PyPortfolioOpt, while less flexible than CVXPY, is much more user-friendly and provides ready-made solutions for portfolio optimization, making it a preferred choice for users focused on portfolio optimization. Pyfolio, a tool developed by Quantopian, is another Python library that focuses on portfolio management and risk analytics. Unlike PyPortfolioOpt, which focuses on portfolio allocation optimization, Pyfolio is more concerned with portfolio performance and risk assessment. Thus, the PyPortfolioOpt module is a powerful tool for optimizing investment portfolios, allowing investors and analysts to achieve a balance between risk and return using advanced quant analysis methods. Thanks to its flexibility and wide range of functions, it opens the door to creating effective asset management strategies tailored to specific goals and constraints.

5. FINDINGS

The developed application is an executable file written in Python. The program can be launched on any device with Python installed and allows a potential investor to automate the process of analyzing stock indices and forming an investment portfolio using economic and mathematical methods of regression analysis, in particular using the Sharpe model.

To begin with, all modules necessary for data processing, calculations, optimization and visualization of results were imported (see Figure 6).

The data for the application is a set of stock indices and unique identifiers of companies listed on stock markets. The user provides this list to the application. To test the application, we downloaded this data from the mobile application of the largest Russian online bank Tinkoff "Tinkoff Investments".

Figure 6. List of connected modules

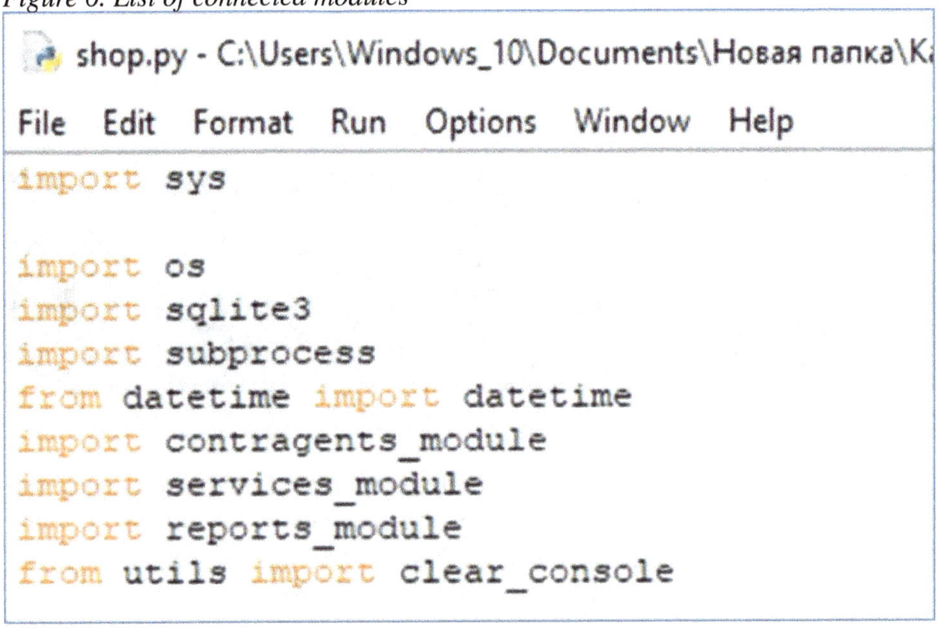

Retrospective data is considered for a time interval that is defined by the user and which will be sufficient to ensure reliable and high-quality analysis.

This data is extracted using an application programming interface that connects to an economic data source and is formed in a Pandas DataFrame, which is a two-dimensional tabular data structure. Each row in this frame corresponds to a trading day, and each column corresponds to a specific exchange rate.

The application processes missing values, that is, various gaps that arise in the data set (they can be associated with non-working days, stock reduction, data errors, etc.)

Next, based on the prepared data, the application calculates the main portfolio indicators for each stock. We select the calculation method (see Figure 7) and the optimization method (see Figure 8)

Figure 7. Selecting a calculation method

```
Calculate the expected profit using the method:
0. Exit
1. Average (MA)
2. The last values have a stronger influence ()
3. CAPM
4. Black-Lettermann
5. Show accumulation
6. Zero the accumulator
Calculation method: 2
```

Figure 8. Selecting an optimization method

```
Maximization of Sharpe coefficient
Minimization of volatility
Minimize risk with the desired income level
Maximize profit if the desired risk
What optimization to use? 1
```

As a result of the calculation, we see (see Figure 9) the value of the average return that an investor can receive by owning shares for a certain period of time, the value of volatility, and the risk index of the share. The higher the volatility, the greater the fluctuations in share prices and, accordingly, the greater the risk.

Figure 9. Calculation results

```
[('MOEX', 37.34), ('CHMF', 27.99), ('MGNT', 25.01), ('NLMK', 9.35),
('LKOH',0.32)]

*************************************
Calculation method: The last values have a stronger influence ()
Optimization method: Maximize the Sharpe ratio
Expected annual return: 100. 59%
Annual volatility: 17.92
Sharpe ratio: 5.5

*************************************
Buy lot for: 100,000
MOEX - Moscow Exchange: 156
CHMF - Severstal:  14
MGNT - Magnet: 3
NLMK - NLMK: 37
There are 4 companies left in the list
1283.00 will remain
if you need to save to a file - press 1 : 0
If you need to accumulate - press 1: 0
Press ENTER to continue…
```

The investment portfolio is optimized using the PyPortfolioOpt module, which offers various optimization methods. In this example, the Sharpe ratio maximization method is selected. This method finds the distribution of shares in the portfolio that provide the maximum risk-adjusted return.

If you select the minimum volatility method, the application will generate a portfolio suitable for risk-averse investors, i.e. with the lowest volatility.

You can also select additional methods that help investors adapt the optimization decision to their desires regarding risk and expected return. You can accumulate the results of different optimization methods.

The included Matplotlib module allows you to present the results visually using graphs and charts. The portfolio composition diagram reflects the percentage distribution for each type of stock in the portfolio (see Figure 10).

The graph of the selected assets shows the dynamics of their pricing (see Figure 11), and the graph of efficient frontiers presents optimal portfolios with different risk-return ratios (see Figure 12).

Figure 10. Percentage of selected assets

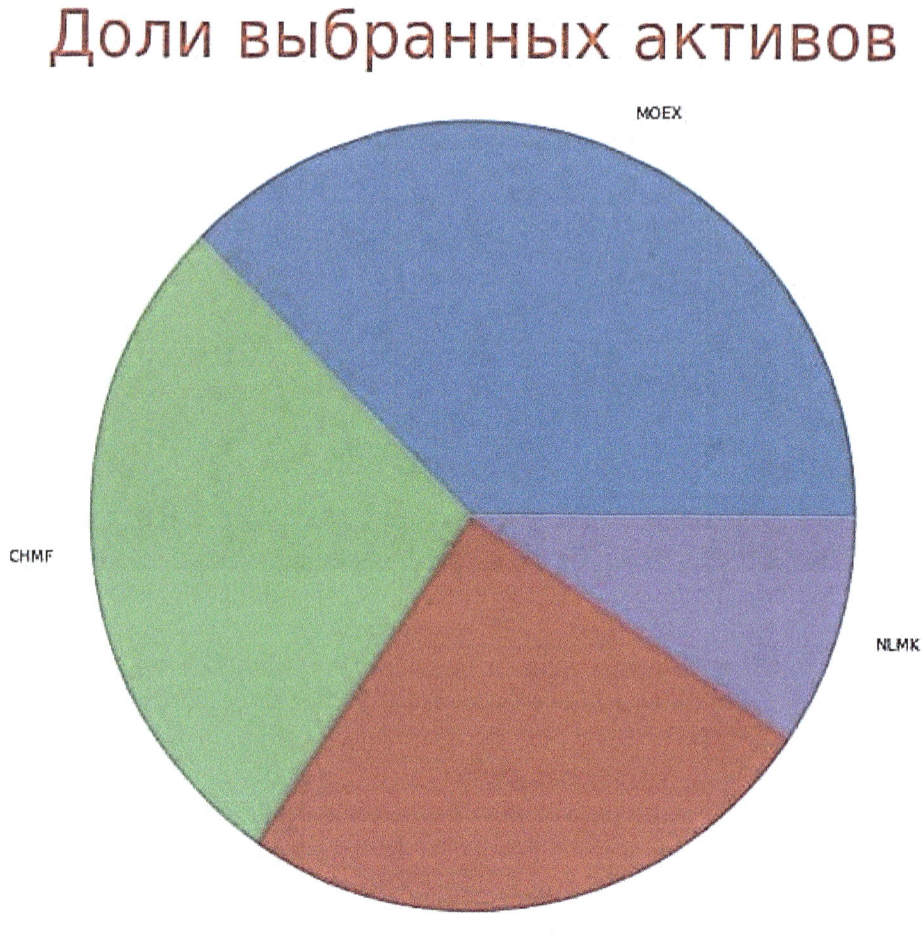

Figure 11. Graph of selected assets

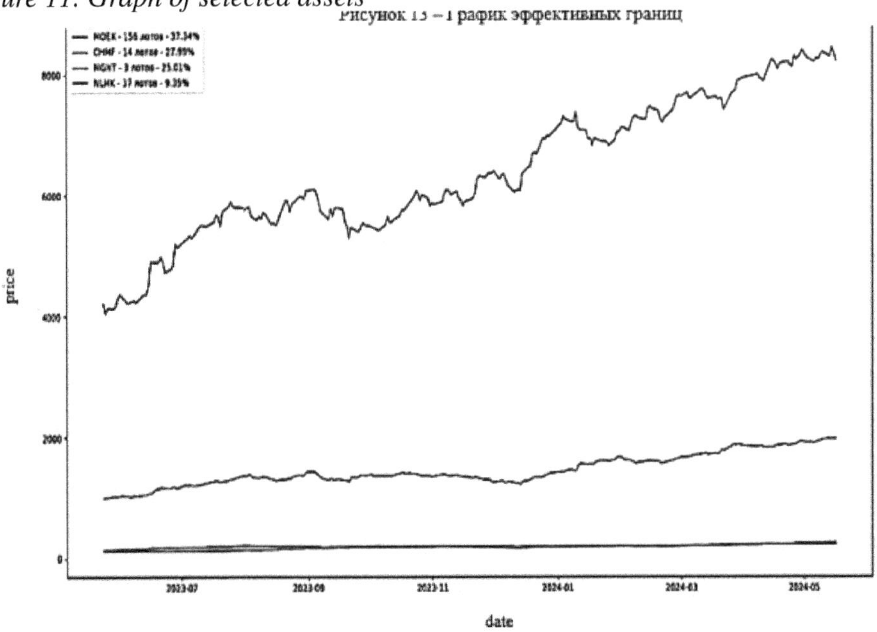

Figure 12. Efficient Frontier Graph

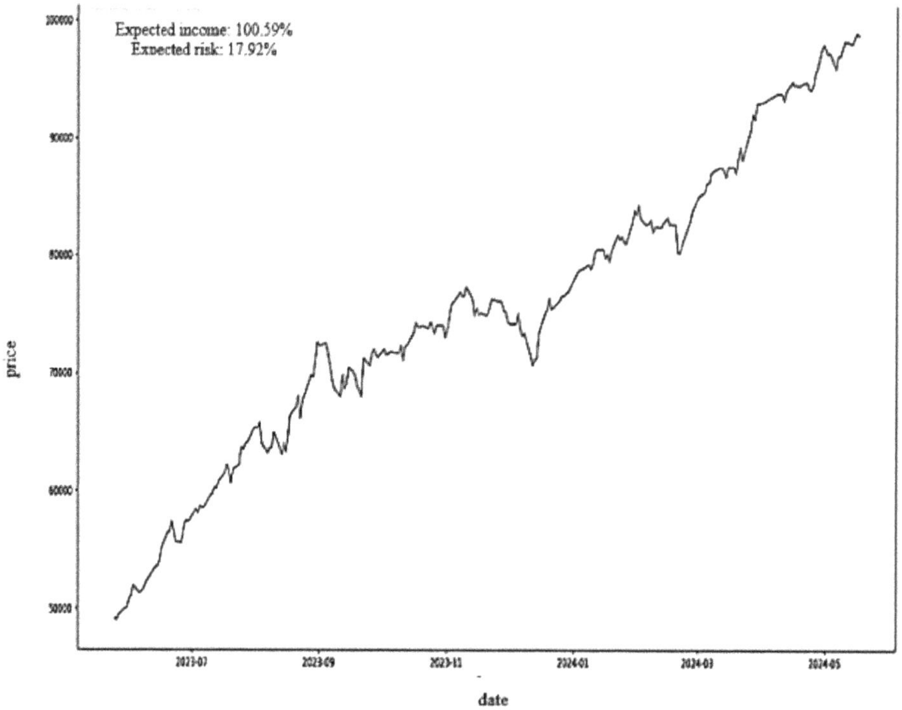

By selecting other values of the investment amount of the set of shares and running the application, we will obtain new results for the formation of the optimal investment portfolio (see Figure 13).

Figure 13. Formation of an optimal portfolio

```
[('MGNT', 10.0), ('SBER',10.0), ('NLMK', 10.0), ('CHMF', 10.0),
('MTSS',10.0), ('LKOH',10.0), ('ETLN',10.0), ('MOEX',
10.0),('MAGN',10.0),('ALRS',8.45),('GMKN',1.55)]

*************************************
Calculation method: The last values have a stronger influence (EMA)
Optimization method: Maximize the Sharpe ratio
Expected annual return: 59.679%
Annual volatility: 14.39
Sharpe ratio: 4.01

***********************************
Buy lot for: 100,000
MGNT - Magnet: 1
SBER - Sber.bank: 30
NLMK - NLMK: 39
CHMF - Severstal:  5
MTSS - MTSS: 32
LKOH - LUKOIL: 1
ETLN - Etalon Group PLC GDR: 100
MOEX - Moscow Exchange: 42
MAGN - Magnitogorsk Iron and Steel Works: 169
ALRS - ALROSA: 104
GMKN - Norilsk Nickel: 10
There are 11 companies left in the list
4678.26 will remain
```

Visualization of the results is presented in Figures 14-15.

Figure 14. Graph of selected assets

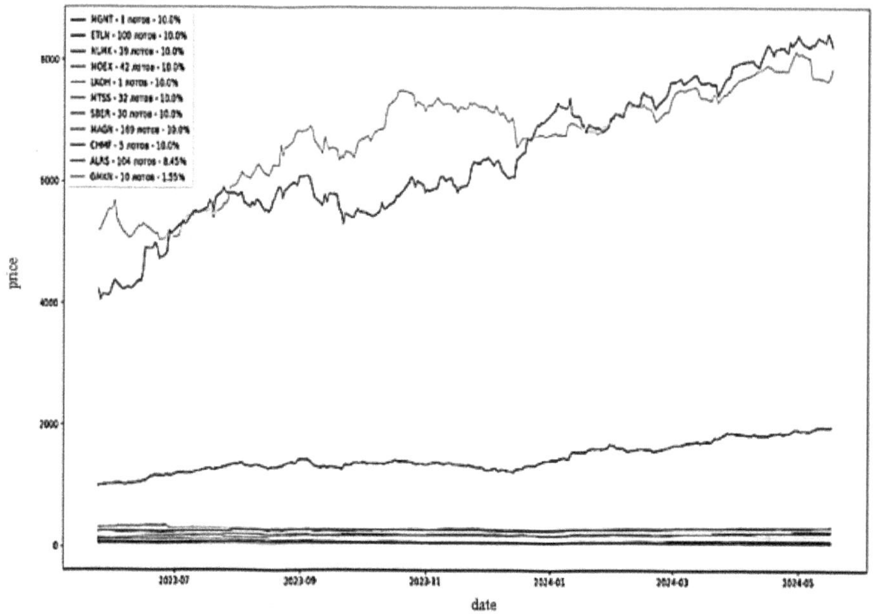

Figure 15. Efficient Frontier Graph

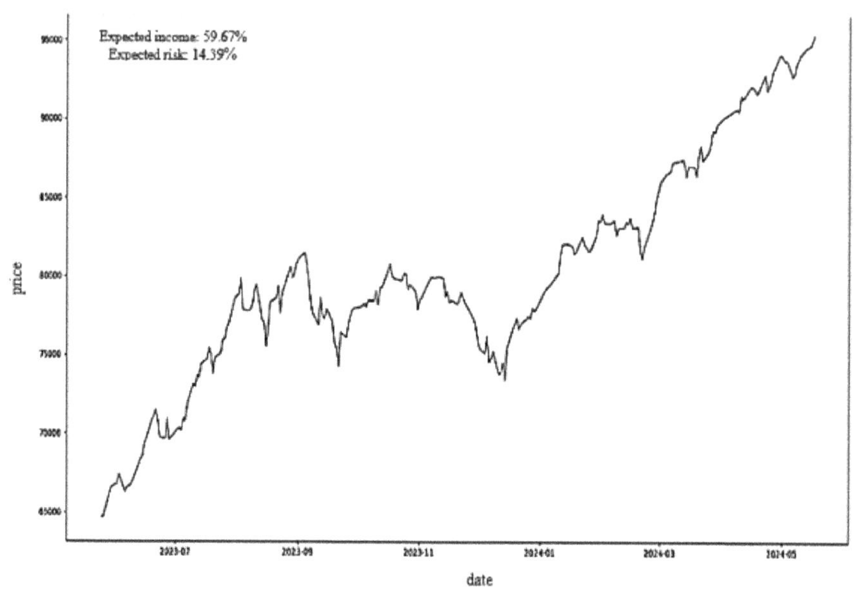

6. SUGGESTIONS

The development of digital technologies, their implementation in all spheres of human activity, including the economy, have led to a transformation of users' attitudes towards bank deposits, investments in securities, etc. Most of the solvent population wants to invest their savings more effectively, although they are not qualified specialists in this area, in the field of the securities market. The developed mobile application can act as an effective tool for forming an investment portfolio and can lay the foundation for new models of hedge funds based on new technologies - robo-advising technologies. This model can be easily adapted for use in other large-scale indices with similar data behavior, also for a better understanding of the situation in the securities market before making investment decisions.

7. IMPLICATIONS

The main features of the developed application include aggregation of financial data, calculation of expected portfolio returns, calculation of variance and standard deviation, portfolio optimization, visualization of results, and ease of use with an intuitive user interface. They help accurately predict stock prices and thus expected profits without requiring specialized knowledge. This study demonstrates a promising opportunity to leverage the capabilities of the Python programming language and Python libraries to define the cone of uncertainty in stock price forecasting. This allows stakeholders to use the results as additional information for their investment decisions. The potential for using this model to improve knowledge in the field of financial portfolio modeling is also considered.

8. SCOPE FOR FURTHER RESEARCH

Investment portfolio optimization is performed using the PyPortfolioOpt module, which offers various optimization methods. In this example, the Sharpe ratio maximization method is selected. This method finds the allocation of stocks in the portfolio that provide the maximum risk-adjusted return.

Additional methods can also be selected that help investors tailor the optimization decision to their desires regarding risk and expected return. The results of different optimization methods can be accumulated. The scipy. optimize and minimize library can be used to solve the problems of optimizing an investment portfolio of securities. This study can be extended by using complex LSTM models, including the window method, multivariate time series forecasting, and other methodologies for forming an investment portfolio, such as the Tobin or Black-Litterman model.

In the modern world, the integration of risk management strategies and the study of new architectures and integration methods are becoming one of the main ones for forecasting and forming an investment portfolio. The development of machine learning technologies leads to the development of more accurate and stable models for the formation of investment portfolios, which is a reliable support for investors in decision-making.

9.CONCLUSION

Developing an application for optimizing an investment portfolio is a complex process that requires deep knowledge of both finance and software development technologies. One of the powerful tools for analyzing the effectiveness of portfolio investments is the Sharpe ratio, due to its simplicity and ability to present complex concepts in an easily interpretable form.

The software implementation of the by_ticker_list.py application is written in the Python programming language. This is a high-level, interpreted language with dynamic typing and expressive syntax, has concise and readable code. Another advantage of the chosen programming language is its cross-platform nature - the independence of the developed application from the installed operating system.

Empirical analysis and validation showed the degree of joint movement between all pairs of stocks. In the correlation matrix, positive correlations indicate that the stocks are moving in the same direction, and negative correlations indicate that they are moving in opposite directions, which is crucial for understanding the diversification potential. This application shows the value of the average return that an investor can receive by owning shares for a certain period of time, the value of volatility, and the risk index of a share. The higher the volatility, the greater the fluctuations in share prices and, accordingly, the greater the risk.

By selecting new values of the investment amount, a set of shares, and launching the application, we will receive new results for forming an optimal investment portfolio. You can also select additional methods that help investors adapt the decision-making on optimization to their desires regarding risk and expected return. You can accumulate the results of different optimization methods. The graph of the selected assets shows the dynamics of their pricing, and the graph of efficient boundaries presents optimal portfolios with different risk-return ratios.

Thus, the development of IT technologies makes it possible to use IT solutions not only by financial professionals, but also allows ordinary people, without being qualified specialists in the field of securities, to invest their savings more effectively.

In our opinion, the developed application can become an effective investment tool for this segment of the population.

REFERENCES

Bhandari, H. N., Rimal, B., Pokhrel, N. R., Rimal, R., Dahal, K. R., & Khatri, R. K. C. (2022). Predicting stock market index using LSTM. *Machine Learning with Applications*, *9*(2), 100320. DOI: 10.1016/j.mlwa.2022.100320

Blanchett, D., & Ratner, H. (2015). Building Efficient Income Portfolios. *Journal of Portfolio Management*, *41*(3), 117–125. DOI: 10.3905/jpm.2015.41.3.117

Calvo, C., Ivorra, C., & Liern, V. (2018). Controlling risk through diversification in portfolio selection with non-historical information. *The Journal of the Operational Research Society*, *69*(10), 1543–1548. DOI: 10.1057/s41274-017-0195-6

Danko, J., & Šoltés, V. (2018). Portfolio creation using graph characteristics. *Investment Management and Financial Innovations*, *15*(1), 180–189. DOI: 10.21511/imfi.15(1).2018.16

Grinblatt, M., & Saxena, K. (2018). When Factors Do Not Span Their Basis Portfolios. *Journal of Financial and Quantitative Analysis*, *53*(6), 2335–2354. DOI: 10.1017/S0022109018000376

Kobets V., Yatsenko V. & Popovych I. (2021) Automated Formation of Insurance Premium for Different Risk Attitude Investment Portfolio Using Robo-Advisor. Communications in Computer and Information Science, ICTERI, Workshops, 3-22. DOI:DOI: 10.1007/978-3-031-14841-5_1

Mangram, M. E. (2013). A Simplified Perspective on Markowitz Portfolio Theory. [MMAR-link]. *Global Journal of Business Research*, *7*(1), 59–70. https://ssrn.com/abstract=2147880

Sharpe, W.F., Alexander, G.J. and Bailey, J.W. (2018). Investments. INFRA-M,1028

Staffini, A. (2022). Stock price forecasting with a deep convolutional generative adversarial network. *Frontiers in Artificial Intelligence*, *5*, 837596. Advance online publication. DOI: 10.3389/frai.2022.837596

Waliszewski, K., & Zięba-Szklarska, M. (2020). Robo-advisors as automated personal financial planners – SWOT analysis. *Journal of Finance and Financial Law*, *3*(27), 155–173. DOI: 10.18778/2391-6478.3.27.09

Zhong, X., & Enke, D. (2019). Predicting the daily return direction of the stock market using hybrid machine learning algorithms. *Financial Innovation*, *5*(24), 24. Advance online publication. DOI: 10.1186/s40854-019-0138-0

Zhou, C., Wu, C., & Wang, Y. (2019). Dynamic portfolio allocation with time-varying jump risk. *Journal of Empirical Finance*, *50*, 113–124. DOI: 10.1016/j.jempfin.2019.01.003

Chapter 5
Review of Monitoring Methods in Energy Storage Systems

Pavel Buchatskiy
https://orcid.org/0000-0002-3161-6567
Adyghe State University, Russia

Semen Teploukhov
Renewable Energy Laboratory, Russia

Stefan Onishchenko
https://orcid.org/0000-0001-9196-673X
Renewable Energy Laboratory, Russia

Sergey Klyushnikov
Renewable Energy Laboratory, Russia

ABSTRACT

Many energy systems are most widely used as elements of power supply systems for critical infrastructure (such as data centers, server infrastructure, etc.) and in distributed energy systems.But they have various disadvantages, the key of which is the inevitable degradation of battery cells, which can lead to premature failure of the entire system and increase operating costs. This problem can be solved through the use of battery management systems (BMS), allowing to optimize the process of operation, thereby extending the overall service life, one of the most important elements of which is a system for monitoring the parameters of the battery cells. In this regard, we considered some modern approaches aimed at organizing the

DOI: 10.4018/979-8-3373-7917-3.ch005

process of monitoring of operating energy storage systems, which are an integral part of distributed energy systems, the functioning of which must be maintained in real time without failures and outages.

1 INTRODUCTION

Energy storage systems can be defined as a number of different technologies that allow the storage and preservation of energy for later use. The current development of the energy industry leads to the need to use energy storage systems, which allow to solve several main tasks (Mitali et al., 2022):

- ensuring autonomy - the possibility of realizing autonomous energy systems, independent of the general network, the functioning of which is ensured through the use of alternative resources, such as renewable energy (Simankov et al., 2023);
- organization of storage facilities used to supply consumers with energy at times of high demand during operation of distributed energy networks dependent on several different generation sources (Buchatskiy et al., 2024);
- use as energy sources in various vehicles that comply with modern environmental standards (Teploukhov et al., 2024);
- organization of backup power supply systems for critical infrastructure (e.g. data centers).

The classification of energy storage systems includes several categories that allow evaluating both energy storage technologies and the application areas of a particular technology. There are many different technologies used to store different types of energy (Koohi-Fayegh & Rosen, 2020; Schischke et al., 2024): electrochemical and battery storage, thermal, thermochemical, pumping, magnetic, chemical and hydrogen storage, flywheels and compressed air storage, as well as more specific types such as biological energy storage. Thus, a short classification (Figure 1) can be made to illustrate the different categories of energy storage devices, providing information on their technical and economic characteristics as well as their respective applications.

Figure 1. Classification of energy storage systems

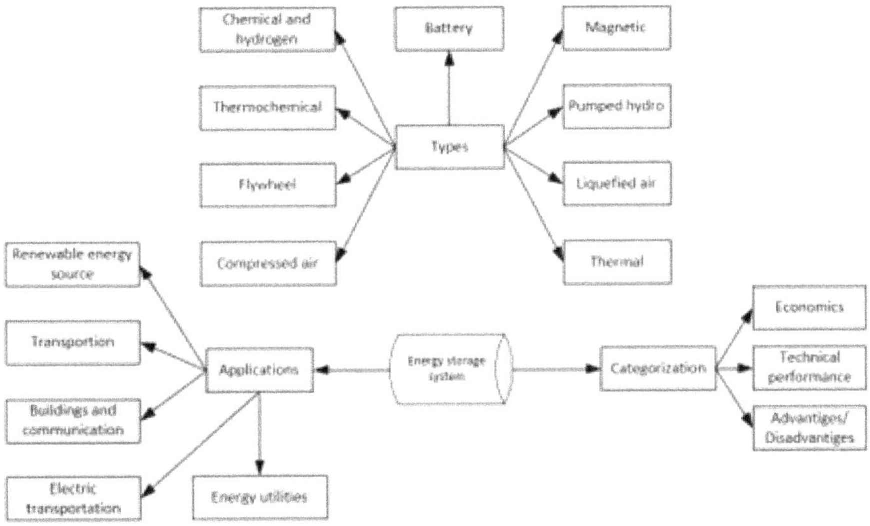

Despite the large number of existing energy storage technologies, not all of them have been widely spread, giving way to more productive and affordable technologies covering more areas of application, as a result of which, the most widespread are electrochemical technologies, which include the most popular lithium-ion and lithium-polymer batteries (Behabtu et al., 2020), facilitated by high levels of specific energy (energy per unit weight) and energy density (energy per unit volume). However, this does not solve the problem of inevitable wear and tear of the used cells and their gradual failure, as a result of which an important aspect is the organization of the management process of such systems (Wang et al., 2021), which allows to extend the service life of the used cells by organizing optimal cycles of charging and discharging of accumulator batteries, and the organization of monitoring systems for such systems allows to determine in advance the elements that need to be replaced, thus avoiding possible failures that lead to rapid failure However, the organization of such control systems is impossible without the implementation of the procedure of quality monitoring of the parameters of the investigated cells, for which it is necessary to use various techniques, a brief review of the main of which will be considered below.

2 METHODS OF MONITORING BATTERY CELL PARAMETERS

Battery management systems (BMS) are systems designed to regulate and optimize the process of battery operation, using electrical, mechanical or other means that allow the implementation of various control methods: charge and discharge control, temperature control, current status, voltage monitoring, etc. (Habib et al., 2023; Vaideeswaran et al., 2019). The key functions of the battery management system are shown in the figure below (Figure 2).

Figure 2. Main functions of the BMS

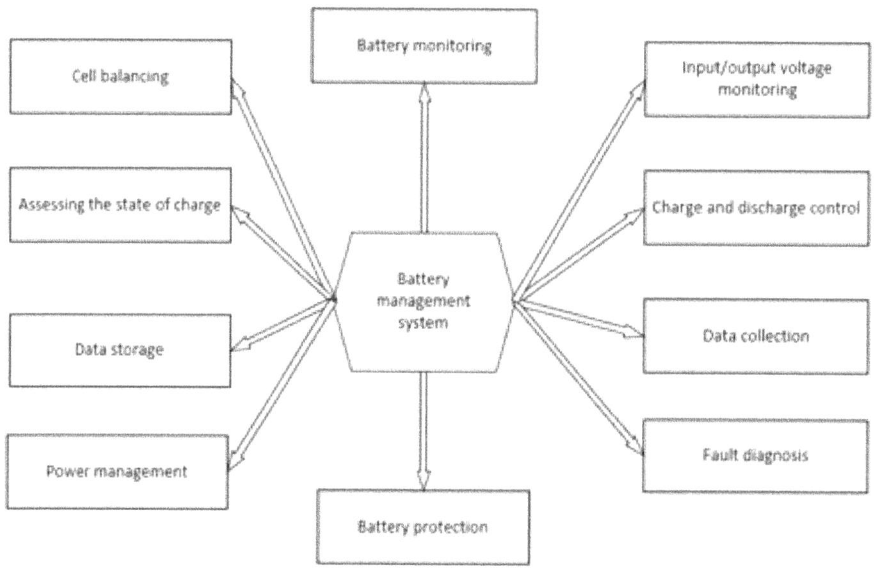

One of the key elements of a BMS is battery monitoring systems, which are part of control systems designed to control charge and discharge currents, monitor temperature, determine cell potential, and track current and voltage levels (Qays et al., 2020). The role of the monitoring system is of key importance in such systems, allowing the determination of all the main characteristics (Nyamathulla & Dhanamjayulu, 2024), so that aspects such as control, safety, balancing and regulation can be realized based on the data obtained from the monitoring of the battery cells. The structural diagram of the control system with integrated cell monitoring module is shown in the figure below (Figure 3).

Figure 3. Structure diagram of cell monitoring

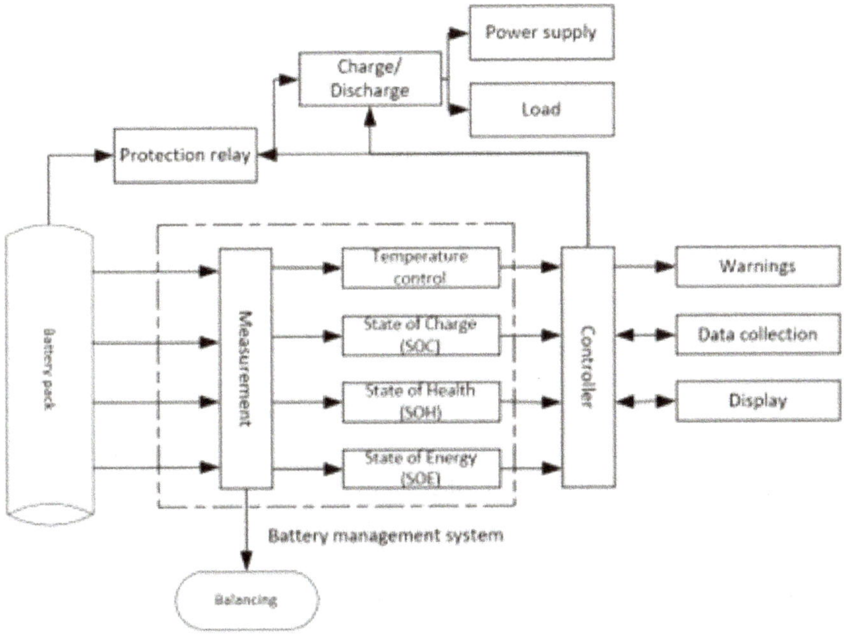

However, it is not always necessary to use a comprehensive management system, in which case it is possible to limit ourselves to measuring only the parameters used to assess the current state of the battery for which the following parameters are measured: state of charge, current capacity, remaining life, state of function, state of performance, state of energy, state of safety and state of temperature (Figure 4) (Snihir et al., 2006).

Figure 4. Battery status assessment system

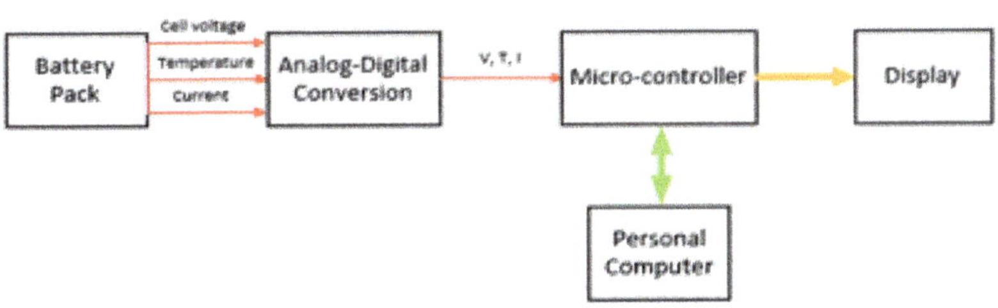

When organizing such a system, the first place is given to the measurement of such parameters as voltage and electric current (Wang et al., 2021), for which additional resistors and analog-to-digital converters are often used to learn these values, which are used later to calculate other characteristics.

One of the important parameters is the State of Health (or SOH), since it is this characteristic that determines the quality of the device performance and directly affects the service life (Yüksek & Alkaya, 2023). The following approaches can be used to estimate SOH:

- measurement of internal resistance is a simple approach, but its realization requires time, which makes it difficult to use in real objects and operating systems;
- internal impedance measurement - service life inevitably affects the internal impedance, both active and reactive. This is the most popular approach, as it provides an accurate indication of current wear;
- measurement of battery charge level - the main disadvantage of this approach is the need for empirical studies to establish the available dependence for the specific type of battery under consideration;
- approaches based on models of devices in use - this approach allows to study devices disconnected or connected to the network, but most models are very complex due to the non-linear nature of the processes involved;
- machine learning-based estimation methods - allow combining experimental and theoretical models. These methods provide high accuracy, but require a significant amount of data for training;
- organization of remote monitoring systems (e.g. based on the Internet of Things) (Burgio et al., 2023).

Classification of approaches to capacity assessment is presented in the figure below (Figure 5).

Figure 5. Classification of approaches to capacity assessment

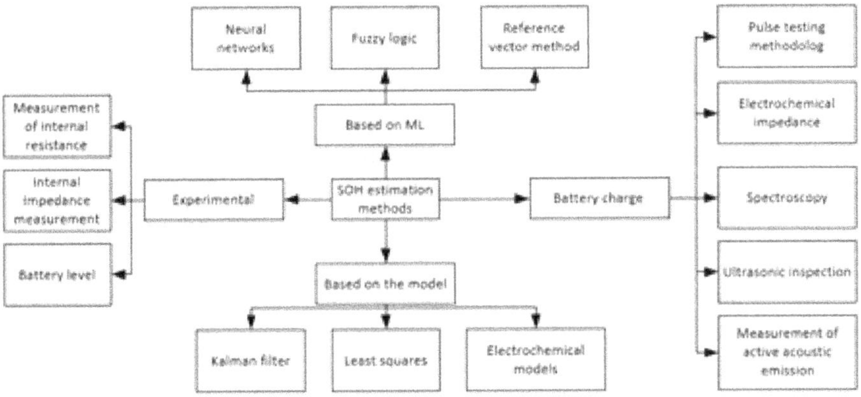

Among the considered approaches to the determination of the remaining battery capacity it is worth highlighting those based on the measurement of internal resistance, as the simplest in implementation and not requiring the construction of complex device models and the availability of large arrays of data on the operation of real systems. To realize the procedure of measuring this value, various approaches are used, among which the following can be distinguished (Schweiger et al., 2010):

- Ohm's Law - is based on measuring the voltage drop across the battery when a load is connected and calculating the internal resistance using the classical formula. This method is easy to implement and allows to quickly assess the current state of the battery, however, it is sensitive to the instability of the load current and does not always give accurate results with high battery capacity;
- Joule's Law - uses the measurement of the heat release of an accumulator when current flows through it. According to the Joule-Lenz law, the internal energy released as heat is directly proportional to the resistance. This method allows to determine not only the total internal resistance, but also to detect local defects, but requires additional temperature control and precise measuring devices;
- AC Resistance - consists of applying a low-signal time current to the battery at a certain frequency and measuring the voltage response. The internal resistance is calculated based on the phase difference between current and voltage. This method is widely used in industry because it provides a quick and

accurate estimate of battery resistance under various operating conditions, but its accuracy depends on the frequency of the signal;

- Electrochemical Impedance Spectroscopy (EIS) is the most accurate and informative. It is based on the analysis of the battery's AC response over a wide frequency range, which allows to study not only the total resistance but also the contribution of individual internal processes such as the ionic conductivity of the electrolyte or the polarization resistance. Although this method requires complex equipment and calculations, it is the gold standard in research and testing of high reliability ac-battery systems.

A more complete classification of battery resistance estimation methods is shown in the figure below (Figure 6) (Waag et al., 2014).

Figure 6. Classification of methods for estimating battery resistance

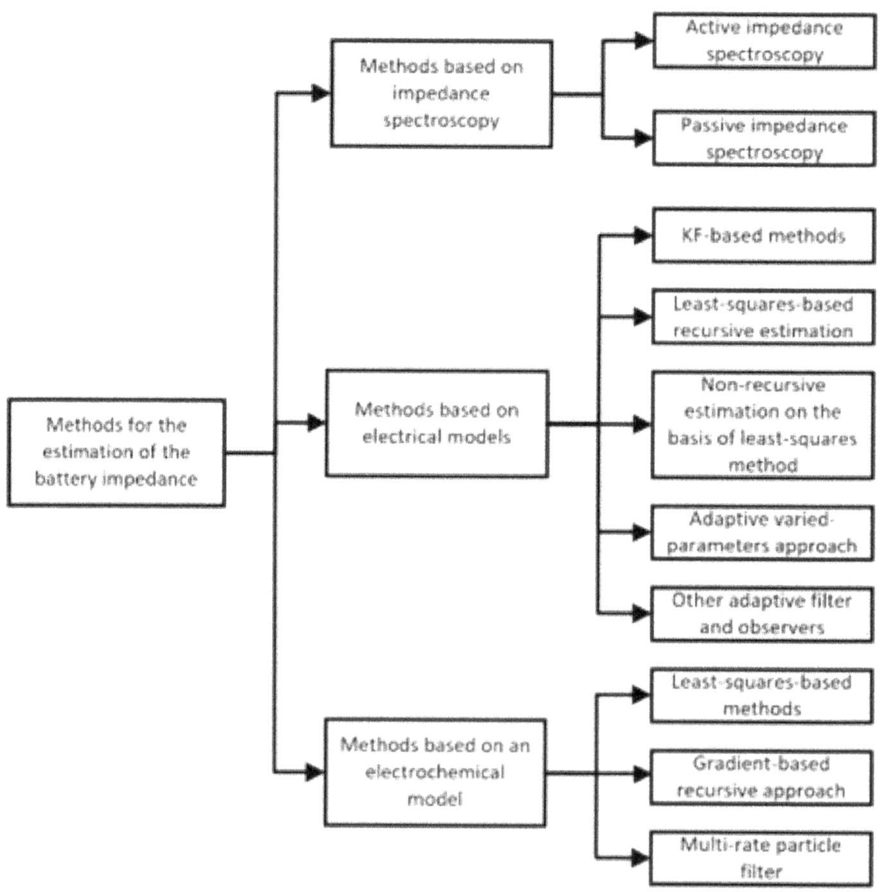

Another important characteristic is the residual life (RUL), the determination of which, under different operating conditions, is of great importance to accurately estimate the battery replacement or maintenance cycle, since it is considered that when the actual capacity of a Li-ion battery drops to 80% of its rated capacity, a critical battery failure is reached and the battery must be replaced. Unlike the previously discussed characteristics, this value can not be measured directly in any way, and its value can only be predicted on the basis of the obtained parameters of the functioning of the battery cell. There are a large number of different techniques for determining this value (Zhao et al., 2023), which can be divided into four broad groups of methods:

- Statistical methods - a group of methods based on the analysis of historical data on battery degradation, often using empirical models such as exponential and step dependencies between the number of charge-discharge cycles and capacity loss. The main advantage of statistical methods is their ease of implementation, but they do not take into account changing operating conditions and may give inaccurate predictions under atypical operating conditions;
- Physics-Based Models - these methods are based on mathematical models describing internal processes in the battery: lithium diffusion, electrolyte resistance, degradation of active materials. Although physicochemical modeling allows to analyze in detail the aging process of the battery, its complexity requires significant computational resources and knowledge in the field of electrochemistry;
- Machine learning methods - these methods use training on large data sets of battery operation data to reveal hidden patterns that allow predicting RUL with high accuracy. For example, recurrent neural networks (RNNs) and LSTM models work effectively with time series data, and the support vector method (SVM) is used for battery state estimation. The disadvantage is the need for large amounts of training data and the difficulty in interpreting the results;
- Hybrid approaches - by combining elements of physical and chemical modeling and machine learning, hybrid methods can improve the accuracy of RUL predictions. For example, you can use a physical model to generate training data and then apply neural network algorithms to adapt the predictions to real-world operating conditions. This approach improves the reliability of predictions and makes the prediction system more versatile.

3 CONCLUSION

This paper briefly reviewed the existing methods of monitoring and determining the key performance characteristics of electrochemical energy storage systems. Monitoring of energy storage systems allows organizing their efficient and careful use, thereby extending their service life, allowing early assessment of elements subject to degradation over time. The main approaches to the organization of monitoring in storage systems have been considered and emphasis has been placed on the existing methods of estimating the current capacity of batteries, their current state and remaining service life. The choice of the approach depends on the type of the used system, capabilities and requirements to the monitoring system, which are determined for each case individually.

REFERENCES

Behabtu, H. A., Messagie, M., Coosemans, T., Berecibar, M., Anlay Fante, K., Kebede, A. A., & Mierlo, J. V. (2020). A Review of Energy Storage Technologies' Application Potentials in Renewable Energy Sources Grid Integration. *Sustainability (Basel), 12*(24), 10511. DOI: 10.3390/su122410511

Buchatskiy, P. Y., Onishchenko, S. V., Teploukhov, S. V., & Kuzmin, K. A. (2024). "Modeling of Wind Generator in the Power Supply System of an Individual Consumer Using SimInTech," 2024 8th International Conference on Information, Control, and Communication Technologies (ICCT), Vladikavkaz, Russian Federation, pp. 1-5, DOI: 10.1109/ICCT62929.2024.10874937

Burgio, A., Cimmino, D., Nappo, A., Smarrazzo, L., & Donatiello, G. (2023). An IoT-Based Solution for Monitoring and Controlling Battery Energy Storage Systems at Residential and Commercial Levels. *Energies, 16*(7), 3140. DOI: 10.3390/en16073140

Habib, A. K. M. A., Hasan, M. K., Issa, G. F., Singh, D., Islam, S., & Ghazal, T. M. (2023). Lithium-Ion Battery Management System for Electric Vehicles: Constraints, Challenges, and Recommendations. *Batteries, 9*(3), 152. DOI: 10.3390/batteries9030152

Koohi-Fayegh, S., & Rosen, M. A. (2020). A review of energy storage types, applications and recent developments. *Journal of Energy Storage, 27*, 101047.

Mitali, J., Dhinakaran, S., & Mohamad, A. A. (2022). Energy storage systems: a review, Energy Storage and Saving, Volume 1, Issue 3, Pages 166-216, ISSN 2772-6835, DOI: 10.1016/j.enss.2022.07.002

Qays, M. O., Buswig, Y., Basri, H., Hossain, M. L., Abu-Siada, A., Rahman, M. M., & Muyeen, S. M. (2020). An Intelligent Controlling Method for Battery Lifetime Increment Using State of Charge Estimation in PV-Battery Hybrid System. *Applied Sciences (Basel, Switzerland), 10*(24), 8799. DOI: 10.3390/app10248799

Schischke, E., Grevé, A., Ehrenstein, U., & Doetsch, C. (2024). Overview of Energy Storage Technologies Besides Batteries. In Passerini, S., Barelli, L., Baumann, M., Peters, J., & Weil, M. (Eds.), *Emerging Battery Technologies to Boost the Clean Energy Transition. The Materials Research Society Series*. Springer., DOI: 10.1007/978-3-031-48359-2_4

Schweiger, H.-G., Obeidi, O., Komesker, O., Raschke, A., Schiemann, M., Zehner, C., Gehnen, M., Keller, M., & Birke, P. (2010). Comparison of Several Methods for Determining the Internal Resistance of Lithium Ion Cells. *Sensors (Basel), 10*(6), 5604–5625. DOI: 10.3390/s100605604 PMID: 22219678

Nyamathulla, S., & Dhanamjayulu, C. (2024). A review of battery energy storage systems and advanced battery management system for different applications: Challenges and recommendations. *Journal of Energy Storage, 86*, 111179.

Simankov, V. S., Buchatskiy, P. Y., Teploukhov, S. V., Cherkasov, A. N., & Onishchenko, S. V. (2023). *"Designing an Intelligent Information and Analytical System for Evaluating Solutions in Renewable Energy Based on Digital Twins," 2023 Seminar on Information Systems Theory and Practice*. ISTP., DOI: 10.1109/ISTP60767.2023.10426987

Snihir, I., Rey, W., Verbitskiy, E., Belfadhel-Ayeb, A., & Notten, P. H. (2006). Battery open-circuit voltage estimation by a method of statistical analysis. *Journal of Power Sources, 159*(2), 1484–1487.

Teploukhov, S. V., Lisov, D. N., Buchatskiy, P. Y., Bychkov, T. Y., & Onishchenko, S. V. (2024). "Development of the Module of Charging Stations Placement for Electric Transport Based on Genetic Algorithm," *2024 XXVII International Conference on Soft Computing and Measurements (SCM)*, Saint Petersburg, Russian Federation, 2024, pp. 123-126, DOI: 10.1109/SCM62608.2024.10554218

Vaideeswaran, V., Bhuvanesh, S., & Devasena, M. (2019). *"Battery Management Systems for Electric Vehicles using Lithium Ion Batteries," 2019 Innovations in Power and Advanced Computing Technologies (i-PACT)*. Vellore., DOI: 10.1109/i-PACT44901.2019.8959965

Waag, W., Fleischer, C., & Sauer, D. U. (2014). Critical review of the methods for monitoring of lithium-ion batteries in electric and hybrid vehicles. *Journal of Power Sources, 258*, 321–339. DOI: 10.1016/j.jpowsour.2014.02.064

Wang, S., Fan, Y., Stroe, D.-I., Fernandez, C., Yu, C., Cao, W., & Chen, Z. (2021). Battery system active control strategies. In *Battery System Modeling* (pp. 313–340). Elsevier., https://doi.org/10.1016/B978-0-323-90472-8.00006-8

Wang, Z., Feng, G., Zhen, D., Gu, F., & Ball, A. (2021). A review on online state of charge and state of health estimation for lithium-ion batteries in electric vehicles. *Energy Reports, 7*, 5141–5161. DOI: 10.1016/j.egyr.2021.08.113

Yüksek, G., & Alkaya, A. (2023). A novel state of health estimation approach based on polynomial model for lithium-ion batteries. *International Journal of Electrochemical Science, 18*(5), 100111.

Zhao, J., Zhu, Y., Zhang, B., Liu, M., Wang, J., Liu, C., & Hao, X. (2023). Review of State Estimation and Remaining Useful Life Prediction Methods for Lithium–Ion Batteries. *Sustainability (Basel), 15*(6), 5014. DOI: 10.3390/su15065014

Chapter 6

Leveraging Gamification, Course, and Individual Characteristics to Boost MOOC Adoption

Paulo Leitão
https://orcid.org/0009-0005-3942-8930
Independent Researcher, Portugal

Diana Sousa
Clan, Portugal

ABSTRACT

MOOCs have transformed global education by offering unprecedented access to knowledge. Yet, engagement and completion rates remain critical challenges. This study presents a conceptual model that examines the interplay between individual learner characteristics and MOOC design features towards course adoption and use, with special emphasis on the role of gamification in sustaining motivation and retention. By considering elements such as learning styles and gamification preferences of its users, MOOC designers can enhance the learning experience and encourage course completion. Beyond the conceptual model, we explore AI as a complementary technology feature, assuming that AI systems can leverage learner styles and MOOC characteristics, in an adaptive fashion, to further enhance engagement and retention. Overall this chapter highlights how integrating Gamified AI-driven innovations, tailored to individual and course characteristics, can strengthen MOOCs as a scalable and sustainable model of education, ensuring their long-term effectiveness in an evolving learning landscape.

DOI: 10.4018/979-8-3373-7917-3.ch006

INTRODUCTION

A Massive Open Online Course (MOOC) is an online interactive step-by-step course aimed at reaching an unlimited number of participants worldwide to create a community of lifelong learners. It's a recent form of global education that has been increasing in popularity over the last years (Ortega-Arranz, Muñoz-Cristóbal, Martínez-Monés, Bote-Lorenzo, & Asensio-Pérez, 2017), delivered in an environment with important differences from previous approaches to online education (Alraimi, Zo, & Ciganek, 2015). MOOC's classes are free and open to all, offering educational content that is generally comprised of short video lectures and supplementary materials. Its input is overseen either with peer-audit and gathering cooperation (e.g. peer assessment and group discussions), or via robotization (e.g. automated assessments and questionnaires), and it usually provides flexibility for learners to study at their own pace.

Coursera, edX, Udacity, Udemy, and Khan Academy are the most well-known MOOC platforms and are each related to exceedingly respected organizations of advanced education (Holdaway & Hawtin, 2013). These players have been conceiving solid partnerships with universities to offer a wider range of courses on different topics. With this strategy, since the creation of this concept, their adoption and number of users have been exponential (Song, Cheung, & Prud'Homme, 2017). Additionally, the integration of gamification elements—such as badges, points, leaderboards, and progress tracking — has also become an important feature in these platforms, designed to enhance the learning experience by capturing learners' attention and engagement (Khalil, Ebner & Admiraal, 2017; Nesterowicz, Bayramova, & Szádeczky, 2024).

MOOCs have changed instructing and learning by offering, through information and communication technologies and connectivity, a practical other option to customary classroom training. While some see MOOCs and traditional classroom courses as complementary paths to achieve higher quality teaching (Li & Yang, 2019), others have argued that open distance learning can, sometimes, be considered a more advanced form of higher education, because it resembles the openness exhibited by individuals in everyday life (Alraimi et al., 2015).

However, regardless of its momentum, the research community also perceives important shortcomings in MOOCs such as the lack of students' motivation, engagement, and interaction (Khalil & Ebner, 2014). The phenomenon is being challenged by its very low completion rates and high attrition (Estrada-Molina & Fuentes-Cancell,2022; Jordan, 2014), which obscures the learning and teaching quality, as well as the overall concept and experience. MOOC success requires continuing usage on the part of students to understand and assimilate the course material properly into their educational pursuits (Khan et al., 2017). Therefore, it is important to discover better approaches to inspire these learners and motivate them to finish the course. Although these high drop-out rates are still being researched (Hagedorn, Renz, Meinel, 2017), previous studies have already recognized a few factors that could clarify the low finish rates: Lack of enough time to follow the course; Different level of the course than expected; Absence of Motivation; Feeling of Isolation and Lack of interactivity; Interest only in a specific part of the course; and, Disappointment with the course (agenda, organization, etc.) (Khalil & Ebner, 2014; Gené, Nuñez & Blanco, 2014). Hagedorn, Renz, and Meinel (2017) also propose that, while some want to achieve a certificate and complete the course as good as possible, others consider the course a success if they just found one piece of information they were searching for. Thusly, to drop out is not always something bad for the participant. This highlights the importance of attending to the individual while studying MOOC adoption and usage, as individual characteristics and preferences might relate to different outcomes. This, in turn, affects how users intend to adopt and actually make use of MOOCs.

The raised matters bring us to this chapter's main focus: how do individual and MOOC characteristics interactively influence MOOC adoption? Are specific users more attracted and influenced to use MOOCs instead of other learning tools due to their inherent characteristics? While previous studies have mainly focused on how specific elements such as technology acceptance variables (e.g., Mendoza, Jung & Kobayishi, 2017) or some targeted MOOC characteristics (e.g., Alraimi et al., 2015) affect the adoption of MOOCs, we believe that the inherent characteristics of MOOCs and individual learner characteristics should also be put into deeper consideration. Seeking to address this gap, our study differentiates itself from previous literature by adopting an integrative perspective to create a conceptual model that combines elements of technology acceptance (UTAUT2), multiple characteristics inherent to the context of MOOCs (Gamification, Openness, and Reputation), and multiple characteristics of individual users (Sex, Age, Learner type, Player type), expanding previous research by deepening the interactive impacts between the various variables (moderation relations).

Futhermore, and in line with the recognition that the main challenges to guarantee engagement in MOOCs are individualized tutoring, interactivity, and feedback (Estrada-Molina & Fuentes-Cancell, 2022), we recognize the impact that Artificial Intelligence (AI) can have towards such factors, tailoring gamification strategies to each learner's individual progress and preferences, thus creating a more personalized and compelling learning journey that can improve student motivation, engagement and attendance rates (Limonova, Santos, Mamede, Filipe, 2023; Vrabie, 2023). With this in mind, we finish our book chapter by exploring the crucial role that AI-driven technologies can have in boosting motivation and engagement through the integration of game-like elements into MOOCs.

To achieve this, we explore how AI-powered adaptive learning systems can dynamically adjust gamified content based on the learner's performance (Fauvel,Yu, Miao, Cui, Song, Zhang, Li & Leung, 2018; Gligorea, Cioca, Oancea, Gorski & Tudorache, 2023; Song, Tang, Liu, Zheng, Wu, Feng & Zhang, 2021), ensuring that the difficulty of tasks and challenges aligns with their current skill level. This not only helps in maintaining a balanced level of challenge and engagement but also encourages continuous progression. Furthermore, AI-driven analytics can track a learner's achievements and milestones (Khalil et al., 2017), providing timely rewards or incentives designed to keep them motivated and committed to completing the course.

We then also explore on how AI can facilitate intelligent feedback and support within gamified MOOCs, with virtual tutors or chatbots that can provide instant feedback on gamified tasks, offer hints or additional resources when learners are struggling, and even suggest new challenges based on their progress (Limonova et al., 2023; Sarshartehrani, Mohammadrezaei, Behravan & Gracanin, 2024; Song et al., 2021). This kind of real-time, personalized support can help sustain learner motivation and reduce the likelihood of dropout by addressing issues as they arise.

In summary, MOOCs represent a transformative approach to education, expanding access and fostering a global community of learners. Despite their potential, significant challenges remain, particularly concerning learner motivation and course completion rates. Through a comprehensive conceptual analysis that explores MOOC and Individual characteristics, we aim to offer insights and practical recommendations for optimizing the MOOC experience, ensuring that these platforms not only attract but also retain learners in their pursuit of knowledge.

THEORETICAL BACKGROUND

MOOCs characteristics

MOOC courses have been sought as a unique opportunity to use technology to implement radical changes in higher education, making it truly universally accessible to everyone with mere access to the internet (Grünewald, Meinel, Totschnig & Willems, 2013; Khan et al., 2017). Additionally, and in a reality characterized by a fast pace of change and a knowledge-based economic model, such online courses can also play an important role in education equity and lifelong learning, creating a new education market that offers flexible, affordable, and fast-track learning opportunities, that present lower entrance barriers due to reduced (to inexistent) financial costs, time limitations, and geographical constraints (Huang, Jew, & Qi, 2023; Santos, 2021).

Despite the massive and heterogeneous growing momentum of both participation and course topics verified in MOOCs (Alraimi et al., 2015, Gené, et al., 2014; Hagedorn et al., 2017;), MOOCs emblematic characteristics have received scarce examination in previous research (Alraimi et al., 2015), with most of the discussions and studies focusing more on the characteristics that distinguish between two types of MOOCs: cMOOC and xMOOC.

While xMOOCs follow a more structured, classroom-like approach with centralized content delivery, regular assessments, and a focus on individual learning and knowledge duplication—often aligned with the cognitive-behaviorist model and offered by prestigious universities—cMOOCs seek to further disrupt the borders of higher education, emphasizing knowledge construction through the pedagogical lenses of connectivism (Dabbagh, Benson, Denham, Joseph, Al-Freih, Zgheib, Fake & Guo, 2016; Saadatdoost, Sim & Mei Hee, 2015). As such, cMOOCs seek to foster long-term engagement by encouraging interaction within a distributed network of peers, learning artifacts, and technologies, creating an exploratory space for new didactic methods that go beyond traditional settings (Dabbagh et al., 2016).

Regardless of pedagogical approach, and within the several specific characteristics of MOOCs, two can be seen as unequivocally vital for adoption: openness and reputation (Alraimi, et al., 2015).

Openness relates to the previously mentioned universal accessibility, deriving from MOOCs absence of cost, affiliation or other differential restrictions (Grünewald et al., 2017). MOOCs openness signify that potential learners can conveniently and timely enroll (adopt) and learn (use) from virtually anywhere and anytime, with very little restraining factors (internet access) (Khan et al., 2017). This characteristic fosters an environment of inclusivity by democratizing education on a global scale, making learning opportunities available to a much broader audience.

Reputation, on the other hand, can be seen as a differentiating factor within the MOOC ecosystem, especially given the ever-growing MOOC offerings (Alraimi et al., 2015). By associating with highly-regarded universities and institutions MOOC platforms can capitalize on their partners credibility to gain a competitive advantage on the adoption decision between diverse offerings (Holdaway & Hawtin, 2013; Alraimi et al., 2015), especially considering that such credibility can cater to job advancement intentions that motivate a large segment of MOOC users (Christensen, Steinmetz, Alcorn, Bennett, Woods & Emanuel, 2013). Hence, the reputation of the partnering institutions can enhance the perceived value of MOOC credentials, increasing their appeal among employers and further encouraging learners to enroll, through the expectation of acquiring both high-quality education and recognized qualifications.

In addition to reputation and openness - and as a strategy to deal with MOOCs high drop-out rates through boosted engagement -, gamification has been implemented on many MOOC platforms (Hagedorn et al., 2017, Nesterowicz, Bayramova, & Szádeczky, 2024), increasingly becoming an inherent feature within the MOOC context, although one that has yet to be deeply explored (Ortega-Arranz et al., 2017).

Despite the absence of a single, universally accepted definition, gamification is largely seen as the inclusion of game elements, mechanics and design techniques in non-game contexts, to influence the motivation or engagement of people to perform certain actions (Deterding, Sicart, Nacke, Ohara, & Dixon, 2011). In the case of MOOCs, this translates into the use of game elements towards the performance of learning activities.

The most typical game elements in MOOCs include the use of points, badges and team leaderboards, although other more complex, and potentially more motivating and engaging elements, such as virtual goods, duels, avatar customizations, or even augmented and virtual reality immersive experiences, are gaining an increased presence in MOOCs (Greene, 2023; Hagedorn et al., 2017; Nesterowicz, Bayramova, & Szádeczky, 2024; Ortega-Arranz et al., 2017). These advanced gamification strategies have the potential to address the greater desire for social interaction and personalization that characterizes modern learners' expectations (Nesterowickz et al., 2024), though challenges may arise due to differing attitudes towards gaming, that are especially prevalent among older adults (Barna & Fodor, 2018).

Regardless of the mechanisms and methodologies adopted, gamification is widely viewed as potentially enhancing learners motivation and the subsequent participation, forum usage and commitment of students towards the course (Gené et al., 2014). However, despite its potential, the true effectiveness of gamification in MOOCs continues to be debated, with its impact varying according to individual learner preferences, course structure, and the specific gamification elements integrated. The key challenge seems to lie in finding the optimal balance between these elements to maintain motivation without overwhelming learners or distracting them from the core educational objectives.

Adoption Models

Among the several theories used by researchers to study MOOC adoption and usage, the Theory of Planned Behavior (TPB), the Technology Acceptance Model (TAM), and the Unified Theory of Acceptance and Use of Technology (UTAUT), surface as the most popular and established models (Song et al., 2017). However, contrasting with the rapid rise of MOOCs, little research examining user's adoption intentions has yet been done within this specific context (Song et al., 2017).

Seeking to address this gap, and for the purpose of the present study, we adopt the UTAUT model as the framework in which to study MOOC adoption. This decision stems from the comprehensiveness of the model, that was built upon several technology acceptance theories (including both TPB and TAM), explaining a larger variance in behavioral intention and actual usage when compared to its predecessors (Venkatesh, Morris, Davis & Davis, 2003).

From the original 4 variables (performance expectancy, effort expectancy, social influence and facilitating conditions) that influence the intention to use a particular technology and its actual usage, the model later evolved to include three additional variables (Hedonic Motivation, Price Value, and Habit), giving rise to the UTAUT2 model (Venkatesh, Thong, & Xu, 2012). Table 1 describes the seven constructs understood as immediate determinants of a person's aim to adopt a given technology in light of the UTAUT2 model.

Table 1. UTAUT2 technology acceptance variables derived from Venkatesh et al. 2012

Construct	Description
Performance expectancy	the degree to which the use of a given technology will provide benefits to consumers in performing certain activities
Effort expectancy	the degree of ease associated with consumers' use of technology
Social influence	the extent to which a consumer perceives that relevant others (family and friends) believe s/he should use a particular technology
Facilitating conditions	consumers' perceptions of resources and support available to perform a behavior
Hedonic motivation	perceived enjoyment in what relates to the fun or pleasure derived from using a given technology
Price value	relates to consumers' cognitive trade-off between the perceived benefits and the monetary cost that they have to bear by purchasing a given technology
Experience and Habit	the extent to which people have acquired automatic behaviors from previous use of technologies that are easily performed due to learning

It should be noted that attending to both individual and social factors, the UTAUT2 model considers that individual variables such as age, gender and experience moderate the relationship between the above-described variables and technology adoption and use (Venkatesh et al., 2012).

Although previous studies have already applied the UTAUT model to MOOC adoption (e.g. Mendoza et al., 2017), to the best of our knowledge the extended version of UTAUT2 has yet to be investigated within the specific context of MOOCs.

For this study, we will adopt the UTAUT2 model of adoption, with the exception of the Price Value variable. It is our belief that given the free nature of most MOOCs, that impose no monetary costs to its adoption and use by consumers, part of the cognitive trade-off element will be compromised, and thus render it invalid.

Individual Characteristics

The different rates of active participation by MOOC users (Willems, Renz, Staubitz, & Meinel, 2014), propose that individual learner characteristics might be an important variable in explaining MOOC adoption and usage. In fact, previous studies have determined that several learner variables play a significant role in MOOC adoption (e.g. Chang, Hung, & Lin, 2015; Mendoza et al., 2017), expanding on the reduced specter of individual characteristics (e.g. age, gender and experience) already incorporated in the UTAUT2 model.

Following suggestions (Mendoza et al., 2017) on the necessity of identifying additional key learner variables that influence the adoption and actual MOOC completion, we propose that an individual's learning style and gaming style affect MOOC adoption and usage by creating different needs that learners expect MOOCs to answer (Hagedorn et al., 2017; Grünewald, et al., 2013).

Recognizing that newer generations have already been "born" in a digital environment, with their learning styles and knowledge acquisition patterns differing significantly from their predecessors (Limonova et al., 2023), and attending to the heterogeneity of MOOC users, it becomes increasingly important to analyze and develop MOOCs that are attractive to different learning styles (Grünewald, et al., 2013). This is crucial not only because diverse learners have varying preferences, but also because adapting to these differences can enhance cognitive engagement and motivation, both of which are key to successful learning outcomes. Forcing users to learn through methods that go against their preferences can result in worse learning experiences and increased desire to abandon the course (Hagedorn et al., 2017).

Seeking to define different learning styles, Kolb (1984) proposes that learning occurs within the framework of two great dimensions: perceiving and processing.

Encompassing the opposite poles of both dimensions, the learning cycle would thus comprise four different phases: concrete experience – associated with feeling the learning in direct ways –, abstract conceptualization – thinking about theories and learnings -, active experimentation – doing and applying the learning through active practices – and reflective observation – associated with reflecting on the different knowledge obtained (Kolb, 1984).

Learner's styles would result from the combination of both dimensions and potential learning phases, resulting in four preferential styles (Kolb, 1984): divergers (concrete experience + reflective observation); assimilators (abstract conceptualization + reflective observation); convergers (abstract conceptualization + reflective observation); and, accommodators (concrete experience + active experimentation).

By applying these learning styles within the MOOC context, we can derive that while the traditional video lectures and textual support resources adequately relate to the preferred learning situations of assimilating learners, only by adopting more active methods will MOOCs be able to fully engage and motivate more converging and accommodating learners (Grünewald et al., 2013).

Gamification elements may be a solution to the above requirement, promoting greater involvement of students by providing a broader set of learning resources tailored to different participant profiles (Gené et al., 2014). However, despite the potentially great impact of gamification on the MOOCs learning process (Gené et al., 2014), we cannot ignore that there will always be a fringe of the population that would not like to see playful content incorporated in MOOCs (Hagedorn et al., 2017). For example, students that are more accustomed and have found success through conventional teaching methods may feel nervous and resistant about transitioning to gamified learning environments (Limonova et al., 2023).

This resistance may be further influenced by individual differences in engagement with game-like elements. According to previous research, different player groups reflect different acceptance levels of gamified content. Specifically, more frequent players tend to exhibit a more positive view about generic playful content, regardless of the specific type involved (Hagedorn et al., 2017). Therefore, player types - as defined by frequency of play - may influence how well participants adopt MOOCs, particularly given the increased incorporation of gamification elements within the MOOC environment (Gené et al., 2014; Hagedorn et al., 2017 Ortega-Arranz et al., 2017).

CONCEPTUAL MODEL

The objective of this chapter is to identify how the interaction between technology acceptance factors, MOOC's contextual characteristics and learner's individual characteristics influence the intention to adopt MOOCs and MOOCs actual usage. To the best of our knowledge, this study proposes a research model that extends previous literature in several ways. First, specific MOOC's characteristics are incorporated with the UTAUT2 framework, studying their direct impact on MOOC's adoption and on technology acceptance variables. Second, we address a key gap in MOOCs literature by adopting a learning style perspective, expanding on how these impact MOOC's adoption. Finally, we explore how the individual characteristics of learners interact with the MOOC's inherent characteristics in influencing MOOCs adoption and actual use.

Previous studies have demonstrated the impact of several technology acceptance variables on MOOC adoption (e.g. Mendoza et al., 2017), with Performance Expectancy and Social Influence surfacing as the most significant factors regarding the intention to adopt MOOCs. Specifically, the users tend to trust that by adopting MOOC as a teaching tool their performance will be enhanced (Mendoza et al., 2017), in alignment with the common motivating factor of job advancement (Christensen et al., 2013). As such we propose that:

> H1 – Performance Expectancy will positively affect behavioral intention moderated by age and gender.

The definition of Effort Expectancy implies that users who find MOOCs easier to use will be more willing to adopt such tools, thus the requirements in terms of skills, effort and time that MOOCs impose on learners will influence the user's decision to adopt such courses (Mendoza et al., 2017).

> H2 – Effort Expectancy will positively affect behavioral intention moderated by age and gender

Regarding Social Influence, users tend to better embrace MOOCs when they see their relevant others (colleagues and friends) trusting such technology, encouraging its adoption by the user (Mendoza et al., 2017).

> H3 – Social Influence will positively affect behavioral intention moderated by age, and gender.

Facilitating conditions have also been shown to directly influence MOOC adoption (Mendoza et al., 2017) with the existence of a favorable set of conditions influencing the final decision to adopt and actually complete the course.

> H4 – Facilitating conditions will positively affect behavioral intention moderated by age and gender.
> H5 – Facilitating conditions will positively affect use behavior moderated by age and gender.

Hedonic Motivation refers to the level of fun or pleasure that might be derived from using a particular technology (Venkatesh et al., 2012), given that mere curiosity has been identified as a common motivating factor for engaging in MOOCs

(Christensen et al., 2013), we hypothesize that the enjoyability of the course content and methodologies will influence MOOC adoption

H6 – Hedonic motivation will positively affect behavioral intention moderated by age and gender.

Reflecting the result of previous experiences (Venkatesh et al., 2012), we propose that individuals who have previously engaged and completed similar learning situations as the ones reflected by the MOOC context will be more willing to enroll and complete MOOC courses.

H7 – Habit will positively affect behavioral intention moderated by age and gender.
H8 – Habit will positively affect use behavior moderated by age and gender.

Attending to the high drop-out rates and low participation (Jordan, 2014; Willem et al., 2014) that currently affect MOOCs, it is important to consider how to move beyond the intention to adopt and the enrollment of students in MOOC courses, as true MOOC success derives from students continuing use of the course material (Khan et al., 2017). Given that behavioral intention displays a strong influence on technology use (Venkatesh et al., 2003), we propose that:

H9 – Behavioral intention will positively affect use behavior.

In what relates to the inherent characteristics of MOOCs, reputation has been identified as a strong predictor of MOOC usage (Alraimi et al., 2015). However, it is possible that, in addition to the direct impact of reputation on MOOC usage, there could also exist an indirect impact of reputation on MOOC adoption through technology acceptance variables. In fact, reputation has been shown to influence perceived enjoyment (Alraimi et al, 2015), a variable similar to Hedonic Motivation, furthermore, it is our belief that the associated credibility that derives from the partnerships with highly regarded institutions can also benefit users that seek job advancement through MOOCs, thus influencing Performance Expectancy. We thus hypothesize that:

H10 – Reputation will positively affect use behavior.
H11 – Reputation will positively affect Hedonic Motivation.
H12 – Reputation will positively affect Performance Expectancy.

Similar to what has been previously described regarding reputation, MOOC openness has also demonstrated a positive influence on MOOC usage, while, simultaneously, directly influencing technology acceptance variables similar to Hedonic Motivation and Performance Expectancy (Alraimi et al., 2015). In addition to these variables, we propose that the flexibility derived from the ability to enroll and learn from virtually anywhere and anytime with very little restraining factors (Khan et al., 2017) provisioned by MOOC openness provides easy and effortless access to this technology. We thus propose that:

H13 – Openness will positively affect use behavior
H14 – Openness will positively affect Hedonic Motivation
H15 – Openness will positively affect Performance Expectancy
H16 – Openness will positively affect Effort Expectancy

Finally, by incorporating game design and techniques, previous studies argue that MOOC participants will be more motivated to both enroll and interact with course materials (Gené et al., 2014; Ortega-Arranz et al., 2017). However, it is worth noticing that a recent study (Hagedorn et al., 2017) has demonstrated that frequent players show more positive attitudes about playful content with high percentages of non-players disagreeing with the introduction of game content on MOOC environment. Thus:

H17 – Gamification will positively affect behavioral intention moderated by player type
H18 – Gamification will positively affect use behavior moderated by player type

Expanding on the impact of such individual variables, it is important to attend to the fact that providing users with learning methods that go against their preferred learning style is believed to result in an overall negative learning experience, diminishing users motivation to engage with course content and resulting in an increased yearning for dropping out of the course (Hagedorn et al., 2017). Reflecting such relationship, Hagedorn and collaborators (2017) have shown that the majority of users that reflect disagreement with playful content (associated with game elements), do so based on a mismatch with their learners type. As such:

H19 – Learning styles will positively influence use behavior.
H20 – Learning styles will moderate the relationship between gamification and behavioral intention
H21 – Learning styles will moderate the relationship between gamification and use behavior

A combination of the UTAUT2 constructs with new MOOC characteristics and Individual characteristics constructs are used as the theoretical support model for the present study, as seen on the figure below.

Figure 1. Research Model

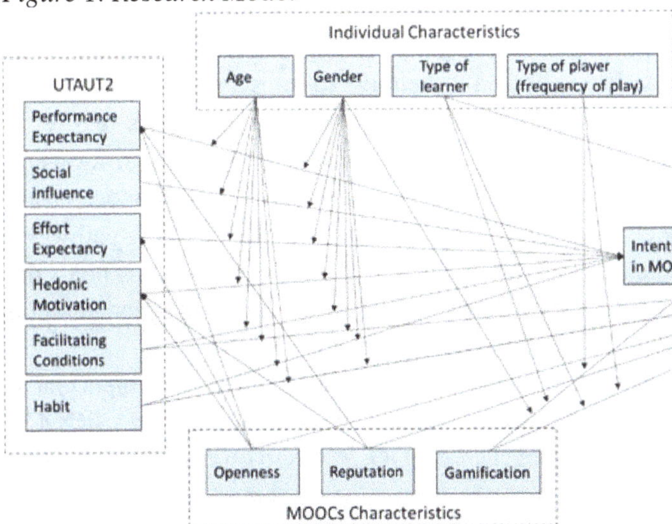

THE IMPACT OF AI IN MOOC ADOPTION

The Role of AI in harnessing Individual Characteristics to Personalize Learning Paths

As our world becomes increasingly driven by technology, it is reshaping the way we live, work, and learn, fueling growing interest in MOOCs, and contributing to higher registration numbers within this sector of the educational landscape. However, and as we have seen until now, it's increasingly important to address individual learning preferences and ensure an engaging learning experience if we want to guarantee that the number of registrations actually translate to a high number of course completion. Otherwise, the challenges faced by MOOCs, particularly in maintaining student motivation, engagement, and interaction (Khalil & Ebner, 2014), will continue to be evident.

Overall, MOOC's learning experience has encountered several difficulties, such as maintaining course quality while accommodating large numbers of students, or overcoming limitations on learner communication. Additionally, another critical weakness that has been pointed out to MOOCs is the lack of practical experience within these courses, which also contributes to higher dropout rates. To address these challenges, effective online learning requires a shift in teaching methods, with a stronger emphasis on personalization to cater to the diverse needs and idiosyncrasies of students (Yu, Miao, Leung & White, 2017).

While such level of personalization might have once be seen as unattainable, recent advancements in AI have allowed these tools to transcend traditional data-centric tasks, with their extensive capabilities and potential synergies (e.g. with gamification) being recognized as a promising solution to transform education and overcome the barriers that hinder MOOC's long-term success (Limonova et al., 2023; Vrabie, 2023). By enabling highly tailored and dynamic learning pathways, and by augmenting personalization, interactivity, and student support, AI holds the potential to foster more fruitful educational experiences, though its implementation requires careful planning and execution (Limonova et al., 2023).

To explore the practical application of AI in enhancing MOOCs, it's important to first revisit the two main formats that define these courses: cMOOCs and xMOOCs, each with distinct pedagogical approaches. cMOOCs, based on connectivism, emphasize interaction within a network of peers, learning resources, and technologies. Participants are encouraged to use various platforms to achieve personal goals, self-organize, and share knowledge. In contrast, xMOOCs are more structured and centralized, focusing on individual learning through video lectures and regular assessments, often linked to cognitive-behaviorist approaches and typically offered by prestigious universities (Dabbagh et al., 2016).

AI can play a significant role in enhancing both formats. As Gligorea and colleagues (2023) explain, AI in e-learning platforms adjusts content and task difficulty using adaptive algorithms, personalizing the learning experience based on performance. In cMOOCs, AI supports learner-driven interactions by tailoring learning paths according to engagement and contributions (Gligorea et al., 2023). Additionally, AI supports peer interactions in cMOOCs by mapping connections, analyzing discussions using Natural Language Processing, and suggesting relevant resources, further fostering collaboration and deeper learning (Conole, 2015). For xMOOCs, AI adapts content to maintain appropriate challenge levels, ensuring the learning process remains engaging and effective (Gligorea et al., 2023). To achieve this, AI uses sentiment analysis to assess student engagement, adapting content based on performance to ensure that learners are appropriately challenged (Conole, 2015). Moreover, AI contributes to forum moderation in xMOOCs by identifying frequently asked questions to which it is able to provide automated responses. Additionally, as Fauvel and colleagues (2018) highlight, AI can assist in analyzing substantial amounts of student-generated data, providing deeper insights into learning processes and facilitating the creation of more engaging, personalized learning experiences for both MOOC formats.

To accommodate for individual differences and characteristics, AI can collect detailed data on student behaviors - such as mouse clicks and participation in forums - using these inputs to identify different learning styles and adjust content accordingly. By identifying patterns of interaction, AI can then categorize students into different engagement types, such as "observers," "solvers," or "collectors," and tailor tasks and pacing to match the level of engagement of each learner (Fauvel et al., 2018). As students' progress, AI continues to analyze responses and interactions, adjusting task complexity to maintain an optimal challenge. Methods like Bayesian Knowledge Tracing are employed to predict the knowledge students have acquired and adjust subsequent tasks accordingly. Hence, if a student demonstrates proficiency in a topic, the system can increase the complexity of future activities, while providing simpler or review-based tasks for the students who struggle (Fauvel et al., 2018).

A complementary method for dynamically adjusting learning tasks and difficulty, relies on performance-based assessments - such as the number of completed tasks or participation in forums -, rather than learning styles. By leveraging these performance data points along the student's learning path, AI can fine-tune course content and task difficulty fostering a more effective learning experience (Fauvel et al., 2018; Gligorea et al., 2023). This personalized approach has been shown to improve academic performance, with some studies indicating increased test scores (Algarra, Bravo, Cevallos & Meneses, 2024; Gligorea et al., 2023). Thus, AI continuously monitors performance, adapting both content and difficulty to match student progress, with AI's ability to provide immediate and nuanced feedback, allowing the assessment process to move beyond grades, focusing instead on growth, learning and continuous improvement (Limonova et al., 2023)

Consubstantiating these ideas with a practical illustration of an actual tool that seeks to fulfill AI's potential in addressing MOOC's limitations, we can highlight the XiaoMu system that employs AI-driven interventions to enhance learning efficiency (Song et al., 2021). Key modules include concept extraction, automated video navigation, neural network-based dropout prediction, machine learning Q&A, reinforcement learning for active questioning, and personalized recommendations. This tool has been tested in concrete experiments with over 100,000 students spread across different courses, showing significant improvements in student engagement and outcomes (Song et al., 2021). These findings underscore AI's potential to personalize online education and improve the learning experience.

The Synergy Between AI and Gamification as a Response to MOOC Challenges

Complementary to the individual contributions of AI, the synergy between AI and gamification is capable of revolutionizing education beyond traditional data analysis, offering a disruptively potent method to amplify student motivation and involvement (Vabrie, 2023; Vaibhav & Gupta, 2014). While the application of game design principles in learning environments is not new, the advent of AI introduces the possibility to develop personalized game elements and rewards based on an individual's data points, preferences and interaction patterns (Modi & Gochait, 2023). This, in turn, makes feasible the creation of a tailor-made gamified environment that can help sustain optimal engagement.

By scrutinizing student behavior, AI can deliver immediate feedback and forecast future performance, using these capabilities to customize the learning journey to individual needs with the aim of maximizing learning outcomes (Abbes, Bennani & Maalel, 2024). When applied to gamification elements, such as scoring systems, competitive rankings, and digital accolades, AI customization allows a persistent cultivation of knowledge acquisition and accomplishment, nurturing a vibrant, inspiring, and more effective educational setting (Khalil et al., 2017).

Furthermore, and to address the challenges posed by feelings of isolation and lack of social interaction - that contributes to a portion of students dropping out of MOOC's (Khalil & Ebner, 2014; Gené et al., 2014; Nesterowicz et al., 2024) - AI can elevate the sense of community by promoting cooperative education. This can be achieved through the assembly of student teams within game-like scenarios, grouping individuals based on their unique tendencies and choices, and customizing gamified team challenges to encourage both personal and collective growth (Limonova et al., 2023; Nesterowicz et al., 2024). Building on the natural student inclination towards competition and success, which enhances learner drive and abilities, such a gamified approach also provides valuable opportunities for learners to cultivate crucial proficiencies like strategic thinking and decision-making within a safe, controlled environment (Greene, 2023).

In addition to the community contributions that such platforms can have, and to tackle the challenges of motivation and misalignment between the level/difficulty of the course and the expectations of students (Khalil & Ebner, 2014; Gené et al., 2014), educators can support themselves in AI-powered feedback mechanisms, to establish an ideal balance of challenges and motivations ultimately promoting engagement and improved learning outcomes (Limonova et al., 2023). One effective way to convey such "real-time" feedback relies on AI tutors in virtual environments, that can further personalize learning by adapting to students' emotions and understanding, mimicking human interaction and thereby boosting engagement and outcomes (Sarshartehrani et al., 2024). AI-guided immersive virtual environments offer dynamic, contextual learning experiences that were previously unattainable in online settings. Research continues to explore the potential of these AI tutors to tailor education by responding to various student factors, underscoring the increasingly pivotal role of technology in education (Sarshartehrani et al., 2024).

In conclusion, the integration of Artificial Intelligence and gamification offers a transformative opportunity to enhance student success in MOOCs by personalizing the learning experience and boosting engagement (Vabrie, 2023; Vaibhav & Gupta, 2014). AI facilitates tailored content delivery and real-time feedback mechanisms (Fauvel et al., 2018; Gligorea et al., 2023; Limonova et al., 2023), while gamification amplifies motivation and engagement (Khalil et al., 2017). Nevertheless, challenges persist, notably in the complexity of development, data quality management, and pedagogical design (Amo, Casany & Alier, 2014). The trajectory of online education indicates a future characterized by hyper-personalization, enhanced interactivity, continuous adaptive learning methodologies (Gligorea et al., 2023; Song et al., 2021), and sophisticated AI-driven tutoring systems (Sarshartehrani et al., 2024), ultimately aimed at improving student outcomes and fostering greater engagement within the MOOC environment (Yu, Miao, Leung & White, 2017).

Nevertheless, while technology holds great promise in moving away from the "one-size-fits-all" education model, it is equally important to ensure that the changes and innovations brought about by these tools do not overshadow the value of quality education or the unique opportunities that MOOCs provide in fostering inclusivity through the promotion of learning and skill development opportunities to a much broader audience.

CONCLUSION

Massive Open Online Courses (MOOCs) have undeniably transformed the educational landscape, providing millions of learners worldwide with unprecedented access to high-quality learning resources. The flexibility, affordability, and broad course offerings of platforms like Coursera, edX, and Udemy have made MOOCs an essential tool in democratizing education (Holdaway & Hawtin, 2013). However, as discussed, despite their widespread popularity, MOOCs continue to face persistent challenges related to low engagement, high dropout rates, and sustaining long-term motivation among learners (Khalil & Ebner, 2014). These issues not only affect the learner experience but also hinder the educational potential MOOCs can offer to individuals and societies.

At the heart of these challenges lies the interaction between learner characteristics and the inherent features of MOOCs. Previous research has shown that individual factors, such as age, gender, learning style, and personality traits, can influence how learners engage with MOOCs and whether they complete the courses (Chang et al., 2015; Mendoza et al., 2017; Willems et al., 2014).Similarly, the specific attributes of MOOCs - such as course content, delivery methods, and gamification - play a critical role in determining learner engagement and retention (Khalil et al., 2017; Nesterowicz et al., 2024). While the open and flexible nature of MOOCs makes them appealing to a wide range of learners, these same features can lead to a lack of structure and accountability, which may foster disengagement and, ultimately, dropout.

This chapter has underscored the need for a more nuanced understanding of how individual learner characteristics and the features of MOOCs interact to influence course outcomes. By combining elements of technology acceptance models with the unique characteristics of MOOCs and learners, we have developed a conceptual model that sheds light on the complex dynamics at play. This model emphasizes that MOOCs must be tailored to the needs and preferences of individual learners to foster a more engaging and motivating experience. As previous research highlights, cultivating student motivation is a crucial factor in determining academic success (Collaço, 2017), and AI-driven personalized approaches can significantly support this process.

One of the most promising ways to address these challenges is through the integration of AI technologies. As explored in this article, AI-driven adaptive learning systems have significant potential to personalize the learner experience in real time (Limonova et al., 2023). These systems can adjust task difficulty, provide personalized feedback, and track learner progress, helping to maintain motivation and engagement throughout courses. In a gamified environment, AI can offer immediate rewards, encouragement, and even virtual tutor strategies that have been shown to improve learner persistence (Khalil et al., 2017). This personalized approach ensures learners are both challenged and supported, enhancing their sense of achievement and progress.

Furthermore, AI can play a pivotal role in addressing MOOC challenges such as isolation and lack of interactivity, which have been identified as a key dropout factor in previous research (Khalil & Ebner, 2014; Gené et al., 2014; Nesterowicz et al., 2024). Through virtual tutors or chatbots, AI can offer instant feedback and guidance, replicating the real-time support typically available in traditional classrooms (Limonova et al., 2023). This interactive, personalized learning experience has the potential to reduce attrition rates, creating a more sustainable and rewarding environment for learners.

Gamification also remains a valuable tool for increasing engagement within MOOCs. By incorporating points, badges, leaderboards, and progress tracking, MOOCs can foster a sense of achievement and competition that motivates learners to stay committed to their studies (Khalil et al., 2017; Nesterowicz et al., 2024). However, and as per the model presented within this chapter, the success of gamification depends on its ability to be tailored to each learner's individual characteristics and preferences. Again, AI enables this level of personalization, ensuring that the gamified experience remains relevant and stimulating. Moving forward, MOOC designers and researchers should continue refining their gamification strategies, incorporating AI to create adaptive, learner-centered experiences that take into account the interaction between MOOC and Learned characteristics in promoting long-term engagement.

While the integration of AI and gamification offers a promising path forward, it is also crucial to address the challenges associated with these technologies. In fact, previous research (Takona, 2024) highlights significant concerns, including ethical issues related to data privacy and algorithmic bias, equity challenges such as the digital divide – i.e. the gap between those who have and those who lack equitable access to technology and internet access to support learning -, and the potential reduction in human interaction – which can further contribute to feelings of isolation and lack of interactivity, thus increasing learner attrition rates. Overcoming these challenges requires careful design and collaboration to ensure that AI integration aligns with core educational values and principles.

Nevertheless, if used correctly, the potential of AI extends beyond just improving general engagement. It also plays a crucial role in fostering inclusivity, particularly for neurodiverse learners. Several applications of AI technologies hold great promise in this transformation. Natural Language Processing and speech recognition tools, for example, can convert speech to text and vice versa, benefiting students with auditory or speech impairments (Chandramma et al., 2025). Additionally, AI technologies can offer descriptive audio for visually impaired students and identify inaccessible elements within educational materials (Chandramma et al., 2025). Overall, adaptive learning systems powered by AI can personalize educational content based on students' learning styles and needs, ensuring that each learner can interact effectively with the material. These advancements make it possible to create more accessible learning environments, ensuring that neurodiverse students, including those with auditory, visual, or cognitive challenges, can thrive within MOOCs.

While the challenges surrounding completion rates and learner motivation are not easily overcome, the integration of AI and gamification offers a promising path forward. By personalizing the learning experience, providing real-time feedback, and offering tailored support, MOOCs can evolve from simple informational tools to dynamic, learner-centered environments. The combination of AI's adaptive capabilities and the motivational power of gamification has the potential to usher in a new era of online education. One where learners are empowered to take control of their educational journey, while receiving the support needed to succeed (Rane, Choudhary & Rane, 2023).

In conclusion, MOOCs represent a powerful tool for expanding access to education on a global scale. However, to fully realize their potential, it is crucial to address the challenges that hinder their effectiveness, particularly in terms of learner engagement and course completion. By understanding the interaction between individual learner characteristics and the inherent features of MOOCs, and by leveraging the potential of AI and gamification, MOOCs can become more personalized, engaging, and motivating. With the right strategies, MOOCs can continue to grow and evolve, fulfilling their promise of providing accessible, high-quality education to learners worldwide.

As the educational landscape continues to evolve, future research on MOOCs must increasingly focus on leveraging AI to cultivate more inclusive and personalized learning experiences. Specifically, it is vital to investigate how AI can better support neurodiverse learners through adaptable interfaces, varied content formats, and tailored feedback mechanisms, thereby addressing their unique educational requirements.

Furthermore, AI offers significant opportunities to transform lifelong learning through the delivery of customized learning experiences and recommendations tailored to individual learners. It is essential to recognize that, according to the World Economic Forum (2025), more than 50% of the global workforce will require reskilling or upskilling by 2030. MOOCs can play a crucial role in supporting individuals undergoing reskilling and those pursuing continuous professional development through upskilling. By integrating AI within MOOC platforms, this technology can ensure that the skills and knowledge being acquired are aligned with each learner's specific needs and career goals. Simultaneously, AI can analyze labor market trends, enabling the personal development plan to be adjusted in response to significant shifts in job requirements or market preferences. By continuously refining learning pathways, AI can not only optimize individual growth but also contribute to broader societal goals, such as reducing skill gaps and promoting greater workforce adaptability in the face of evolving industries.

Finally, it is crucial to examine the impact of AI on the emotional well-being and mental health of learners within online educational environments. Studies exploring AI's ability to identify indicators of emotional distress and offer appropriate support can be instrumental in fostering a stronger sense of belonging and enhanced social interaction, both of which are critical for learner retention and academic success. By thoughtfully integrating these AI-driven innovations, MOOCs hold the potential to revolutionize education even further, making it more accessible, inclusive, and ultimately more effective for a wide spectrum of learners across the globe.

REFERENCES

Abbes, F., Bennani, S., & Maalel, A. (2024). Generative AI and gamification for personalized learning: Literature review and future challenges. *SN Computer Science*, *5*(8), 1–12. DOI: 10.1007/s42979-024-03491-z

Algarra, E. J. D., Bravo, C. B., Cevallos, M. B. M., & Meneses, E. L. (2024). Pedagogical and Technical Analyses of Massive Open Online Courses on Artificial Intelligence. *Applied Sciences (Basel, Switzerland)*, *14*(3), 1051. DOI: 10.3390/app14031051

Alraimi, K. M., Zo, H., & Ciganek, A. P. (2015). Understanding the MOOCs continuance: The role of openness and reputation. *Computers & Education*, *80*, 28–38. https://doi.or/10.1016/j.compedu.2014.08.006. DOI: 10.1016/j.compedu.2014.08.006

Amo, D., Casany, M. J., & Alier, M. (2014). Approaches for quality in pedagogical and design fundamentals in MOOCS. *Teoría de la Educación. Educación y Cultura en la Sociedad de la Información*, *15*(1), 70–89.

Barna, B., & Fodor, S. (2018). An empirical study on the use of gamification on IT courses at higher education. In M.Auer, D. Guralnick, & I. Simonics (Eds.), *Teaching and learning in a digital world: Proceedings of the 2017 International Conference on Interactive Collaborative Learning. Advances in Intelligent Systems and Computing, 715,* pp.684-692. DOI: 10.1007/978-3-319-73210-7_80

Chandramma, R., Babu, K. S., Ranjith, K. S., Sinha, A. K., Neerugatti, V., & Reddy, D. S. (2025, January). Enhancing E-learning Accessibility through AI (Artificial Intelligence) and Inclusive Design. In *2025 6th International Conference on Mobile Computing and Sustainable Informatics (ICMCSI)* (pp. 1466-1471). IEEE. DOI: 10.1109/ICMCSI64620.2025.10883148

Chang, R. I., Hung, Y. H., & Lin, C. F. (2015). Survey of learning experiences and influence of learning style preferences on user intentions regarding MOOCs. *British Journal of Educational Technology*, *46*(3), 528–541. DOI: 10.1111/bjet.12275

Christensen, G., Steinmetz, A., Alcorn, B., Bennett, A., Woods, D., & Emanuel, E. J. (2013). The MOOC Phenomenon: Who Takes Massive Open Online Courses and Why? SSRN *Electronic Journal*. Retrieved March 10, 2025 from https://papers.ssrn.com/sol3/papers.cfm?abstract_id=2350964 DOI: 10.2139/ssrn.2350964

Collaço, C. M. (2017). Increasing Student Engagement in Higher Education. *Journal of Higher Education Theory and Practice*, *17*(4), 40–47.

Conole, G. (2015). MOOCs as disruptive technologies: strategies for enhancing the learner experience and quality of MOOCs. *Revista de Educación a Distancia (RED)*, (39). Retrieved March 11, 2025 from https://revistas.um.es/red/article/view/234221

Dabbagh, N., Benson, A. D., Denham, A., Joseph, R., Al-Freih, M., Zgheib, G., Fake, H., & Guo, Z. (2016). Massive Open Online Courses. In: N., Dabbagh, A.D., Benson, A., Denham, R., Joseph, M., Al-Freih, G., Zgheib, H., Fake, & Z., Guo (Eds). *Learning Technologies and Globalization: Pedagogical Frameworks and Applications, 9–13.* DOI: 10.1007/978-3-319-22963-8_2

Deterding, S., Sicart, M., Nacke, L., Ohara, K., & Dixon, D. (2011). Gamification. using game-design elements in non-gaming contexts. *Proceedings of the 2011 annual conference extended abstracts on Human factors in computing systems* - CHI EA 11, New York, USA, 2425-2428. DOI: 10.1145/1979742.1979575

Estrada-Molina, O., & Fuentes-Cancell, D. (2022). Engagement and desertion in MOOCs: Systematic review. *Comunicar, 30*(70), 111–124. DOI: 10.3916/C70-2022-09

Fauvel, S., Yu, H., Miao, C., Cui, L., Song, H., Zhang, L., Li, X., & Leung, C. (2018). Artificial Intelligence Powered MOOCs: A Brief Survey. *2018 IEEE International Conference on Agents (ICA)*, 56-61. DOI: 10.1109/AGENTS.2018.8460059

Gené, O. B., Núñez, M. M., & Blanco, Á. F. (2014). Gamification in MOOC: challenges, opportunities and proposals for advancing MOOC model. *Proceedings of the Second International Conference on Technological Ecosystems for Enhancing Multiculturality* - TEEM 14. DOI: 10.1145/2669711.2669902

Gligorea, I., Cioca, M., Oancea, R., Gorski, A. T., Gorski, H., & Tudorache, P. (2023). Adaptive learning using artificial intelligence in e-learning: A literature review. *Education Sciences, 13*(12), 1216. DOI: 10.3390/educsci13121216

Greene, R. T. (2023). *The future of instructional design: Engaging students through gamified, personalized, and flexible learning with AI and partnerships.* e-Learning Industry. Retrieved on March 25, 2025 from: https://elearningindustry.com/future-of-instructional-design-engaging-students-through-gamified-personalized-flexible-learning-with-ai-and-partnerships

Grünewald, F., Meinel, C., Totschnig, M., & Willems, C. (2013). Designing MOOCs for the Support of Multiple Learning Styles. *Lecture Notes in Computer Science Scaling up Learning for Sustained Impact*, 371-382. DOI: 10.1007/978-3-642-40814-4_29

Hagedorn, C., Renz, J., & Meinel, C. (2017). Introducing digital game-based learning in MOOCs: What do the learners want and need? *IEEE Global Engineering Education Conference* (EDUCON). DOI: 10.1109/EDUCON.2017.7942987

Holdaway, X., & Hawtin, N. (2013). Major players in the MOOC Universe. Retrieved March 2, 2025 from: http://chronicle.com/article/The-Major-Players-in-the-MOOC/138817/

Huang, H., Jew, L. J. J., & Qi, D. (2023). Take a MOOC and then drop: A systematic review of MOOC engagement pattern and dropout factor. *Heliyon*, *9*(4), e15220. DOI: 10.1016/j.heliyon.2023.e15220 PMID: 37123924

Jordan, K. (2014). Initial trends in enrollment and completion of massive open online courses. *International Review of Research in Open and Distance Learning*, *15*(1), 133–160. DOI: 10.19173/irrodl.v15i1.1651

Khalil, H., & Ebner, M. (2014). MOOCs Completion Rates and Possible Methods to Improve Retention - A Literature Review. *In Proceedings of World Conference on Educational Multimedia*, Hypermedia and Telecommunications, 1236-1244.

Khalil, M., Ebner, M., & Admiraal, W. (2017, October). *How can gamification improve MOOC student engagement?* Paper presented at the European Conference on Game Based Learning, Graz, Austria.

Khan, I. U., Hameed, Z., Yu, Y., Islam, T., Sheikh, Z., & Khan, S. U. (2017). Predicting the acceptance of MOOCs in a developing country: Application of task-technology fit model, social motivation, and self-determination theory. *Telematics and Informatics*. Advance online publication. DOI: 10.1016/j.tele.2017.09.009

Kolb, D. (1984). *Experiential Learning: Experience As The Source Of Learning And Development*. Prentice-Hall.

Li, T., & Yang, N. (2019). Comparing MOOCs with traditional courses for quality teaching in higher education. *DEStech Transactions on Social Science, Education and Human Science*. DOI: 10.12783/dtssehs/meit2018/27637

Limonova, V., Santos, A., Mamede, H.M., & Filipe, V. (2023). The Research Context of Artificial Intelligence and Gamification to Improve Student Engagement and Attendance in Higher Education. *RE@D –Revista de Educação à Distância e Elearning,* Vol.6, Nº2, 1-17.

Mendoza, G., Jung, I., & Kobayashi, S. (2017). A Review of Empirical Studies on MOOC Adoption: Applying the Unified Theory of Acceptance and Use of Technology. *International Journal for Educational Media and Technology, 11*(1), 15–24.

Modi, T., & Gochhait, S. (2023). Impact of Artificial Intelligence on Gamification: Current Applications. *2023 International Conference on Innovative Data Communication Technologies and Applications (ICIDCA),* pp. 287-290. DOI: 10.1109/ICIDCA56705.2023.10099771

Nesterowicz, K., Bayramova, U., & Szádeczky, T. (2024). Gamification in Massive Open Online Courses After the Pandemic. In L.A., Tomei and D.D. Carbonara (Eds.) *Exploring Technology-Infused Education in the Post-Pandemic Era* (pp. 489-513), IGI Global DOI: 10.4018/979-8-3693-2885-9.ch017

Ortega-Arranz, A., Muñoz-Cristóbal, J. A., Martínez-Monés, A., Bote-Lorenzo, M. L., & Asensio-Pérez, J. I. (2017). How Gamification Is Being Implemented in MOOCs? A Systematic Literature Review. In Lavoué, É., Drachsler, H., Verbert, K., Broisin, J., & Pérez-Sanagustín, M. (Eds.), Lecture Notes in Computer Science: Vol. 10474. *Data Driven Approaches in Digital Education. EC-TEL 2017.* Springer. DOI: 10.1007/978-3-319-66610-5_40

Rane, N., Choudhary, S., & Rane, J. (2023, November). *Education 4.0 and 5.0: Integrating Artificial Intelligence (AI) for Personalized and Adaptive Learning. SSRN.* DOI: 10.2139/ssrn.4638365

Saadatdoost, R., Sim, A. T. H., Jafarkarimi, H., & Mei Hee, J. (2015). Exploring MOOC from education and Information Systems perspectives: A short literature review. *Educational Review, 67*(4), 505–518. DOI: 10.1080/00131911.2015.1058748

Santos, C. L. (2021). Formação Complementar. In Leitão, P. (Ed.), *Transição entre o mundo académico e o mundo laboral:Um mundo de oportunidades* (pp. 59–74). Editora Rh.

Sarshartehrani, F., Mohammadrezaei, E., Behravan, M., & Gracanin, D. (2024, June). Enhancing e-learning experience through embodied AI tutors in immersive virtual environments: A multifaceted approach for personalized educational adaptation. In *International Conference on Human-Computer Interaction* (pp. 272–287). Cham: Springer Nature Switzerland. DOI: 10.1007/978-3-031-60609-0_20

Song, Z., Tang, J., Liu, T. X., Zheng, W., Wu, L., Feng, W., & Zhang, J. (2021). XiaoMu: An AI-driven assistant for MOOCs. *Science China. Information Sciences, 64*(6), 164101. DOI: 10.1007/s11432-019-9939-2

Song, Z. X., Cheung, M. F., & Prud'Homme, S. (2017). *Theoretical Frameworks and Research Methods in the Study of MOOC/e-Learning Behaviors: A Theoretical and Empirical Review. New Ecology for Education — Communication X Learning.* Springer., DOI: 10.1007/978-981-10-4346-8_5

Takona, J. (2024). AI in Education: Shaping the Future of Teaching and Learning. *International Journal of Current Educational Studies*, *3*(2), 1–25. DOI: 10.46328/ijces.121

Vaibhav, A., & Gupta, P. (2014, December). Gamification of MOOCs for increasing user engagement. In *2014 IEEE International Conference on MOOC, Innovation and Technology in Education (MITE)* (pp. 290–295). IEEE. DOI: 10.1109/MITE.2014.7020290

Venkatesh, V., Morris, M. G., Davis, G. B., & Davis, F. D. (2003). User Acceptance of Information Technology: *Toward a Unified View. Management Information Systems Quarterly*, *27*(3), 425–478. DOI: 10.2307/30036540

Venkatesh, V., Thong, J. Y. L., & Xu, X. (2012). Consumer Acceptance and Use of Information Technology: Extending the Unified Theory of Acceptance and Use of Technology. *Management Information Systems Quarterly*, *36*(1), 157–178. DOI: 10.2307/41410412

Vrabie, C. (2023). Education 3.0 – AI and gamification tools for increasing student engagement and knowledge retention. *Lecture Notes in Business Information Processing*, *495*, 74–87. DOI: 10.1007/978-3-031-43590-4_5

Willems, C., Renz, J., Staubitz, T., & Meinel, C. (2014), Reflections on Enrollment Numbers and Success Rates at the openHPI MOOC Platform, *Proceeding of the Second MOOC European Stakeholders Summit (EMOOCs2014),* Lausanne, Switzerland, 101-106

World Economic Forum. (2025). *The Future of Jobs Report 2025.* Geneva: World Economic Forum. Retrieved March 28, 2025 from https://reports.weforum.org/docs/WEF_Future_of_Jobs_Report_2025.pdf

Yu, H., Miao, C., Leung, C., & White, T. J. (2017). Towards AI-powered personalization in MOOC learning. *npj Science of Learning, 2(1)*, 15. DOI: 10.1038/s41539-017-0016-3

Chapter 7
Trade vis-à-vis Human Rights of Farmers Under TRIPS

Ananya Pandey
https://orcid.org/0000-0002-2419-6314
Christ University, Bangalore, India

Achyuta Nanda Mishra
https://orcid.org/0000-0002-4113-0974
Christ University, Bangalore, India

ABSTRACT

The Agreement on Trade-Related Aspects of Intellectual Property Rights (TRIPS) plays a crucial role in shaping the global intellectual property system that significantly impacts farmers' rights. In agriculture, TRIPS grants intellectual property protection in the form of patents and plant variety protection. Such protection conflicts with the farmer's rights to save, sow and exchange the seeds. This raises human rights concerns related to farmers such as access to genetic resources and their cultural rights, food security, livelihood that are enshrined under various international human rights instruments such as International Covenant on Economic, Social and Cultural Rights (ICESCR) and Universal Declaration on Human Rights (UDHR). This intersection between TRIPS and the human rights of farmers requires a more inclusive approach that balances human rights with innovation. In light of the above perspective, this chapter analyzes farmers' human rights integration under the domain of intellectual property rights protected under TRIPS.

DOI: 10.4018/979-8-3373-7917-3.ch007

INTRODUCTION

The most extensive multilateral agreement on intellectual property is the World Trade Organization (WTO) administered Agreement on Trade-Related Aspects of Intellectual Property Rights (TRIPS). It came into force on 1 January 1995. It is essential for promoting trade in innovation and knowledge, settling intellectual property trade disputes, and allowing WTO members to pursue their domestic policy goals. It frames the intellectual property system in terms of innovation, and technology transfer. The Agreement is a formal acknowledgement of setting minimum standards for protecting intellectual property by the member states.

However, the stringent intellectual property protection of the Agreement has been a subject of primary concern particularly in sectors like agriculture, pharmaceuticals, and biotechnology and their impact on access to resources and traditional knowledge. The agriculture sector is significantly impacted by the protection of Intellectual Property Rights (IPRs), particularly patents and plant variety protection (Wright & Pardey, 2006). Article 27.3 (b) of the TRIPS Agreement mandates that plant and animal inventions, as well as plant variations, can be protected by patents, an effective sui generis system, or a combination of the two (WTO, n.d.). This has led to an increase in the control of corporate interest over the seeds. It also results in restricting farmers' trading and seed-saving practices and raising concerns about food security, biodiversity and farmers' rights. The expansion of IPRs over the agricultural sector in the form of patent and plant variety protection, has led to the erosion of traditional farming practices that resulted in the marginalization of small-scale farmers, particularly in the Global South (Borowiak, 2004). The farmers are made to be utterly dependent on the patented seeds, which restricts their autonomy for seed selection, establishes monopolization, reduces the expansion of biodiversity and threatens traditional agricultural practices (Kruft, 2001). This adversely affects the human rights of farmers' such as the right to seed, food security and livelihood. TRIPS have led to concerns over biopiracy, where corporates exploit traditional agricultural knowledge without proper benefit sharing with the indigenous communities (Imran et al., 2021). This results in a threat to their food security and weakens the agriculture system. The TRIPS framework has profoundly affected, farmers' rights particularly in the Global South, where the impact of stringent intellectual property law has limited the seed sovereignty (Rogers, 2010). The Agreement majorly prioritizes the corporate interest of the industrialized nations over the societal welfare that exacerbates inequalities between the Global North and the Global South. In light of the above perspective, the chapter critically examines the TRIPS Agreement from a human rights perspective, particularly in relation to patents, plant variety protection and farmers rights. The potential for alternative models such as open-access seed culture and community seed banks, will also be explored as measures to counterbalance the impact of TRIPS on farmers' rights.

TRIPS AND AGRICULTURE

The *Punta del Este* Declaration of 1986 launched a series of negotiations on "trade-related aspects of intellectual property rights" as part of the Uruguay Round mandate for international trade negotiations (Sauvé, 2017). These negotiations eventually resulted in the signing of the TRIPS Agreement, an essential component of the single initiative that formed the legal basis for the newly established WTO in 1995 (Watal et al., 2015). The TRIPS Agreement specifies minimum criteria for the protection of IPRs in each of the identified categories of intellectual property, implying that WTO Members may provide higher but not lower levels of intellectual property protection (Correa & Yusuf, 2016). The TRIPS Agreement provides these standards primarily by combining many multinational intellectual property accords (Mitchell & Voon, 2009). The Agreement strongly aims to harmonize trade liberalization with advancing innovations in the form of IPRs.

In the agriculture sector, the patent and plant variety protection accorded to the plant varieties under the Agreement (Article 27.3(b)), has diverse approaches across several nations. Where the developed countries favor stringent patent protection, the developing and underdeveloped countries (such as India and Africa) on the other hand, advocate for flexible and farmer-friendly practices.

Article 27.3 (b) of the Agreement reads as follows:

"(3) Members may also exclude from patentability:

(b) plants and animals other than micro-organisms, and essentially biological processes for the production of plants or animals other than non-biological and microbiological processes. However, Members shall provide for the protection of plant varieties either by patents or by an effective sui generis system or by any combination thereof. The provisions of this subparagraph shall be reviewed four years after the date of entry into force of the WTO Agreement." (WTO, n.d.)

This provision exclusively stated that the member nations shall give protection to plant varieties in the form of patents, plant variety protection or a combination of the protection measures. This provision has significant implications for plant research, biodiversity, farmer rights, and food security (Aggarwal & Satpute, 2013). Article 27.3 TRIPS covers plant breeders, although underdeveloped nations encounter challenges in accessing this protection. Patents cannot protect traditional knowledge systems and communal innovations made by indigenous peoples over lengthy periods due to the need for a recognized inventor (Breining-Kaufmann, 2005). Today, scientific plant breeding relies on community-based and traditional knowledge from farmer varieties. However, TRIPS do not recognize traditional knowledge as it is not intended to conserve it. Significantly, after the adoption of the Agreement, it sparked a contentious debate related to biopiracy, access to genetic resources, protection of traditional knowledge, in countries with rich biodiversity (Antons, 2010). TRIPS, promoted by business interests in the North, aimed to maximize profits. This would not have been possible without legal access to indigenous knowledge, techniques, and resources from the South (Nordin, 2011). The Agreement is being used to legally transfer biodiversity and traditional knowledge from the South to the North, as the North has financial resources, legal muscle, and technical superiority while the South lacks these resources (Adhikari, 2005).

With regard to the genetically modified seeds and plants, the corporations that spent a considerable amount of money in patenting those ‘own them’ (Fuglie & MacDonald, 2023). The others are supposed to pay the cost of patented seed and plant to the patent-holding corporations for their usage. Since the majority of the world's genetic diversity resides in the South, traditional national corporations patent genetically modified crops and plants without compensation to the traditional farming community for their efforts in the original research (Dhar & Anuradha, 2004). Such practices also lead to the monopolization of seeds, leading to loss of biodiversity. In this perspective, the Agreement prioritizes corporate interest over the traditional knowledge of the farmers (Wynberg, 2023). Furthermore, the inclusion of such practices by the corporations as promoted by the Agreements, negatively impacts farmers' rights such as access to seeds and knowledge, which directly affects their right to livelihood, food security and environment (Nabuuma et al., 2022).

FARMERS RIGHTS UNDER INTERNATIONAL LAW

Farmers provide the food that the world relies on, feeding the planet. The first legally binding international agreement to formally acknowledge the role that farmers and local and indigenous communities play in the development and conservation of plant genetic resources for food and agriculture (PGRFA) is the International Treaty on Plant Genetic Resources (International Treaty), 2001 (Lawson, 2015). Regarding plant genetic resources for food and agriculture, Article 9 of the International Treaty calls for recognizing, realizing, and advancing farmers' rights (Prip, 2024). The protection, advancement, and realization of farmers' rights can be achieved through a several policies and procedures (Stringer et al., 2020). Farmers and agricultural communities can carry out their responsibilities as PGRFA creators and stewards and continue to feed the globe for future generations if these rights are promoted and realized (Bjørnstad, 2004).

Human Rights Dimension of Farmers Rights

Farmer's Rights are deeply rooted in human rights, encompassing the fundamental freedom related to food security, IPRs and sustainable livelihoods (Winter, 2010). The protection of farmers' rights not only ensures their economic and societal well-being but also steps towards the contribution of global food security, expansion of biodiversity and rural development (Food and Agriculture Organization (FAO), 2017). The core human rights approach to farmers' rights can be reflected in the right to food, livelihood, work, participation and cultural heritage (Apodaca, 2012). All these dimensions are reflected under several international instruments such as the United Declaration on Human Rights (UDHR), the International Covenant on Economic, Social and Cultural Rights (ICESCR) and the United Nations Declaration on the Rights of Peasants and Other People Working in Rural Areas (UNDROP). The United Nations Committee on Economic, Social and Cultural Rights (CESCR) mandates the protection of human rights over trade agreements.

One of the most critical aspects of farmers' rights is the right to food and livelihood, which is in close association with access, use, saving and exchange of seeds (Adhikari et al., 2021). Furthermore, the right to work is another aspect that complicates the dependence of small-holder farmers for their livelihood (Rodgers, 2021). In order to sustain productivity, ownership rights are essential, i.e. farmers' right to own the land and agricultural productivity. Additionally, the right to participation is another crucial dimension that helps the farmers to raise their voice against the seed monopolization and participate in the policy-making process that significantly affects their communities (Haugen, 2020). Also, cultural rights are crucial to farmers since traditional and indigenous knowledge is a part of their cultural heritage. Traditional knowledge includes both the content of knowledge and traditional cultural representations, such as particular signs and symbols (Correa, 2001). In a narrow sense, traditional knowledge refers to knowledge in general, specifically knowledge coming from intellectual effort in a traditional setting, which encompasses know-how, practices, skills, and innovations (Aguilar, 2001). It appears in various contexts, including agricultural, scientific, technological, ecological, and medical knowledge, as well as biodiversity-related information. Upholding the human rights dimensions of farmers' rights is crucial to ensure food security, social justice and sustainable development (Cullet, 2004; Haugen et al., 2011).

INTERSECTION BETWEEN TRIPS AND FARMERS RIGHTS

The intersection between TRIPS and farmers' rights raises multiple contradictions concerning plant variety protection and its potential to impact the traditional knowledge of farmers and indigenous communities. This leads to hindrances in the human rights of farmers, such as the right to food, livelihood, culture and food security.

Human Right to Food and Agriculture

Food is a fundamental commodity that plays a significant role in society's culture and policies. It is a cultural statement, biological necessity, and a source of commerce, due to increased productivity and trade opportunities (Assan, 2023). However, modern trade regulations and multilateral trade agreements such as TRIPS have the potential to conflict between the culture and human rights related to food and agriculture (Wang et al., 2023). The TRIPS Agreement is of concern in terms of the right to food, particularly its possible influence on farmers' rights. Where under Article 11 of the ICESCR, the state parties are authorized to make individual or collaborative efforts to enhance food production, preservation, and distribution practices by utilizing all available technical and scientific knowledge, educating people about nutrition principles, and creating or reorganizing agricultural systems to maximize the use and development of natural resources (Alston, 1984), TRIPS on the other hand, enhances plant variety protection that limits farmers' ability to conserve, exchange, and replant seeds (Cullet, 1999). Farmers who use high-yield or unique seed varieties from plant breeders need to acquire permission from the breeders before reusing them. Granting monopoly rights to breeders (in line with TRIPS) worsens the crisis of malnutrition and food insecurity (Perrin, 1994). The stringent intellectual property protection imposed by TRIPS on seeds and genetically modified plants raises concerns over impact on food security and the ability of small farmers to sustain their livelihood (Amantae et al., 2024). This practice limits farmers' rights to freely exchange, save and sow the seeds, which results in the plant breeders and corporations monopolizing seeds. This contradicts the ICESCR's emphasis on equitable access to food (Oguamanam, 2006).

Impact on Farmers in Global South

Historically, farmers relied upon their traditional knowledge to sustain their livelihood and enhance food security, but the rise of patented seeds and genetically modified crops, which are subject to control by multinational corporations and agribusinesses, has created a system of dependency on those seeds and plants (Azadi et al., 2016). This creates an excessive financial burden on the smallholder farmers who are bound to buy the seeds yearly. With the gradual implementation of IPRs over specific plant varieties, control over genetic and biological resources and associated information held by commercial enterprises, states, and individual farmers has grown more controversial (Blankeney, 2004). The implementation of

IPRs in agriculture brings up particular issues regarding farmers' control over their resources and expertise. At the expense of small and subsistence farmers, IPRs generally helps agribusinesses control seeds and associated information (Oguamanam, 2014). Also, many issues pertaining to research and development are exacerbated by the implementation and expansion of IPRs in agriculture. There are concerns that the biotechnology industry's "over-patentability" could hinder rather than foster innovation in the public and commercial sectors (Dludlu, 2014).

It is important to consider the implementation of IPRs in agriculture in a broader context, which includes how it affects biodiversity management. The long-term viability of agricultural systems primarily depends on biodiversity, and agro-biodiversity in particular (Srivastava, 2006). Agro-biodiversity is particularly significant since it directly affects human nutrition. Since biological resources are the primary input to agricultural production systems and the majority of current agricultural products have evolved through the selection and gathering of plant and animal species, agriculture and biodiversity management are closely linked (Scherr & McNeely, 2008).

Farmers' Human Rights Violation under TRIPS

Since major businesses and multinational corporations began to obtain global control of the seed market, seeds have also been about big-time generating money. The standard has shifted from variety to monopolization (Mahoney, 2018). As the new system of seed distribution, monopoly control based on property rights gradually replaced sharing. Seeds have become a worldwide commodity in the service of industrial farming and large businesses, with less or no regard paid to local adaptation to the distinctive methods, habitats, and needs of family farms (Louwaars, 2007). After implementing the TRIPS Agreement and the privatization of the seed system, farmers and the local communities have become vulnerable to access, use, save and exchange the seeds freely (Matthews, 2003). This practice undermines food security, disrupts traditional agricultural practices and increases economic vulnerabilities for the farmers that directly affect their livelihood (Yazdanpanah et al., 2021). The Agreement also has several adverse socio-economic and cultural consequences. For instance, biopiracy i.e., the unlawful takeover of agricultural and indigenous communities' expertise and genetic resources by individuals or institutions seeking exclusive monopoly control via patents or plant variety protection (Shiva, 2001).

TRIPS have a significantly disproportionate impact on the indigenous communities and farmers. The traditional knowledge accumulated over generations of inheritance is important for enhancing and promoting sustainability in farming (Melash et al., 2023). However, TRIPS provides strong intellectual property protection to the seeds and plants in the form of patents and plant variety protection. It disregards the customary rights of the indigenous communities (Upadhyay et al., n.d.). The patenting of seeds under TRIPS forces the farmers to buy patented seeds at high prices. This restricts them from saving and reusing their seeds as they did under the traditional practice. It further exacerbates farmers' economic hardship and vulnerability and increases the dominance of seed corporations over the seed system (Peschard, 2022). This clearly illustrates how the TRIPS Agreement undermines the cultural and economic heritage of farmers and the indigenous communities (Foster, 1998). Additionally, the right to food security, i.e., where the people are free to define their own seeds and have full control and access to their products is also not highlighted in the Agreement (Cullet, 2004; Haugen et al., 2011). The compliance of the patent regime under the TRIPS for the genetic resources by the corporate entities is crucial to be considered. The increasing demand for the genetically modified crops has also eroded biodiversity by increasing the monopolization of seeds (Jacobsen et al., 2013).

Furthermore, biopiracy can impact indigenous populations in a variety of ways. Without adequate recompense or incentive for traditional knowledge of natural resources, the abrupt increase in commercial value of the species producing the active component may render it expensive for the native people (Mackey & Liang, 2012). One of the most well-known examples of biopiracy is W.R. Grace's 1994 patent for a neem tree seed extract used in Neemex, an antifungal spray (Bhukta, 2020). Although the business claimed that its patent resulted from a novel idea, neem extracts have been used by rural farmers in India for over 2,000 years as insect repellents, soaps, and contraceptives (Oxfam, 2025). After years of activism and farmer opposition, the Environmental Protection Organization (EPO) revoked the patent in 2000 due to a "lack of novelty and innovative step." While the neem patent was revoked, it remains a challenge to regulate against biopiracy because the term has no legal definition and restrictions vary by location (Winterbottom, 2021). This uncertainty allows for innumerable examples of firms patenting everything from gene sequences to crop varieties to human cell lines without adequately compensating the countries and communities of origin (Merz & Cho, 2005).

CASE STUDIES: EVALUATING THE REAL-WORLD IMPACT

Plant breeding and seed manufacturing became independent operations from farming in the twentieth century, and industrial kinds increasingly displaced farmers varieties. This occurred across several decades in Europe and North America, aided by new technologies such as hybrid vehicles. It gained popularity in Asia, Africa, and Latin America after the 1960s, when so-called development programs emphasized 'high-yielding' crops and the use of chemical inputs. In the last 20 years, the globe has seen a new situation in which an aggressive wave of seed regulations has been released, frequently in the name of trade liberalization, with the goal of prohibiting practically all actions carried out by farmers using their seeds. Some of the case studies related to the impact of trade liberalization under TRIPS affecting farmers and the indigenous communities are described below.

Africa

The involvement of international agreements with the domestic legal frameworks, including multinational corporations and smallholder farmers, has become contentious with the governance of IPRs and biosafety in Africa (Peschar et al., 2023). The reconciliation of seed sovereignty with the mandatory obligation laid under TRIPS is one of the main issues in this landscape. The African Region faces significant pressure from foreign investors and domestic agricultural needs to adopt the framework mandated by the TRIPS Agreement under Article 27.3(b), i.e., adoption of a patent or an effective sui-generis system or a combination of both in order to protect plant life (Oloko, 2016). Foreign governments are flooding Africa with new public-private partnerships and development initiatives to privatize seed markets, remove the public sector from plant breeders, and transform farmers into dealers in the global agroindustry supply chain (Mangeni, 2019). Furthermore, international seed businesses and private organizations are attempting to assist African governments in rewriting legislation to make trade and investment "fair" and "responsible," which means beneficial to corporate interests (Grain, 2020). As a result, African farmers and civil society organizations oppose new seed legislation daily.

The autonomy of smallholder farmers, which limits their ability to save, exchange, sow and reuse the seeds, is affected by the heavy reliance on genetically modified crops and hybrid seeds protected under the strict intellectual property regime (Grain, 2020). This, in turn, affects the farmers' right to livelihood and food security. One of the examples in the African region is from Mozambique, where the farmers resisted stringent intellectual property protection by developing a local seed system (Smalley, 2013). The G8 New Alliance for Food Security and Nutrition is driving the new Seed Law in Mozambique, to create new opportunities for African agribusiness (De Schutter, 2015). Typically, this involves pushing cash crop development to fulfill market expectations rather than addressing local people' food needs. It also entails pushing expensive hybrid seeds that require agrochemicals and serve just the aim of producing commodities for global markets (Calicioglu et al., 2019). Farmers in Mozambique have resorted to strengthening their own peasant seed systems in this circumstance. Working with Brazilian peasant farmer movements since 2012 has given them exposure to the country's seed system setup expertise (Borras Jr, 2023). The goal of this partnership was to teach Mozambican farmers how to choose and grow seeds that they believe are essential to have on their fields in significant quantities. The Brazilian, Mozambican, and South African governments were supposed to have supported its expansion due to its success. However, only the Brazilian government had provided the funds and resources to assist the farmers' seed initiative as the enlarged program was about to begin (Milhorance, 2017).

Furthermore, in the case of *Malawian Civil Society Organizations versus The Government of Malawi* (2020), a group of civil societies in Malawi challenged the adoption of the Plant Variety Protection Act, which was enforced after the UPOV Convention, 1991 (Hunga et al., 2023). The petitioners in this case argue that the Act disproportionately favors the multinational agribusinesses and restricts the rights of small-holder farmers to save, share and exchange seeds, which are essential for the growth of food security and biodiversity (Gondwe, 2023). The conflict between international trade and the rights of the farmers under customary law was highlighted. The Court held that there is a need for more balanced regulations to overcome the inadequate safeguards for small-holder farmers under the intellectual property regime (Prip, 2024). Another case that upheld the rights of farmers over trade regulations is *The Kenya Seed Company versus Small-Holder Farmer Association*, 2019. This case highlights the challenges of implementing the plant variety protection law against traditional agricultural practices. The seed company sued the small-holder farmers for the unauthorized reproduction of their registered maize crop (Munyi & De Jonge, 2015). The petitioners argued that such acts of the farmers violated the country's' plant breeder's rights legislation. However, the respondents contended over their constitutional right to food security and livelihood. The Kenya High Court ruled for a prompt legislative reform that protects farmers' rights under the intellectual property framework (Peschard & Randeria, 2020).

India

As a member of the World Trade Organization (WTO), India was supposed to harmonize its relationship and compliance with the TRIPS Agreement (Sundaram, 2014). This led to the enactment of the Protection of Plant Varieties and Farmers Rights Act, 2001. The enforcement of the Act has raised several concerns about biopiracy and corporate governance over the seeds resulting in the depletion of traditional knowledge and biodiversity (Banerjee, 2019). One such issue was highlighted in the patenting of genetically modified seeds, particularly in the case of bacterium *Bacillus thuringiensis* (Bt) Cotton (Abbas, 2018).

The Indian farmers movement along with the organization Navdanya, led by social activist, Dr. Vandana Shiva, has been at the forefront of resisting the dominance of corporate seeds over the interest of the farming community (Futo & Nasehi, 2023). The case of *Monsanto LLC Pvt Ltd versus Nuziveedu Seeds*, 2019, is one of the most significant cases related to the genetically modified seed patent in India (Padmaja, 2020). In this case Monsanto claimed a patent over the Bt Cotton seed that was genetically modified to resist the bollworms. The State Government implemented price control measures to guarantee that farmers could continue to purchase genetically modified seeds (Chawla, 2018). The trait value (royalty) charged for using genetically modified technology was drastically lowered, and the Minimum Support Price (MSP) for cotton seeds was set. Nuziveedu refused to provide Monsanto the larger royalty agreed upon in their contract and followed the government-imposed pricing (Chawla, 2019). In 2015, Monsanto dissolved the agreement, claiming that the conditions were binding regardless of changes in local regulations. In Monsanto v. Nuziveedu, the Supreme Court upheld the necessity of interpreting Section 3(j) in a sophisticated manner. In contrast to naturally occurring biological processes, Monsanto's Bt cotton technique included artificial genetic manipulation (Singh, 2018). As a result, it may be argued that the procedure was exempt from Section 3(j), which mainly targets naturally occurring plants and animals.

United States of America (USA)

The increment in the several multinational corporations that dominate the seed industry in the USA has resulted in increased privatization of the agriculture sector (Partelow et al., 2019). The Patent Models for agriculture biotechnology and control over the production of genetically modified seeds and crops is very strong in the region (Abdul Aziz et al., 2022). Before purchasing any patented seed in the USA- such as Monsanto's Bt Cotton or Soybean seeds- the farmers must sign an agreement that prohibits them from saving those seeds (Glover, 2007). This leads to an excessive financial burden on the farmers, as they are forced to buy a new set of patented seeds for every new season. As a result of the Agreement, Monsanto, i.e., a giant agricultural company, filed infringement cases against several farmers for unlawfully using and saving the seeds (Murphy et al., 2015).

Furthermore, one of the most controversial steps by the US government was the enforcement of the Monsanto Protection Act of 2013 (Blass, 2011). This Act legally saved the biotech industry from challenges related to genetically modified crops and seeds. The Act granted the US Department of Agriculture (USDA), the authority to continue selling and cultivating the genetically modified crops, even if it was undergoing some legal dispute (Bratspeis, 2012). The legislation was widely criticized for favoring multinational corporations. The Act negatively impacted the farmers not only in the US region but globally (Giuliani & Macchi, 2014). The Act furthered the reliance of farmers on genetically modified seeds and crops by giving control to biotech companies over the seeds (Kruft, 2001). The growing monopolization of the seed market and the rising price of the seeds made the conditions vulnerable for small and independent farmers (Shiva & Crompton, 1998). This practice highlighted the uneven influence of multinational corporations in shaping agricultural policies that significantly affect farmers' right to livelihood, food security and dependence (Viana et al., 2022).

CHALLENGES IN VUCA AND BANI CONTEXT FOR FARMERS' RIGHTS UNDER TRIPS

The evolving landscape in the field of agriculture, which can be characterized by VUCA (Volatile, Uncertain, Complex, Ambiguous) and BANI (Brittle, Anxious, Non-linear, Incomprehensible) framework, often faces significant challenges for farmers under TRIPS. It is crucial to view these challenges from a human rights perspective. The VUCA perspective shows the volatility in agriculture with shifting trade practices and rapid technological advancements. This threatens the traditional farming practices of the Indigenous communities and the seed sovereignty of the farmers. This increases the economic instability for farmers who rely on their traditional knowledge system. Further, the export restriction under TRIPS disrupts the agricultural supply chain, restricting access to advanced seeds and crops. The *uncertainty* of human rights imperatives surrounding the global governance framework on TRIPS and seed sovereignty is creating legal disruptions that leave farmers vulnerable to exploitative patents and monopolistic control over genetically modified resources by multinational giants of the seed system. Further, the *complexity* perspective involves the balancing of farmers' human rights, such as the right to livelihood, food security, decision-making and culture, with that of stringent intellectual property protection provided under the TRIPS Agreement. Lastly, the *ambiguity* in TRIPS concerning the protection of traditional knowledge and the limitation of the sui-generis system results in non-consistent enforcement that affects smallholder farmers, particularly in the global South, as discussed above.

The BANI framework provides a lens to understand the fragility of farmers' rights in the era of globalization, where economic intellect prevails over societal welfare. The *brittle* agricultural system, highly dependent on a global seed system and market, faced challenges during the COVID-19 pandemic that resulted in a surge in the prices of biotechnology seeds and disruption in the supply chain. This brittleness was further heightened by the giant multinational corporations who took over the monopoly of the seed system, severely limiting the farmer's autonomy over their seeds. The *anxiety* exacerbated by the stringent intellectual property protection under the TRIPS raises concerns over unequal access to technology and the marginalization of traditional farming communities. Smallholder farmers who cannot understand the patenting of technologies, i.e. genetically modified seeds and crops, are forced to purchase the seeds for agribusinesses that, broaden the economic and technological gap. The *non-linear* perspective of BANI includes the lack of clear regulatory standards and guidelines that create disproportionate shifts in the agricultural market by restricting farmers from being involved in any of the decision-making processes. Lastly, the incomprehensibility of the rapid evolution of the technological framework restricts the farmers from navigating their rights under the TRIPS Agreement. The complexity of strong intellectual property protection under TRIPS makes it difficult for the farmers to claim their rights over the seeds. To overcome these challenges, a substantial revision to the TRIPS Agreement is mandatory that includes strengthening TRIPS flexibilities (parallel imports and compulsory licensing), availability of open access seed banks and implementing farmer-centric legal protection that ensures equal participation in a digital agrarian economy. Therefore, a commitment to a human rights perspective under TRIPS can create an inclusive future for farmers that eliminates the increasing volatility and incomprehensibility in this tech-driven world.

REFORMING TRIPS TO ADDRESS THE HUMAN RIGHTS OF FARMERS

In the Global South, the TRIPS Agreement disproportionately benefits the multinational corporations and agribusinesses over the welfare and interest of small-holder farmers. As discussed above, the stringent patent protection over the seeds provided in the Agreement limits farmers' rights to save, sow, and exchange seeds, which is essential for the livelihood of farmers, global food security and the enhancement of biodiversity. Therefore, reforming TRIPS is mandatory to align with the human rights imperatives and is important in order to create a balanced and sustainable agricultural system (Walker, 2001). *Firstly,* the involvement of provisions as outlined under Article 9 of the ITPGRFA that recognize farmers rights is crucial to be determined. This can be achieved by giving farmers the exclusive right to farmers to save, sow and exchange seeds (Prip, 2024). Certain flexibilities to developing countries should be provided to adopt an effective sui-generis system that aligns with human rights commitments. *Secondly,* a human rights-based approach to TRIPS must also address the right to access essential agricultural technologies, particularly in the case of genetically modified crops and seeds. Additionally, reforms that acknowledge the contribution of traditional knowledge and Indigenous communities should receive compensation for their efforts with the establishment of proper benefit-sharing mechanisms (Elvin-Lewis, 2007). *Lastly,* the participation of farmers' rights can be strengthened with the global intellectual property policy decision making process, which is equally essential to demonstrate the decision-making ability and participation of farmers. Without such implementation, TRIPS shall continue to exacerbate inequalities that undermine farmer's rights to livelihood, food security and cultural rights (Guida, 2025). Therefore, aligning TRIPS with the human rights instruments such as UNDROP is essential for an inclusive agricultural system. The reforms that balance farmers' rights with corporate interests can benefit present and future generations.

CONCLUSION

Agriculture, which progressed from subsistence cultivation to barter and trade, has significantly impacted the culture, traditions, and history of human civilizations worldwide. Farmers' access to seeds is threatened by intellectual property and seed marketing legislation. Farmers and commercial seed systems share fluid boundaries. Commercial seed regimes sometimes restrict or prevent peasants from using, saving, exchanging, and selling farm-saved seeds (Elfstrand et al., 2011). The intersection of IPRs mentioned under TRIPS and farmers' rights as human rights pose a significant challenge in the global agricultural system. TRIPS result in monopolization of the agricultural sector's seed system, limiting farmers' rights to save, sow and exchange seeds. The rigid enforcement of intellectual property protection under the Agreement affects small-holder farmers who rely heavily on traditional knowledge and seeds-saving practices. Disputes are exacerbated in rural countries in the Global South, where farmers' seed systems are the primary source of seeds. Sui generis plant variety protection schemes tailored to local needs are more effective in ensuring the right to seeds as outlined in UNDROP. Some jurisdictions in the Global North advocate for adopting the UPOV 1991 model as the exclusive paradigm for regulating intellectual property rights over plant varieties. To address problems between farmers' rights to seeds and intellectual property and seed marketing rules, it is crucial to increase the protection of the right to seeds in international law, including recognition of farmers' rights to seeds.

Addressing these challenges, requires an inclusive approach for farmers that is fundamental to balancing the interests of farmers with that of corporations. Mechanisms such as exemption for farmers to save and reuse the seeds can result in the harmonization of trade with human rights. Furthermore, an integrated policy-making process with an inclusive and diverse approach is essential to safeguard innovation and farmers' livelihood. A more just and equitable approach to the TRIPS framework would strengthen human rights and contribute to long-term food security, environment sustainability and economic stability in rural areas.

REFERENCES

Abbas, M. S. T. (2018). Genetically engineered (modified) crops (*Bacillus thuringiensis* crops) and the world controversy on their safety. *Egyptian Journal of Biological Pest Control, 28*(1), 1–12. DOI: 10.1186/s41938-018-0051-2

Abdul Aziz, M., Brini, F., Rouached, H., & Masmoudi, K. (2022). Genetically engineered crops for sustainably enhanced food production systems. *Frontiers in Plant Science, 13*, 1027828. DOI: 10.3389/fpls.2022.1027828 PMID: 36426158

Adhikari, K., Bikundo, E., Chacko, X., Chapman, S., Humphries, F., Johnson, H., Keast, E., Lawson, C., Malbon, J., Robinson, D., Rourke, M., Sanderson, J., & Tranter, K. (2021). What should farmers' rights look like? The possible substance of a right. *Agronomy (Basel), 11*(2), 367. DOI: 10.3390/agronomy11020367

Adhikari, R. (2005). Emerging issues relating to conflicts between TRIPS and biodiversity: development implications for South Asia. *South Asian Yearbook of Trade and Development*, 261-288.

Aggarwal, V., & Satpute, A. (2013). Role of TRIPS in the Indian agriculture sector: Balancing traditional knowledge and biotechnology. *AgEcon Search.* https://www.aares.org.au/publications/AgEcon-search/

Aguilar, G. (2001). Access to genetic resources and protection of traditional knowledge in the territories of indigenous peoples. *Environmental Science & Policy, 4*(4-5), 241–256. DOI: 10.1016/S1462-9011(01)00028-4

Alston, P. (1984). International law and the human right to food. In *The Right to Food* (pp. 9–68). Brill Nijhoff. DOI: 10.1163/9789004482302_004

Antons, C. (2010). The role of traditional knowledge and access to genetic resources in biodiversity conservation in Southeast Asia. *Biodiversity and Conservation, 19*(4), 1189–1204. DOI: 10.1007/s10531-010-9816-y

Apodaca, C. (2012). The right to food. In *the Handbook of human rights* (pp. 349-358). Routledge.

Assan, N. (2023). Socio-cultural, economic, and environmental implications for innovation in sustainable food in Africa. *Frontiers in Sustainable Food Systems, 7*, 1192422. DOI: 10.3389/fsufs.2023.1192422

Azadi, H., Samiee, A., Mahmoudi, H., Jouzi, Z., Rafiaani Khachak, P., De Maeyer, P., & Witlox, F. (2016). Genetically modified crops and small-scale farmers: Main opportunities and challenges. *Critical Reviews in Biotechnology, 36*(3), 434–446. PMID: 25566797

Banerjee, M. (2019). Biopiracy in India: Seed diversity and the scramble for knowledge. *Phytomedicine, 53*, 296–301. DOI: 10.1016/j.phymed.2018.10.017 PMID: 30448389

Bhukta, A. (2020). Bio-piracy. In *Legal Protection for Traditional Knowledge: Towards a New Law for Indigenous Intellectual Property* (pp. 45–72). Emerald Publishing Limited. DOI: 10.1108/978-1-80043-063-120200004

Bjørnstad, S. I. B. (2004). *Breakthrough for" the South"?: an analysis of the recognition of Farmers' Rights in the International Treaty on Plant Genetic Resources for Food and Agriculture* (Master's thesis).

Blas, E. A. (2011). Monsanto Co. v. Geertson Seed Farms: Why the National Environmental Policy Act of 1969 Fails to Protect the Environment from Current Biotechnology. *Sustainable Development Law Journal, 14*, 35.

Borowiak, C. (2004). Farmers' rights: Intellectual property regimes and the struggle over seeds. *Politics & Society, 32*(4), 511–543. DOI: 10.1177/0032329204269979

Borras, S. M.Jr. (2023). La Via Campesina–transforming agrarian and knowledge politics, and co-constructing a field: A laudatio. *The Journal of Peasant Studies, 50*(2), 691–724. DOI: 10.1080/03066150.2023.2176760

Breining-Kaufmann, C. (2005). The right to food and trade in agriculture. *Human rights and international trade*, 341-381.

Calicioglu, O., Flammini, A., Bracco, S., Bellù, L., & Sims, R. (2019). The future challenges of food and agriculture: An integrated analysis of trends and solutions. *Sustainability (Basel), 11*(1), 222. DOI: 10.3390/su11010222

Chawla, V. (2018). Monsanto's Loss of Patent Over Bt Cotton is a Victory for Farmers and Indian Seed Companies. *Economic and Political Weekly, 53*, 25.

Chawla, V. (2019). Indian Supreme Court on patentability of genetically modified life forms—A missed opportunity? *Journal of Intellectual Property Law & Practice, 14*(5), 343–345. DOI: 10.1093/jiplp/jpz032

Correa, C. M. (2001). Traditional knowledge and intellectual property. *Geneva: The Quaker United Nations Office (QUNO), 17.*

Correa, C. M., & Yusuf, A. A. (Eds.). (2016). *Intellectual property and international trade: The TRIPS agreement: The TRIPS agreement*. Kluwer Law International BV.

Cullet, P. (1999). Revision of the TRIPS Agreement concerning the protection of plant varieties: Lessons from India concerning the development of a sui generis system. *The Journal of World Intellectual Property*, *2*(4), 617–656. DOI: 10.1111/j.1747-1796.1999.tb00081.x

Cullet, P. (2004). *Food security and intellectual property rights in developing countries*. RIBios (Réseau interdisciplinaire biosécurité): Institut universitaire d'études du développement (IUED).

De Schutter, O. (2015). The New alliance for food security and nutrition in Africa.

Dhar, B., & Anuradha, R. V. (2004). Access, benefit-sharing and intellectual property rights. *The Journal of World Intellectual Property*, *7*(5), 597–639. DOI: 10.1111/j.1747-1796.2004.tb00221.x

Dludlu, C. P. (2014). *A comparative study of the patentability standards with respect to pharmaceutical inventions in the United Kingdom and South Africa* (Doctoral dissertation, Queen Mary University of London).

Elfstrand, S., Malmer, P., & Skagerfält, B. (2011). Strengthening agricultural biodiversity for smallholder livelihoods. *What knowledge is needed to overcome constraints and release potentials*.

Elvin-Lewis, M. (2007). Evolving Concepts Related to Achieving Benefit Sharing For Custodians Of Traditional Knowledge. https://utoronto.scholaris.ca/items/83a11d88-303f-47b9-8e54-e7205581a868

Food and Agriculture Organization (FAO). (2017). Farmers Rights. https://openknowledge.fao.org/server/api/core/bitstreams/c6767e7b-331d-421f-b936-f1922a3bc54d/content

Foster, M. (1998). The Human Genome Diversity Project and the patenting of life: Indigenous peoples cry out. *Canterbury L. Rev.*, *7*, 343.

Fuglie, K., & MacDonald, J. M. (2023). Expanded intellectual property protections for crop seeds increase innovation and market power for companies. *Amber Waves: The Economics of Food, Farming, Natural Resources, and Rural America, 2023*.

Futo, T., & Nasehi, E. E. (2023). The Seeds of Vandana Shiva (2021): A Credible Actor of Change. *Kadın/Woman 2000, 24*(1), 77-96.

Giuliani, E., & Macchi, C. (2014). Multinational corporations' economic and human rights impacts on developing countries: A review and research agenda. *Cambridge Journal of Economics, 38*(2), 479–517. DOI: 10.1093/cje/bet060

Glover, D. (2007). Monsanto and smallholder farmers: A case study in CSR. *Third World Quarterly, 28*(4), 851–867. DOI: 10.1080/01436590701336739

GONDWE. K. J. (2023). *INVESTIGATING THE NGO's-LOCAL GOVERNMENT RELATIONS IN GOVERNANCE SECTOR IN MALAWI: CASE STUDY OF BALAKA DISTRICT COUNCIL* (Doctoral dissertation, UNIVERSITY OF MALAWI).

GRAIN, A. (2020). *The real seeds producers: small-scale farmers save, use, share and enhance the seed diversity of the crops that feed africa.*

Guida, A. (2025). *From policy to plate: international trade law's role in advancing sustainable food governance and supporting dignified food access.* CERIDAP.

Haugen, H. M. (2020). The UN Declaration on Peasants' Rights (UNDROP): Is Article 19 on seed rights adequately balancing intellectual property rights and the right to food? *The Journal of World Intellectual Property, 23*(3-4), 288–309. DOI: 10.1111/jwip.12152

Haugen, H. M., Muller, M. R., & Narasimhan, S. M. (2011). Food security and intellectual property rights: Finding the linkages. *Intellectual property and human development: Current trends and future scenarios, 103*, 108.

Hunga, H. G., Chiwaula, L., Mulwafu, W., & Katundu, M. (2023). The seed sector development in low-income countries: Lessons from the Malawi seed sector policy process. *Frontiers in Sustainable Food Systems, 7*, 891116. DOI: 10.3389/fsufs.2023.891116

Imran, Y., Wijekoon, N., Gonawala, L., Chiang, Y. C., & De Silva, K. R. D. (2021). Biopiracy: Abolish corporate hijacking of indigenous medicinal entities. *TheScientificWorldJournal, 2021*(1), 8898842. DOI: 10.1155/2021/8898842 PMID: 33679261

Jacobsen, S. E., Sørensen, M., Pedersen, S. M., & Weiner, J. (2013). Feeding the world: Genetically modified crops versus agricultural biodiversity. *Agronomy for Sustainable Development, 33*(4), 651–662. DOI: 10.1007/s13593-013-0138-9

Kruft, D. (2001). Impacts of genetically-modified crops and seeds on farmers. *The Agricultural Law Resource and Reference Center*, 1-11.

Lawson, C. (2015). Implementing Farmers' Rights: Finding Meaning and Purpose for the International Treaty on Plant Genetic Resources for Food and Agriculture Commitments? *Available at SSRN* 2626668.

Louwaars, N. (2007). *Seeds of Confusion: The Impact of Policies on Seed Systems.* Wageningen University and Research.

Mackey, T. K., & Liang, B. A. (2012). Integrating biodiversity management and indigenous biopiracy protection to promote environmental justice and global health. *American Journal of Public Health, 102*(6), 1091–1095. DOI: 10.2105/AJPH.2011.300408 PMID: 22515858

Mahoney, S. E. (2018). Owning the World's Seed Supply: How Seed Industry Mergers Threaten Global Food Security. *Geo. Envtl. L. Rev., 31*, 563.

Mangeni, B. (2019). The role of public-private partnerships (PPPs) in ensuring technology access for farmers in sub-Saharan Africa. *African Journal of Food, Agriculture, Nutrition and Development, 19*(1), 14137–14155. DOI: 10.18697/ajfand.84.BLFB1018

Matthews, D. (2003). *Globalising intellectual property rights: the TRIPS Agreement.* Routledge. DOI: 10.4324/9780203165683

Melash, A. A., Bogale, A. A., Migbaru, A. T., Chakilu, G. G., Percze, A., Ábrahám, É. B., & Mengistu, D. K. (2023). Indigenous agricultural knowledge: A neglected human based resource for sustainable crop protection and production. *Heliyon, 9*(1), e12978. DOI: 10.1016/j.heliyon.2023.e12978 PMID: 36711305

Merz, J. F., & Cho, M. K. (2005). What are gene patents and why are people worried about them? *Public Health Genomics, 8*(4), 203–208. DOI: 10.1159/000087956 PMID: 16244473

Milhorance, C. (2017). The Brazil–Mozambique experience in rural development: bringing conflict back into transfer studies. In *Public Policy Transfer* (pp. 125–148). Edward Elgar Publishing. DOI: 10.4337/9781785368042.00013

Mitchell, A. D., & Voon, T. (2009). TRIPS', in Daniel Bethlehem, and others (eds), *The Oxford Handbook of International Trade Law* (2009; online edn, Oxford Academic, 18 Sept. 2012), , accessed 7 Mar. 2025.DOI: 10.1093/oxfordhb/9780199231928.013.0008

Munyi, P., & De Jonge, B. (2015). Seed systems support in Kenya: Consideration for an integrated seed sector development approach. *Journal of Sustainable Development, 8*(2), 161. DOI: 10.5539/jsd.v8n2p161

Murphy, L., Noga, K., & Rose, M. (2015). Seeking Pure Fields: The Case Against Preemption of State Bans on Genetically Engineered Crops. *University of San Francisco Law Review. University of San Francisco. School of Law, 49*(3), 3.

Nabuuma, D., Reimers, C., Hoang, K. T., Stomph, T., Swaans, K., & Raneri, J. E. (2022). Impact of seed system interventions on food and nutrition security in low- and middle-income countries: A scoping review. *Global Food Security*, *33*, 100638. DOI: 10.1016/j.gfs.2022.100638

Nordin, R. (2011). Protection of indigenous knowledge: Mismatch between the convention on biological diversity and TRIPS agreement. *Jurnal Undang-undang dan Masyarakat, 15*, 179.

Oguamanam, C. (2006). Intellectual Property Rights in Plant Genetic Resources: Farmers' Rights and Food Security of Indigenous and Local Communities. *The Drake Journal of Agricultural Law*, *11*, 273.

Oguamanam, C. (2014). Farmers' rights and the intellectual property dynamic in agriculture. *Sage Handbook of Intellectual Property*, 238-257.

Oloko, T. O. (2016). An examination of Article 27 of the TRIPS Agreement in relation to the provisions on patentable subject matter under the PDA in Nigeria. *Commonwealth Law Bulletin*, *42*(2), 236–260. DOI: 10.1080/03050718.2016.1187072

Oxfam. (2025, January 20). Takers not Makers: The unjust poverty and unearned wealth of colonialism. *Oxfam.* https://www.kedv.org.tr/public/uploads/files/Oxfam%202025%20Takers%20Not%20Makers%20(%C3%9Cretenler%20De%C4%9Fil%20%C3%9Czerine%20Konanlar)%20EN%20Report.pdf

Padmaja, S. S. (2020). Intellectual Property, Competition, and Regulatory Policies: A Case of Bt Cotton Seed Industry in India. *Competition Commission of India Journal on Competition Law and Policy*, 79-101.

Partelow, S., Abson, D. J., Schlüter, A., Fernández-Giménez, M., Von Wehrden, H., & Collier, N. (2019). Privatizing the commons: New approaches need broader evaluative criteria for sustainability. *The International Journal of the Commons, 13*(1), 747. DOI: 10.18352/ijc.938

Perrin, R. K. (1994). Intellectual property rights in agricultural development.

Peschard, K., & Randeria, S. (2020). 'Keeping seeds in our hands': The rise of seed activism. *The Journal of Peasant Studies*, *47*(4), 613–647. DOI: 10.1080/03066150.2020.1753705

Peschard, K. E. (2022). *Seed activism: Patent politics and litigation in the global south.* MIT Press. DOI: 10.7551/mitpress/14484.001.0001

Prip, C. (2024). Farmers' Rights in the Plant Treaty: Interrelations and recent interactions with other international regimes and processes. *Development Studies Research, 11*(1), 2357095. DOI: 10.1080/21665095.2024.2357095

Rodgers, G. (2021). Work and conditions of work. In *Critical Issues in Human Rights and Development* (pp. 282–299). Edward Elgar Publishing.

Rogers, J. T. (2010). The encroachment of intellectual property protections on the rights of farmers. *The Drake Journal of Agricultural Law, 15*, 149.

Sauvé, P. (2017). A First Look at Investment in the Final Act of the Uruguay Round. In *Globalization and International Investment* (pp. 241–252). Routledge.

Scherr, S. J., & McNeely, J. A. (2008). Biodiversity conservation and agricultural sustainability: Towards a new paradigm of 'ecoagriculture landscapes. *Philosophical Transactions of the Royal Society of London. Series B, Biological Sciences, 363*(1491), 477–494. DOI: 10.1098/rstb.2007.2165 PMID: 17652072

Shiva, V. (2001). *Stolen harvest: The hijacking of the global food supply*. Zed Books.

Shiva, V., & Crompton, T. (1998). Monopoly and monoculture: Trends in Indian seed industry. *Economic and Political Weekly*, •••, A137–A151.

Singh, K. K. (2018). Bayer-Monsanto Merger and India's IP Approach to Agricultural Biotechnology: Navigating through a complex web of law and policy. *Asian Biotechnology and Development Review*, •••, 20.

Smalley, R. (2013). Plantations, contract farming and commercial farming areas in Africa: A comparative review. *Future Agricultures Working Paper, 55*, 73.

Srivastava, U. (2006). Intellectual property rights and agro-biodiversity. *Indian Journal of Plant Genetic Resources, 19*(01), 1–11.

Stringer, L. C., Fraser, E. D., Harris, D., Lyon, C., Pereira, L., Ward, C. F., & Simelton, E. (2020). Adaptation and development pathways for different types of farmers. *Environmental Science & Policy, 104*, 174–189. DOI: 10.1016/j.envsci.2019.10.007

Sundaram, J. (2014). India's trade-related aspects of Intellectual Property Rights compliant pharmaceutical patent laws: What lessons for India and other developing countries? *Information & Communications Technology Law, 23*(1), 1–30. DOI: 10.1080/13600834.2014.891310

Upadhyay, M., Neupane, S., & Dhaubadel, S. (n.d.). *Agreement on TRIPS: Legal Framework for Plant Variety Protection*. https://www.fao.org/4/ae896e/ae896e08.pdf

Viana, C. M., Freire, D., Abrantes, P., Rocha, J., & Pereira, P. (2022). Agricultural land systems importance for supporting food security and sustainable development goals: A systematic review. *The Science of the Total Environment*, *806*, 150718. DOI: 10.1016/j.scitotenv.2021.150718 PMID: 34606855

Walker, S. (2001). *The TRIPS agreement, sustainable development and the public interest: discussion paper* (No. 41). IUCN.

Wang, X., Ma, L., Yan, S., Chen, X., & Growe, A. (2023). Trade for food security: The stability of global agricultural trade networks. *Foods*, *12*(2), 271. DOI: 10.3390/foods12020271 PMID: 36673361

Watal, J., Taubman, A., & Cottier, T. (2015). Revisiting the TRIPS Negotiations. In *WTO, The making of the TRIPS agreement: Personal insights from the Uruguay round negotiations. pp- 3-13.* WTO.

Winter, L. (2010). Cultivating farmers' rights: Reconciling food security, indigenous agriculture, and TRIPS. *Vanderbilt Journal of Transnational Law*, *43*, 223.

Winterbottom, A. (2021). Becoming "traditional" a transnational history of neem and biopiracy discourse. *Osiris*, *36*(1), 262–283. DOI: 10.1086/713897

Wright, B. D., & Pardey, P. G. (2006). The evolving rights to intellectual property protection in the agricultural biosciences. *International Journal of Technology and Globalisation*, *2*(1-2), 12–29. DOI: 10.1504/IJTG.2006.009124

WTO. (n.d.). TRIPS and the Conventions related to it. https://www.wto.org/english/tratop_e/trips_e/ta_docs_e/1_tripsandconventions_e.pdf

Wynberg, R. (2023). Biopiracy: Crying wolf or a lever for equity and conservation? *Research Policy*, *52*(2), 104674. DOI: 10.1016/j.respol.2022.104674

Yazdanpanah, M., Tajeri Moghadam, M., Savari, M., Zobeidi, T., Sieber, S., & Löhr, K. (2021). The impact of livelihood assets on the food security of farmers in Southern Iran during the COVID-19 pandemic. *International Journal of Environmental Research and Public Health*, *18*(10), 5310. DOI: 10.3390/ijerph18105310 PMID: 34067638

KEY TERMS AND DEFINITIONS

TRIPS: The TRIPS Agreement, which came into effect on 1 January 1995, is to date the most comprehensive multilateral agreement on intellectual property. It sets out the minimum standards for the protection of intellectual property for the member states in the form of patents, copyrights, plant variety protection, trademarks and geographical indications.

Intellectual Property Rights: Intellectual Property Rights are legal protections granted to creators and inventors, allowing them exclusive rights to use and benefit from their creations, such as inventions, literary and artistic works, designs, and symbols, for a specific period.

Genetically Modified Seeds: Genetically modified (GM) seeds, also known as GMO seeds, are seeds whose DNA has been altered using genetic engineering techniques to introduce specific traits, such as pest resistance or herbicide tolerance.

Human Rights: Human rights are inherent, universal, and inalienable rights that all people possess simply by virtue of being human, regardless of their nationality, ethnicity, religion, or any other status. They are foundational, including the right to life, liberty, and freedom from discrimination.

Plant Variety Protection: Plant Variety Protection is a form of intellectual property right that legally protects new plant varieties, giving breeders exclusive rights to use, sell, and distribute their creations for a set period, encouraging innovation and investment in plant breeding.

Biotechnology for Plants: Biotechnology for Plants utilizes scientific techniques to modify and manipulate plant cells or tissues for beneficial purposes, encompassing techniques like genetic engineering and tissue culture to enhance traits like yield, quality, and resistance to adverse conditions.

Chapter 8
Enhancing Cognitive Skills Through Smart Classroom Technologies in the Vuca-Bani World: How Student Participation Channels the Impact and Learning Motivation Shapes This Relationship

Indrajit A. Doddanavar
https://orcid.org/0000-0003-1203-1747
Jain College of Engineering and Research, Belagavi, India

Sweeta Agrawal
https://orcid.org/0000-0002-9576-5668
Ramadevi Women's University, Bhubaneswar, India

Shubhangi Rote
https://orcid.org/0009-0005-4678-997X
Padmaraj Ariga Bharatesh College of Business Administration, Belagavi, India

Amit Subramanyam
https://orcid.org/0009-0005-2846-1700
Rani Chanamma University, Belagavi, India

DOI: 10.4018/979-8-3373-7917-3.ch008

ABSTRACT

The smart classroom technologies with skill development in a VUCA-BANI context is what this study focuses on. The research uses a sample of 550 students of UG and above in India. It assesses the impacts of IoT, AR/VR, Gamification, and AI on cognitive skills through the mediating effects of student participation and moderating effects of learning motivation. Structural equation modeling indicates that IoT, AR/VR, and GAM have substantial direct and indirect impact on cognitive skill development, whereas the impacts of AI are nearly fully mediated through students' participation. The study suggests that student participation is a vital mediator and learning motivation an important moderator. IPMA suggests that learning motivation and student participation have the strongest impact in developing cognitive skills. This study helps to understand the impact of smart education on the formation of basic cognitive skills and provides guidance on the development and implementation of strategic educational policies within the context of an overly complicated and uncertain world

INTRODUCTION

In this modern world defined by volatility, uncertainty, complexity, and ambiguity (VUCA) as well as brittleness, anxiety, non-linearity, and incomprehensibility (BANI), educational systems globally face immense pressure to change systems that have served them for decades (Bennett & Lemoine, 2014; Cascio & Montealegre, 2016). Issues such as outdated framework of classroom learning fails to address the needs of modern learners wherein technology, interactivity and personalization are at the core of the learning experience. These traditional frameworks are less likely to engage learners constructively or develop the requisite skills to be successful in a technologically driven society (Khlaif, 2018; Alrousan et al., 2022).

Technologies such as AI focuses on building crucial critical cognitive skills such as problem solving, adaptability, and critical thinking. In today's student environment, such skills are the foundation for a holistic skill set. These skills are not merely supplementary components of the learning process; they form the foundation for preparing students to excel in an ever-evolving environment. However, the reliance on traditional educational practices often falls short in achieving these objectives, revealing an urgent need for innovative teaching strategies and tools that align with the demands of contemporary learners (AlGerafi et al., 2023; Lawson et al., 2024).

Smart classroom technologies are considered as the new frontier in innovation in teaching and providing novel approaches towards making teaching and learning more effective. New and more engaging ways of immersing students in the education world using the Internet of Things (IoT), Augmented Reality and Virtual Reality (AR/VR), gamification, and Artificial Intelligence have been made possible. These technologies transform the learning setting by providing captivating, fully immersive experiences that replace traditional techniques and nurture the development of vital cognitive skills (Gargrish et al., 2022). These technologies integrated specially designed spaces in classroom environments for students allow teachers to create self-directed, comprehensive and multi-dimensional learning environments that engage more diverse learners.

The shift using ordinary instructional technologies to smart technologies in classrooms represents more than just an improvement in instructional practice. These technologies transform knowledge acquisition and retention but also engage students more actively by fostering learning experience application which makes them resilient, adaptable, and skillful in the complex interdependent world (Collins and Halverson, 2018; Fischer et al., 2020). In this way, learners change from being mere knowledge passive recipients to becoming engaged learners who take part in their education.

Regardless, the effectiveness of smart classroom technologies goes beyond their mere existence. These resources are most useful when utilized properly, which, in turn, pivots on how motivated students are to learn along with their active participation in the classroom. Students who are driven by the desire to learn and have access to technology will, in fact, have the most to gain from cognitive advancements offered by these innovations. These factors greatly affect the outcomes but are rarely accurately captured, which impacts the ability of instructors and policymakers to leverage the smart classroom technologies to their maximum degree. A lack of understanding regarding the extent of student participation and motivation affects the technology – cognitive development link will cause the benefits of these innovations to not be exploited as intended.

In the context of the VUCA-BANI world, where adaptability and resilience are indispensable, this chapter explores the intersection of smart classroom technologies, student participation, and learning motivation. It examines how these elements collectively contribute to enhancing cognitive skills while addressing broader concerns related to educational equity, accessibility, and innovation. The focus of this investigation is on trying to merge new technological development with the practical teaching problem of achieving meaningful learning, with the goal of understanding how to make the best use of these tools in meeting the challenges of a constantly changing world.

The objectives of this study are as follows:

- To determine the extent to which smart classroom technologies influence the development of students' cognitive skills.

- Examine the relationships between student engagement, motivation, and their use of the technologies.
- Formulate and disseminate practical suggestions for teachers, government officials, and business people dealing with smart classroom technologies.

The research seeks to answer the following critical questions:

- Which smart classroom technologies have positive contribution towards the development of cognitive skills?
- Does a student's involvement in these processes help their cognitive development in utilizing those technologies?
- To what extent does the degree of learning motivation affect the relationship between the technology and their cognitive benefits?

By addressing these questions, this chapter aims to highlight the transformative potential of smart classroom technologies and their role in cultivating essential cognitive skills. There is a need for incorporation of more advanced technologies together with the application of active participation and motivational techniques so that learners can adapt and succeed in very dynamic harsh environments. Through this exploration, the chapter advocates for a holistic approach to education—one that ensures equity, innovation, and meaningful learning outcomes for all.

SIGNIFICANCE OF THE STUDY

The significance of this study lies in its potential to bridge critical gaps in the understanding and implementation of smart classroom technologies, particularly within the context of modern education. The advent of technology and changing views on education make this research important because it highlights how IoT, AR/VR, gamification, and AI can be used to build cognitive skills and undergo the educational changes IoT enabled productivity promises. These techniques can provide a resolution to the challenges posed by the traditional approaches.

Understanding the student participation and learning motivation as mediating factors increases the construction of knowledge which goes beyond the mere application of smart devices in pedagogical practices. It examines the context in which the learner operates and how the learner engages with and in the deployed smart technologies to work for learning.

Participation and motivation are not only secondary elements but also act as a stimulus for development and transformation. Therefore, the study demonstrates how other technologies, beyond the devices themselves, can facilitate the development of skills such as critical thinking, problem solving, creativity, adaptability, and many more essential for dealing with uncertainties in today's complex and unpredictable world.

Taking into consideration the VUCA-BANI world, this study is focused on resilience, adaptability, and innovative thinking. Educational practices to prepare learners to navigate uncertainties and complexities within the context of VUCA-BANI are baffling. This study also sheds light on the transformative potential of smart classroom technologies which enables example solutions to equip Stakeholders with the tools possible whom are prejudice to the prospect of enhancing learning through classroom adoption technologies. This transformation will only occur if the tools are not just adopted but integrated within the framework in a manner that now maximizes the effectiveness of the tools.

Policymakers, technology developers, and even educators will greatly benefit from this study since its findings cater to various stakeholders. While educators stand to benefit from evidence-based techniques for enhancing motivation and engagement through the use of technology, policymakers will be able to develop smart classroom policies as a result of fact driven policy recommendations. Finally, technology developers will benefit from suggestions on classroom technology that account for the preferences of students and teachers.

Ultimately, this study contributes to the growing body of knowledge on educational innovation, serving as a foundation for further research in this field. It emphasizes the importance of a holistic approach to integrating technology, participation, and motivation in education, ensuring that learners are equipped not only with knowledge but with the cognitive tools required to excel in a rapidly changing world.

RESEARCH PROBLEM

The modern world, sometimes coined as VUCA-BANI (Volatile, Uncertain, Complex, Ambiguous, Brittle, Anxious, Non-Linear, and Incomprehensible), brings new problems that impose difficulty for educators all around the globe. As technology develops, problems keep getting more difficult to solve. Smart classroom technologies could offer a potential solution, however, their effectiveness in developing basic cognitive skills, especially in the context of higher education, is highly contested.

This study addresses several critical research problems:

1. There exists a cognitive skills gap regarding how smart classroom technologies IoT, Augered Reality and Virtual Reality (AR/VR), Gamification (GAM), and Artificial Intelligence (AI) is understood to influence on the students in higher education.

2. In regard to other learning outcomes these technologies may impact, the role of student participation as a mediator is also seldom captured.
3. There has not been sufficient exploration how learning motivation may act as a moderator in the relationship between student participation and cognitive skill development in technology enhanced learning environments.
4. There is a lack of practical research regarding the significance and effectiveness of various smart classroom technologies at developing cognitive skills in learners, particularly for a VUCA-BANI world.

To help solve these questions, this research aims to inform best practices on the deployment of smart classroom technologies and the enhancement of cognitive skill development in contemporary classrooms. These results will assist towards the creation of approaches that are grounded in evidence for the use of technology in higher education, thus improving the readiness of students to face modern challenges.

LITERTAURE REVIEW

The adoption of modern technologies in educational environments has changed the approaches associated with nurturing cognitive skills and involvement of learners. Modern educational technologies impact learning outcomes through Constructivist, Cognitivist, Behaviorist theories, Technology Acceptance Models (TAM and UTAUT), and Cognitive Load Theory (CLT). These theories are framed within the context of VUCA-BANI (Volatility, Uncertainty, Complexity, Ambiguity, and Brittleness, Anxiety, Non-Linearity, and Incomprehensibility) and mark a significant boundary for the discussion on how smart technologies enable learners to cope with the difficult times by enhancing their adaptive abilities.

VUCA-BANI World and Smart Technologies

Relying on the traditional teaching approaches and methods in VUCA-BANI world might not be efficient in solving the problems of complex and fast-paced learning environments. Advanced technologies enhance learning outcomes through flexibility and adaptability that enables students to navigate through them more effectively.

Table 1 provides a structured analysis of how emerging educational technologies address volatility, uncertainty, complexity, and ambiguity while fostering adaptability, resilience, and intelligence in modern learning environments.

Table 1. Smart Education in the VUCA-BANI World

Concept/Framework	Description	Role of Smart Technologies
VUCA: Volatile, Uncertain, Complex, Ambiguous	VUCA represents environments characterized by constant change, unpredictability, complexity, and unclear outcomes.	Cutting-edge technologies including AI, IoT, and AR/VR provide smart learning environments that can easily be pivoted to fit the needs of the learner and other factors. These new technologies help learners reason and think critically in irrational circumstances.
BANI: Brittle, Anxious, Nonlinear, Incomprehensible	BANI refers to systems that are fragile and prone to disruption, creating anxious, nonlinear, and difficult-to-understand experiences.	Smart education technologies alleviate anxiety associated with learning by offering a predictable and organized lesson outline. AR/VR and AI engage students actively for better understanding of difficult subjects, therefore making them less overwhelming.

Constructivism and Smart Education

Vygotsky (1978) and Piaget (1973) claim that individuals, acquire knowledge through interactions in their environment. Learning is more than passively absorbing information; rather, as it happens, the learner actively participates. This means that a learner can assimilate new experiences to their existing mental structure. In smart education, AR, VR, and IoT allow students not only to passively watch but also to actively take part in the learning, thus engaging them.

AR/VR technology removes the boundaries from a traditional classroom by enabling learners to interact with complex content in the real world. For instance, learners can immerse themselves in 3D models of the human body, experience historical events, or conduct science experiments virtually (Bailey & Bailenson, 2017; Ahn et al., 2022). Such activities are further enhanced by Vygotsky's (1978) social constructivism that puts forward social relations as the core of cognitive learning. These ideas are executed with the help of IoT devices through interactions, feedback, data uploading and sharing real time information, facilitates collaboration in the learning process (Li et al., 2024; Hamad et al., 2024). Students are motivated to participate in productive activities through a reinforcement system which helps to achieve the desired behavior (Kurni & KG, 2024; Mora, 2020). Furthermore, AI applications can boost engagement and learning by offering instant feedback and rewards.

Cognitivism and Cognitive Skill Development

Cognitivism deals with mental activities in the learning process such as attention, memory, and mental problem solving (Sweller, 1988). This theory contrasts with behaviorism by emphasizing the internal processing of information, rather than merely external behaviors. According to Bruner (1966), learning involves structuring knowledge and organizing information to facilitate understanding and retention.

Therefore, in smart classrooms, technologies like AI powered adaptive learning systems with AR/VR encourages the engaging of deeper cognitive processes making it easier to understand complex problems. Baker et al. (2019) urge that AI systems help in creating learning content tailored to a student's pace, style, and preferences thereby providing balanced levels of cognitive load. Furthermore, AR/VR helps in sustaining focused and enhanced interactive learning, thereby, aiding in their understanding and memory of difficult topics (Gargrish et al., 2022, AlGerafi et al., 2023). These technologies help in presenting and processing the information better, which is exactly what is needed to nurture cognitive development.

Behaviorism and Motivation in Smart Classrooms

The behavior of students learning in the classroom has been considered and researched by behaviorists like Skinner (1953) from the perspective of cognitive and psychomotor domain activities. In this case, students' actions and level of engagement in a classroom during a lesson depend on the rewards and disciplinary measures that they receive. The use of gamification in smart classrooms is connected with the use of behaviorist principles. Adding points, badges, and the use of leaderboards facilitates participation in the exercises and keeps them actively engaged (Deterding et al., 2011).

Gamified systems in smart classrooms encourages students to compete to achieve certain milestones and, as a result, increase the level of engagement with the material learning. The approaches were developed on the principle of using reward system so as to compel the learners to engage in more and more positive behaviors (Kurni & KG, 2024; Mora, 2020). Additionally, AI applications that provide instant feedback to answers submitted and give rewards boost engagement and learning in a balanced manner.

Technology Acceptance Models: TAM and UTAUT

The models TAM and UTAUT evaluate the levels of acceptance and usage of educational technology. The Technology Acceptance Model (TAM), developed by Davis (1989), argues that acceptance of a technology is a function of its perceived ease of use and perceived usefulness. This model was improved by Venkatesh et al. (2003) who added additional parameters to the UTAUT model i.e. performance expectancy, effort expectancy, social influence, and facilitating conditions.

The technology acceptance models cover the acceptance and integration of the Internet of Things, Artificial Intelligence, Augmented Reality/ Virtual Reality, and gamification which incorporate both the student and teacher roles. Alharbi et al. (2021) and Khan et al. (2022) pointed out that the adoption of smart technologies by teachers is largely a function of perceived ease of use and perceived usefulness of the technologies.

Additionally, the adoption of such technologies also depends on the facilitating conditions which include adequate training, technical assistance, and the required infrastructure (Venkatesh et al., 2003). The technology acceptance models also emphasize the role of social influence that peers, administrators, and policymakers have on acceptance (Taylor & Todd, 1995).

Cognitive Load Theory and Smart Technologies

CLT, or the Cognitive Load Theory, was first proposed by Sweller (1988) and it explains that the most effective learning happens when cognitive resources are best utilized. CLT emphasizes that instructional methods should reduce unnecessary cognitive load to allow for more effective processing of learning materials. Technologies such as AR/VR and AI can provide interactive learning experiences which are engaging and meaningful to students, in turn reducing cognitive load.

AR/VR allows students to engage with complex, abstract concepts in a tangible, visual way, reducing the extraneous cognitive load often experienced in traditional learning environments. This radically diminishes the extraneous cognitive load that is often associated with traditional pedagogy. The study by Dai et al. (2023) showed that using VR as part of learning enhanced student comprehension of complex scientific subjects. More importantly, adaptive AI powered learning systems would be configured so that more homework would be automatically assigned as the student progressed, and the system would keep track of the learner's advancement. This would bring balanced cognitive workload and improved learning outcomes (Baker et al., 2019).

Through this literature review, the application of intelligent technologies in education is analyzed from the perspective of different theories. Constructivism, cognitivism, and Behaviorism elucidate the relationships between the use of AR/VR technologies, AI, IoT, and gamification and the development of cognitive abilities and interests of students.

The Technology Acceptance Models as well as UTAUT address the integration of these technologies, whereas techniques of cognitive load theory focus on how to enhance learning by reducing cognition overload. These theories are in consonance within the design and functioning of smart education systems to improve learning and engagement of students with new technologies in the changing world. The complete literature frame work is depicted in the Figure 1.

Figure 1. Literature review

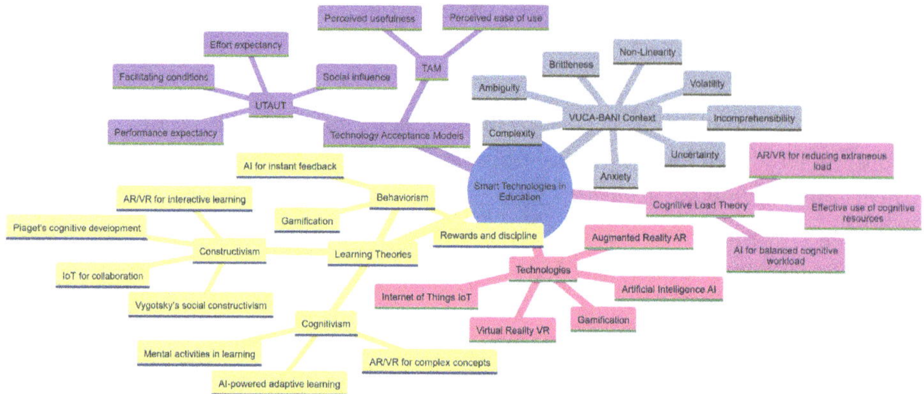

CONCEPTUAL FRAMEWORK AND HYPOTHESIS DEVELOPMENT

The function and influence of technology have become more relevant than ever with the development of smart education systems (Zawacki-Richter et al., 2019). Smart education technologies include the Internet of Things (IoT), Artificial Intelligence (AI), gamification, and Augmented Reality/Virtual Reality (AR/VR) that help in the achievement of cognitive development objectives (Bond et al., 2020). Such advancement is crucial in solving the challenges of a volatile, uncertain, complex, and ambiguous (VUCA) world (Bennett & Lemoine, 2014) or brittle, anxious, non-linear, and incomprehensible (BANI) environment which try to explain modern learning systems' ecosystems' intricacies (Clegg et al., 2021). Huang et al. (2021) argue that integrated technologies in smart classrooms can improve learning by changing learner's perceptions, engaging, and motivating them to devise solutions in a challenging world.

This study seeks to establish the connections these smart technologies have with other important aspects like the degree of engagement of the students and their interest in acquiring cognitive skills. The study seeks to clarify how new technologies in education pass through a VUCA-BANI ecosystem while rendering a learning experience (Selwyn, 2020). From the literature review presented (Table 1 and Figure 1), the hypotheses which were formulated in this study are meant to account for these relationships as well as propose solutions on how educational technology can be used to improve learning in the world today.

Research Hypotheses.

A main area of this study is the examination of the effects that smart classroom technologies have on student engagement. In a world as fast-paced as VUCA-BANI, there is a necessity for engagement and adaptability, and so, these technologies should positively contribute to deeper learning (Clegg et al., 2021; Bennett & Lemoine, 2014). The hypotheses that follow are designed to assess how actively participating students engage in their lessons with the aid of various smart technologies (Al-Fuqaha et al., 2015).

The use of AI adaptive tools foster cognitive engagement by personalizing teaching to specific learners. In an ever changing and uncertain educational context, AI fosters cognitive flexibility along with problem-solving and recalling of information by customizing experiences to fit the learners' strengths and weaknesses. Also, the use of AI personalized intelligent tutoring systems along with machine learning algorithms enables the analysis of performance data from students in order to give constructive comments and feedback that strengthen cognitive development (Zawacki-Richter et al., 2019).

H1: AI-powered learning in smart classrooms enhances cognitive skill development.

AI enhances student engagement by providing interactive and adaptable learning experiences. AI makes participation more motivating and encourages students to stay involved by appealing to their individual learning styles. In addition, AI powered chatbots, virtual assistants, and tailored suggestions enable students to actively interact with the learning content which helps foster ongoing interaction and engagement (Selwyn, 2020).

H2: AI integration in smart classrooms positively influences student participation.

Innovative technologies, such as AR and VR, create engaging learning opportunities which foster deep understanding and learning retention. These technologies build cognitive processing, spatial reasoning, and analysis skills by enabling students to actively engage with challenging topics. Furthermore, VR simulations coupled with AR overlays foster a safe, interactive, hands-on environment for students to try out real-life situations, thereby enhancing their cognitive skills (Makransky & Petersen, 2021).

H3: The use of AR/VR in smart classrooms enhances cognitive skill development.

With AR and VR technologies, students can now interact with learning materials in a more sophisticated manner. This, in turn, increase motivation, curiosity, and willingness to participate in class activities. Those with cognitive overload are able to relieve some of the burden as they learn more effectively and efficiently through AR/VR, making it easier for students with differing learning paradigms to learn (Makransky & Petersen, 2021).

H4: The use of AR/VR technology in smart classrooms increases student participation.

Gamification incorporates playful learning techniques, such as point scoring, rewarding, and setting challenges, which improves problem-solving, decision making, and memory skills. It boosts the effectiveness of learning in today's advanced educational environment. Furthermore, leaderboards, progress indicators, and target-based challenges enable critical and strategic thinking, thus increasing cognitive growth (Deterding et al., 2011).

H5: Gamification in smart classrooms positively influences cognitive skill development.

Integrating features like achievement systems and leaderboards has been shown to increase gamification's motivational and engaging capabilities. Students are motivated to participate and engage interactively during and after the learning process. In addition, reward systems and competitive features promote a feeling of achievement and collaboration among peers which enhances participation in class activities (Hamari et al., 2014).

H6: Gamification enhances student participation in smart classrooms.

The integration of IoT in classrooms allows instantaneous data capturing, updating, and learning, which spearheads cognitive skill growth. Smart learning ecosystems cultivate critical, metacognitive, and autonomous reasoning through the implementation of interactive learning paradigms. Student achievement is further enhanced with IoT's integration with sensors, wearables, and other connected devices as they provide real-time information (Hwang et al., 2020).

H7: IoT integration in smart classrooms enhances cognitive skill development.

The Internet of Things enhances interaction between peers in real-time, enhances feedback, and facilitates personalized learning which makes learning more participatory and engaging. In the changing educational environment, tools based on the Internet of Things allow students to participate dynamically and flexibly. The Internet of Things devices helps students to work collaboratively, contribute to discussions and access course materials in a more organized and interactive way (Hwang et al., 2020).

H8: IoT-driven smart classrooms increase student participation.

Motivated students are likely to interact deeply with learning materials which translates to improved outcomes such as critical thinking and problem solving. Motivated students demonstrate a remarkable improvement in cognitive development and knowledge retention, especially in novel educational contexts. Students who are intrinsically motivated are likely to be ready to learn new ideas, use knowledge, and engage in activities that need higher-order thinking (Chi & Wylie, 2014).

H9: Learning motivation positively impacts cognitive skill development.

Engaged students are more likely to develop active higher-order cognitive skills needed to solve complex problems. Smart classrooms enable students to take part in collaborative learning as well as critical thinking which improves cognitive flexibility. Learning motivated by participation enables more profound conversations, knowledge transfer between learners, and deep thinking, which supports the development of advanced cognitive skills.

H10: Student participation in smart classrooms positively influences cognitive skill development.

Mediating Hypotheses.

These hypotheses analyze student participation as an intervening variable in the association between smart classroom technologies and cognitive skill development. In the contemporary VUCA-BANI world which requires cognitive resilience, critical thinking, and problem-solving, it is proposed that student participation is the intervening variable which captures the value these technologies offer and transforms it into cognitive outcomes (Huang et al., 2021).

The following hypotheses analyze student participation as a mediating variable in the influence of each technology on cognitive skill development. Participation serves as the direct effect of the interactive and personalized learning experience these technologies offer.

With the aid of AI, learning technologies provide personalized instructions which accommodate the level and interest of the learner, thus ensuring active participation. This, in turn, helps in developing cognitive skills like remembering, critical thinking, and problem-solving. Therefore, student participation is defined as a mediating variable in the relationship between the adoption of AI and the development of cognitive skills.

H11: AI significantly impacts cognitive skill development through student participation.

The integration of IoT tools fosters real time collaboration and personalized learning experiences due to advanced connectivity. Students develop cognitive skills by dealing with uncertainty, analyzing data, and solving problems while using the technologies. This assumption proposes that the cognitive skills developed by IoT tools depend on the students' active engagement in the learning process.

H12: The use of IoT tools significantly impacts cognitive skill development through student participation.

By integrating AR/VR technologies, learners are immersed in engaging environments that require active participation for completion, while enabling learners to virtually discover intricate concepts and scenarios. Tools of this nature foster deep engagement to encourage hands-on learning to enhance cognitive skills, including problem-solving and higher-order thinking skills. This hypothesis posits focus on the link between student participation in AR/VR applications and cognitive skill development.

H13: AR/VR significantly impacts cognitive skill development through student participation.

With the use of rewards and competition, gamification makes students more motivated and engaged, which improves their cognitive skills, such as problem solving and strategic thinking. Gamification motivates participation through interactive teaching, and participation positively correlates with the development of cognitive skills. This hypothesis suggests that student participation modifies the positive effect of gamification on cognitive skills.

H14: Gamification significantly impacts cognitive skill development through student participation.

Moderating Hypotheses.

This hypothesis looks at how motivation to learn, influences the connection between student participation and cognitive skill development. Motivation has a great influence on student participation and the amount of effort that they put into work; therefore, a clear strategy must be put in place to ensure that the rewards and outcomes of technology-based learning are fully maximized.

Motivated students are more likely to engage in learning activities, persist through problems, and employ better thinking skills. Motivated learners help transform active student engagement into better cognitive results, which is increasingly important in today's world where change and creativity are needed. This suggests that learning motivation influences the engagement-participation-cognitive skill development relationship by guaranteeing that students who have motivation achieve more pseudo-intellectual development relative to their exposure in participation.

H15: Learning motivation positively moderates the relationship between student participation and cognitive skill development.

Figure 2. Proposed model - Smart Classroom Technologies in the VUCA-BANI World

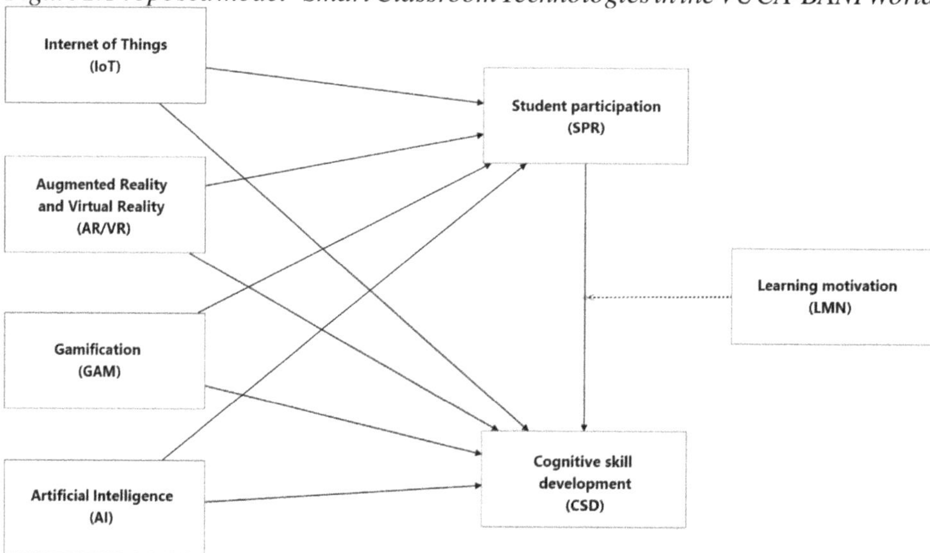

Figure 2 aims to explain the impact of smart classroom technologies on the development of cognitive skills through the mediating effects of student participation and learning motivation. In the context of VUCA-BANI, these hypotheses seek to understand how these technologies not only increase student engagement but also help nurture the cognitive resilience needed to succeed in a fast-changing world.

RESEARCH METHODOLOGY

This study employs a quantitative approach and measures the effect of smart classroom technologies on cognitive skill development with student participation and learning motivation as intervening variables. A stratified and systematic sampling method was employed to proportionately capture students from different disciplines and levels of study in India. In total, the final sample collected was 550.

Figure 3 summarizes model of the methodology and study processes of the research. It illustrates the stepwise approach of the research study, where participation and motivation of students form an integral part of the study concerning smart classroom technologies.

Figure 3. Research Process Flowchart

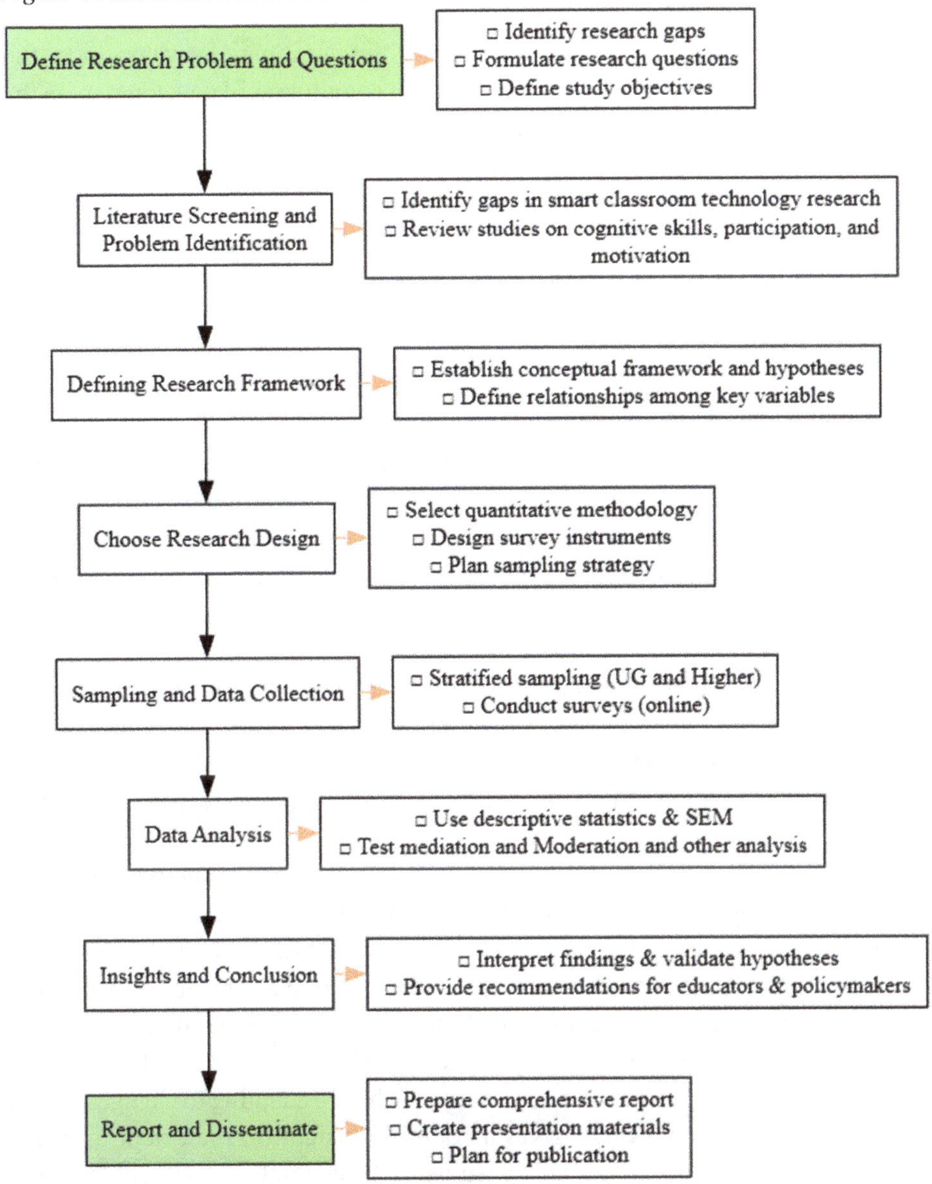

Sampling Frame and Proportions

The sampling frame consists of students from various academic levels, institutions, and geographic locations, ensuring diverse representation.

Questionnaire Development for Smart Education Research

The primary focus behind the development of this questionnaire was to assess the impact of technology enhanced learning tools on cognitive skill development in the VUCA and BANI world (Executive Academy, 2025; Stephan Grabmeier, n.d.). This tool was designed to gather data from students regarding their exposure to emerging educational technologies like AI, IoT, AR/VR, and GAM and its subsequent effect on SPR, LMN, and CSD (eSchool News, 2024; Multishoring, 2025; Nsflow, n.d.).

All constructs were operationalized using multiple Likert-scale items, where respondents could express their agreement or disagreement on a 7-point scale (1 = Strongly Disagree to 7 = Strongly Agree). The thematic sections of the questionnaire were aligned with key research variables which ensured coherence in the questionnaire. The first section focused on technology tools (IoT, AR/VR, GAM, and AI), investigating how easy these tools were to use, if they were engaging, and if students actually learned. The second section assessed student participation in digital learning environments, while the third section observed cognitive skill development as a result of technology. Learning motivation was also assessed as an important factor influencing participation in smart learning environments (Yin, 2018; Wang et al., 2021).

A demographic section was added for respondents to provide their age, gender, academic level, field of study, geographic location, and previous encounters with educational technology. The respondents' access to devices and the internet was also analyzed in relation to the availability of digital learning resources (Masters of ERP, n.d.).

The questionnaire (Table 2) was subjected to a pilot study and pre-testing so that the wording hypothetically would facilitate enhance understanding while maintaining high reliability and construct validity. For assessing internal consistency and convergent validity, the statistical procedure of Cronbach's Alpha, Composite Reliabilities (CR) and Average Variance Extracted (AVE) were utilized.

Conducting the survey through digital means makes it more accessible to a wide range of users, which is important with learners who are adapting to changes within the education system. This tool gives a complete insight on the extent to which technology-enhanced pedagogies foster students' behavioral engagement, learning motivation, cognitive processes, which assists teachers in developing flexible instructional settings in the context of rapid technological advancement (Di Giacomo et al., 2017; UCW, 2024).

The questionnaire that it used for the survey

Table 2. Questionnaire deployed

		1 Strongly Disagree	2 Disagree	3 Somewhat Disagree	4 Neutral	5 Somewhat Agree	6 Agree	7 Strongly Agree
IoT Tools								
	1. IoT tools in my learning environment are easy to use							
	2. IoT devices enhance my learning experience							
	3. IoT integration helps me understand complex concepts							
	4. IoT tools provide real-time feedback for my learning							
AR/VR								
	1. AR/VR technology makes learning more engaging							
	2. I can easily navigate AR/VR learning environments							
	3. AR/VR helps me visualize abstract concepts							
	4. AR/VR experiences improve my practical understanding							
Gamification								
	1. Gaming elements make learning more enjoyable							
	2. I feel motivated by achievement badges and rewards							
	3. Competitive elements enhance my learning engagement							
	4. Gamified challenges help me practice course content							
	5. Level progression helps track my learning progress							
AI								
	1. AI-powered recommendations are relevant to my learning							
	2. AI assistance helps me identify knowledge gaps							
	3. AI-based feedback improves my learning outcomes							
	4. AI tools provide personalized learning experiences							
Student Participation								
	1. I actively participate in technology-enhanced learning activities							
	2. I regularly contribute to online discussions							
	3. I collaborate effectively with peers using digital tools							
	4. I complete all interactive learning tasks							
	5. I seek help when needed through digital platforms							
	6. I engage with multimedia learning content							
Cognitive Skill Development								
	1. Technology helps improve my critical thinking							
	2. I can better analyze problems using digital tools							

continued on following page

Table 2. Continued

		1 Strongly Disagree	2 Disagree	3 Somewhat Disagree	4 Neutral	5 Somewhat Agree	6 Agree	7 Strongly Agree
	3. My problem-solving skills have improved							
	4. I can apply learned concepts in new situations							
	5. My decision-making abilities have enhanced							
Learning Motivation								
	1. Technology increases my interest in learning							
	2. I feel more confident using technology for learning							
	3. I am motivated to explore additional learning resources							
	4. I maintain consistent engagement in online learning							

Data Collection

Data was collected through a structured survey questionnaire designed to measure the main aspects of the study. The questionnaire aimed at capturing the:

1. The extent to which students know about and use smart classroom technologies.
2. The degree of active learning and motivation of the students within the learning activity.
3. Perceptions regarding the impact of these technologies on cognitive skill development.

To achieve optimal coverage as well as inclusivity, offline and online surveys were used in combination. This combination allowed for a wider scope and reduced the chance of sampling errors.

Inclusion Criteria

The study includes participants who meet the following criteria to ensure relevance and reliability in analysing technology adoption in education:

1. **Academic Enrollment:** Participants must be currently enrolled in undergraduate, or above programs.
2. **Field of Study:** Respondents should belong to Science, Commerce, Humanities, or related disciplines.
3. **Institution Type:** Students from government, private, deemed universities, and autonomous institutions are included.

4. **Geographic Location:** Participants must be from urban, semi-urban, or rural areas in India.
5. **Age Group:** Only students aged 18 years and above are considered.
6. **Technology Exposure:** Respondents should have some level of experience with educational technology, ranging from no experience to more than three years.
7. **Device Ownership & Internet Access:** Participants must own or have access to digital devices (smartphones, laptops, tablets, etc.) and have some form of internet connectivity.
8. **Willingness to Participate:** Only students who voluntarily consent to participate in the study are included.

Data Analysis Techniques

The study goals were accomplished while investigating the interconnections between variables with the aid of sophisticated statistical methods. The sophisticated methods utilized also analyzed descriptive statistics to prepare the demographic and academic data of the participants. Technology's automation of cognitive skill development was mediated by student engagement and motivation, thus the SEM model tested the mediation effects of student participation and interest. Accurate analysis and data processing were executed through SPSS and SmartPLS, which facilitated the interpretation of advanced models and provided more complex results. The use of SEM also improved the understanding of the observed patterns, relationships, and mediating variables in regard to the objectives of the study.

Consideration of Ethics

The rights and privacy of the subjects were kept confidential in this research, so ethical procedures were set in place throughout the research. Consent guarantees were provided through anonymity as well as a withdraw option. Permission was granted to various educational authorities to collect such sensitive data within established protocols.

DATA ANALYSIS AND INTERPRETATION

Data Preparation and Screening

Preparing and screening data are some of the most important processes in a research study, making certain that the data is pristine, trustworthy, and ready for analysis. This study focuses on the impact of modern technologies such as IoT, AR and VR, Gamification, and AI on student engagement, motivation to learn, and the acquisition of different competencies. Prior to the statistical analysis, some missing values needed to be filled, outliers defined and dealt with, normality assessed, internal consistency and validity checked, and all prerequisites for the following procedures had to be completed.

Dealing with absent values is one of the most concerning matters during data preprocessing as data absence can cause skewed results and lower statistical efficacy (Little & Rubin, 2019). The first phase required determining the missing portions and evaluating their distribution. For total random loss of data (MCAR) methods as either listwise or pairwise deletion were preferred (Enders, 2010). However, if the data was missing in a systematic way (not missing completely at random), more sophisticated algorithms such as the use of mean imputation, regression imputation, and multiple imputations were employed in order to maintain the integrity of the data set (van Buuren, 2018). Given that certain aspects such as motivational factors in learning and development of cognitive skills are constrained to student's self-report responses, absent values for these variables can alter the outcomes in expectation (Peugh & Enders, 2004). Thus, my aim was to ensure that missing data did not affect the outcome by employing appropriate imputation techniques that corresponded to the degree and type of absence (Graham, 2009).

Analyzing and handling outliers was the next important task. Outliers can interfere with statistical relationships which can lead to incorrect conclusions (Aguinis et al., 2013). Extreme values were identified and Boxplots and Z-scores were utilized to flag any observations that were over ±3 standard deviations (Tabachnick & Fidell, 2019). With regards to the students' responses to technology-based learning tools, it was crucial to understand whether extreme values were the result of genuine participation, motivation, and skill development or simple data entry mistakes (Osborne & Overbay, 2004). If data entry mistakes had resulted in an outlier value, then those values were either adjusted or deleted. However, if extreme values were valid responses, statistical techniques such as Winsorization (trimming extreme values) or data transformations (e.g., log transformation) were applied to reduce their undue influence on the analysis (Ghosh & Vogt, 2012; Kwak & Kim, 2017).

Normality assessment was another key component of data screening. Many statistical methods assume a normal distribution of variables; thus, histograms, Q-Q plots, skewness, kurtosis, and the Shapiro-Wilk test were used to check for normality (Ghasemi & Zahediasl, 2012). Technology-related variables such as IoT, AR/VR, AI, and Gamification were particularly examined for skewness, as strong preferences toward or against these technologies could result in a non-normal distribution (Hair et al., 2019). Similarly, psychological constructs such as student participation, learning motivation, and cognitive skill development were assessed to determine if they followed a normal distribution (Field, 2017). If variables were found to be significantly skewed, appropriate transformations such as log, square root, or Box-Cox transformations were applied to approximate normality (Osborne, 2010).

Reliability and internal consistency were also evaluated to ensure that multi-item scales used in the study produced consistent and stable results (DeVellis, 2016). Cronbach's alpha was calculated for each scale, with a threshold of 0.70 indicating acceptable reliability (Taber, 2018). If any scale exhibited low reliability, item-total correlations were examined, and poorly performing items were removed or modified (Streiner, 2003). Given the subjective nature of constructs such as learning motivation and cognitive skill development, ensuring high internal consistency was crucial to maintaining measurement reliability (Drost, 2011).

Construct validity was established by performing exploratory factor analysis (EFA) (Thompson, 2004). The data sanity check, Kaiser-Meyer-Olkin (KMO) test, and Bartlett's test of sphericity were done to find out whether the data were eligible for factor analysis (Williams et al., 2010). Bartlett's test being significant along with KMO value being more than 0.60 factorability was confirmed (Tabachnick & Fidell, 2019). Principal Component Analysis (PCA) with varimax rotation was performed to extract meaningful factors while ensuring that each variable was loaded onto its expected construct (Costello & Osborne, 2005). Items with weak or cross-loadings were adjusted from the construct to maintain construct validity (Hair et al., 2019). This step proved critical to make sure student participation, learning motivation, and cognitive skill development accurately represented their intended constructs (Messick, 1995).

Concerns for multicollinearity were the leading issue which is the use of two or more independent variables. The high correlation creates uncertainty in regression models (Alin, 2010). Assessment of multicollinearity was done by calculating the Variance Inflation Factor (VIF) where any $VIF < 10$ is deemed acceptable for the study (Hair et al., 2014). Other than that, tolerances above 0.10 were checked confirm that sufficient independence among the variables exists (O'Brien, 2007). In cases where high multicollinearity was found, improving model stability was feasibly done by merging correlated variables into composite indices or displacing redundant variables (Dormann et al., 2013).

Before proceeding with deeper analyses, six other assumption tests were conducted. Scatterplots were used to test for linearity, checking that independent and dependent variables' relationships were correctly fitted (Tabachnick & Fidell, 2019). The examination of residual plots to check if error variance was constant with respect to the different amounts of the independent variables tested confirmed that homoscedasticity was present (Osborne & Waters, 2002). The independence of residuals was assessed employing the Durbin-Watson statistic, verifying whether residuals were autocorrelated (Field, 2017). All of these assumptions were necessary to perform valid statistical tests like correlation, regression, and SEM analysis (Kline, 2016).

After completing all of these data cleaning procedures, the dataset was considered complete and ready for analysis. Working with closed missing values, dealt outliers, performed tests for normality, established reliability, and met most of the statistical assumptions provided an opportunity to explore the effects of IoT, AR/VR, AI, and Gamification on students' participation, learning motivation, and cognitive skill development in a more precise and valid manner. These preparatory processes increased the methodological rigor of the study, ensuring that the results and conclusions drawn will not only be accurate but also meaningful.

Demographic Variables

Table 3. Demographic Variables

Variable	Option 1	Option 2	Option 3	Option 4	Option 5
Age (Years)	18-20 (5.6%)	21-23 (28.5%)	24-26 (23.5%)	27-29 (27.1%)	30+ (15.3%)
Gender	Male (64.9%)	Female (35.1%)			
Current Academic Level	Undergraduate (11.3%)	Postgraduate (35.3%)	Doctoral (31.1%)	Other (22.4%)	
Field of Study	Science (12.5%)	Commerce (36.7%)	Humanities (32.9%)	Other (17.8%)	
Year of Study	First Year (6.5%)	Second Year (30.5%)	Third Year (24.5%)	Fourth Year (29.5%)	Other (8.9%)
Type of Institution	Government (11.8%)	Private (36.4%)	Deemed University (31.6%)	Autonomous (20.2%)	
Geographic Location	Urban (11.8%)	Semi-urban (36.4%)	Rural (51.8%)		
Experience with EdTech	No experience (5.8%)	<1 year (28.5%)	1-2 years (23.3%)	2-3 years (27.5%)	3+ years (14.9%)
Device Ownership	Smartphone (6.2%)	Laptop (29.8%)	Desktop (23.1%)	Tablet (30.9%)	Other (10.0%)
Internet Access	Home broadband (4.7%)	Mobile data (25.5%)	Both (23.8%)	Other (46.0%)	

From the Table 3 the following is the demography of the respondent of this study,

Age Distribution: The age distribution of the respondents is categorized into five groups: 18-20 years, 21-23 years, 24-26 years, 27-29 years, and 30 years and above. The largest proportion of respondents, 28.5% (n=157), belong to the 21-23 years age group, followed by 27.1% (n=149) in the 27-29 years category. The 24-26 years group comprises 23.5% (n=129) of the sample, while 15.3% (n=84) fall into the 30 years and above category. The smallest proportion of respondents, 5.6% (n=31), are in the 18-20 years age range. These findings suggest that the majority of the participants are in their early to late twenties, which aligns with higher education enrollment trends.

Gender Representation: Gender distribution among the respondents indicates that the majority, 64.9% (n=357), identify as male, while 35.1% (n=193) identify as female. No respondents identified as other gender categories. The gender distribution provides insight into the representation of male and female students in higher education and may help in understanding whether gender influences perceptions and adoption of educational technology.

Current Academic Level: The academic level of respondents is divided into four categories: Undergraduate, Postgraduate, Doctoral, and Other. The largest proportion of respondents, 35.3% (n=194), are postgraduate students, followed closely by 31.1% (n=171) at the doctoral level. Additionally, 22.4% (n=123) fall under the other category, which may include diploma or certification students, while 11.3% (n=62) are undergraduate students. This indicates that a significant portion of the sample consists of postgraduate and doctoral-level students, which may reflect a higher level of engagement with advanced learning technologies.

Field of Study: The study fields of respondents are categorized as Science, Commerce, Humanities, and Other. A substantial number of participants, 36.7% (n=202), are from the Commerce field, while 32.9% (n=181) belong to Science. The Humanities field represents 17.8% (n=98) of the sample, whereas 12.5% (n=69) fall into the Other category. The predominance of commerce and science students suggests a strong inclination toward technology-driven learning in these domains.

Year of Study: Participants belong to various academic years: First Year, Second Year, Third Year, Fourth Year, and Other. The Second Year has the highest representation at 30.5% (n=168), followed by Fourth Year students at 29.5% (n=162). Third Year students constitute 24.5% (n=135), while 8.9% (n=49) belong to the Other category, which may include part-time or research-based students. The lowest proportion, 6.5% (n=36), are in their First Year. This distribution suggests that students in their second and fourth years are more engaged with educational technology.

Type of Institution: The respondents come from different types of institutions: Government, Private, Deemed University, and Autonomous. The highest percentage, 36.4% (n=200), are from Private institutions, followed by 31.6% (n=174) from Deemed Universities. Autonomous institutions account for 20.2% (n=111), while Government institutions represent the smallest share at 11.8% (n=65). This distribution indicates a stronger representation of students from private and deemed universities.

Geographic Location: The geographic distribution of respondents is divided into Urban, Semi-Urban, and Rural categories. The largest segment, 51.8% (n=285), belongs to the Rural category, followed by 36.4% (n=200) from Semi-Urban areas. Only 11.8% (n=65) of the respondents are from Urban areas. This data suggests that a significant portion of the sample comprises students from non-urban areas, which may have implications for access to and familiarity with digital learning tools.

Previous Experience with Educational Technology: Respondents' prior experience with educational technology is classified into five categories: No Experience, Less than 1 year, 1-2 years, 2-3 years, and More than 3 years. The largest percentage, 28.5% (n=157), have less than 1 year of experience, while 27.5% (n=151) have 2-3 years of experience. A total of 23.3% (n=128) have 1-2 years of experience, and 14.9% (n=82) have been using educational technology for more than 3 years. Only 5.8% (n=32) reported having no prior experience with educational technology. This distribution suggests that most students have some level of exposure to digital learning tools.

Device Ownership: Participants described having numerous digital devices, classified as a smartphone, laptop, desktop, tablet, and other. The most commonly owned device is a Tablet, reported by 30.9% (n=170) of the respondents. Laptops are owned by 29.8% (n=164), followed by Desktops at 23.1% (n=127). Smartphones are owned by 6.2% (n=34) of the respondents, while Other devices account for 10.0% (n=55). This data illustrates the significant ownership of laptops and tablets within the student population, which are essential for participating in digital learning environments.

Internet Access: The participants have different degrees of access to the internet classified as home broadband, mobile data, both, and other. The highest proportion, 46.0% (n=253), reported having access to both home broadband and mobile data. Mobile data only is used by 25.5% (n=140) of respondents, while home broadband only is used by 23.8% (n=131). A smaller segment, 4.7% (n=26), reported using other sources of internet access. The data indicates that although a good number of students have internet access from multiple sources, a large proportion does not, with a sole dependence on mobile data and that could limit students from digitally learning effectively.

The demographic analysis reveals key insights into the student population's engagement with educational technology. The majority of respondents are in their early to late twenties, with a higher representation of male students (Selwyn, 2016). Most participants belong to postgraduate and doctoral programs, with Commerce and Science being the most common fields of study (Henderson et al., 2017). Students in their second and fourth years are more prominently represented, and a large proportion of respondents come from private institutions and rural areas (Castañeda & Selwyn, 2018).

Regarding technological readiness, most students have prior experience with educational technology, with a significant proportion having 1-3 years of exposure.

Device ownership is high, particularly for laptops and tablets, and most students have access to multiple internet sources. These insights are crucial for understanding the extent to which students are equipped to engage with emerging learning tools such as IoT, AR/VR, AI, and gamification-based education.

Factor Analysis for Construct Validity: Data Analysis and Interpretation

Identifying unobserved elements within a dataset is easily achievable with factor analysis as it allows collection of correlated variables into a meaningful construct (Thompson, 2004). This analysis serves to prove the construct validity of the measurement scale designed to evaluate the students' engagement with educational technology (Messick, 1995). The Varimax rotation Principal Components Method (PCA) was utilized to extract factor(s) and assess their contribution to the determined variance in the dataset (Costello & Osborne, 2005). Other preliminary requirements for conducting factor analysis such as the Kaiser-Meyer-Olkin (KMO) Measure of Sampling Adequacy and Bartlett's Test of Sphericity were performed to check whether the dataset is suitable for factor analysis (Williams et al., 2010).

Table 4. KMO and Bartlett's Test

Kaiser-Meyer-Olkin Measure of Sampling Adequacy.		**0.882**
Bartlett's Test of Sphericity	Approx. Chi-Square	8305.329
	Df	528
	Sig.	.000

With reference to the Table 4, the KMO measure was 0.882 which exceeds the threshold of 0.6. This demonstrates that the sample size is sufficient, and therefore the dataset is suitable for factor analysis (Kaiser, 1974). The KMO measure also indicates a high value suggesting a substantial proportion of variance in common among the factors which makes them adequate for factor extraction (Tabachnick & Fidell, 2019). In addition, Bartlett's test showed a KMO value of ($\chi^2 = 8305.329$, df $= 528$, $p < 0.001$) which sufficiently proves the correlation matrix is not an identity matrix and is able to undergo factor analysis (Hair et al., 2014).

Figure 4. Scree Plot

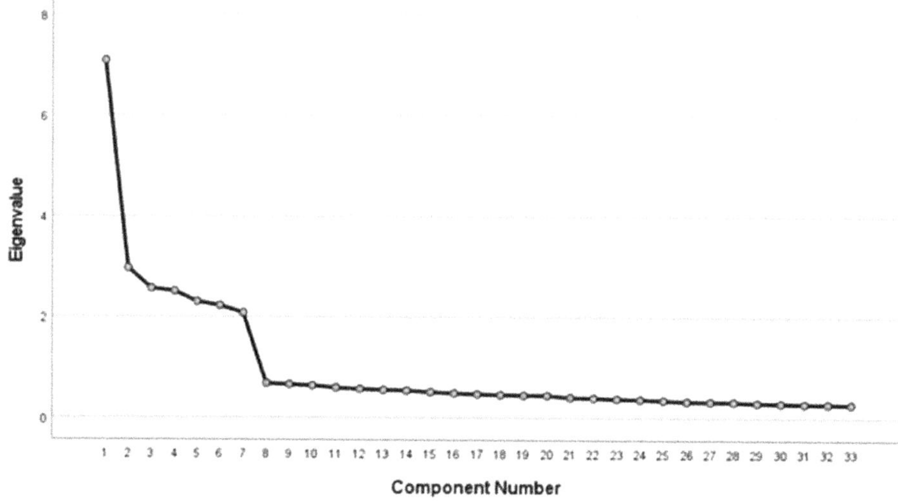

Table 5. Total Variance Explained

Component	Initial Eigenvalues		
	Total	% of Variance	Cumulative %
1	7.088	21.478	21.478
2	2.963	8.980	30.458
3	2.560	7.758	38.216
4	2.504	7.588	45.803
5	2.301	6.974	52.777
6	2.221	6.731	59.508
7	2.077	6.295	65.803

Following these preliminary tests, an eigenvalue criterion was applied to determine the number of factors to retain. According to Kaiser's criterion, factors with eigenvalues greater than 1 should be retained for further interpretation (Kaiser, 1960). As mentioned in the Table 5, the total variance explained table revealed that seven components met this criterion, accounting for a cumulative variance of 65.803%. This means that the extracted factors collectively explain 65.803% of the total variance in the dataset, which is considered an adequate level of explanation in social science research (Hair et al., 2014). The first factor had the highest eigenvalue of 7.088, contributing 21.478% of the total variance. The subsequent factors explained progressively smaller portions of the variance, with the second factor contributing 8.98%, the third factor 7.758%, the fourth factor 7.588%, and so on. The scree plot (Figure 4), a graphical representation of eigenvalues, confirmed the selection of these seven factors as it showed a clear "elbow" at the seventh component, indicating the point where the explained variance starts to level off (Cattell, 1966; Ledesma & Valero-Mora, 2007).

Table 6. Rotated Component Matrix (EFA)

	Component						
	1	2	3	4	5	6	7
SPR3	.830						
SPR5	.827						
SPR2	.820						
SPR4	.814						
SPR1	.809						
SPR6	.804						
CSD2		.758					
CSD5		.757					
CSD4		.749					
CSD6		.742					
CSD3		.735					
CSD1		.675					
GAM4			.771				
GAM5			.767				
GAM3			.761				
GAM1			.748				
GAM2			.743				
LMN1				.831			
LMN4				.831			
LMN3				.825			
LMN2				.823			
IoT1					.857		
IoT4					.834		
IoT3					.824		
IoT2					.806		
AI2						.782	

continued on following page

Table 6. Continued

	Component						
	1	**2**	**3**	**4**	**5**	**6**	**7**
AI4						.782	
AI3						.779	
AI1						.776	
ARVR4							.792
ARVR3							.785
ARVR2							.775
ARVR1							.774

Extraction Method: Principal Component Analysis.
Rotation Method: Varimax with Kaiser Normalization.
a. Rotation converged in 5 iterations.

To further interpret the extracted factors, a rotated component matrix was examined. Varimax rotation was applied to achieve a clearer and more meaningful pattern of factor loadings by maximizing the variance of squared loadings for each factor (Abdi, 2003). The results revealed that each factor had strong loadings (Table 6) for distinct sets of variables, supporting the construct validity of the measurement scale (Messick, 1995). The first factor was identified as Student Participation, comprising six variables with high loadings (ranging from 0.804 to 0.830). This factor reflects the extent to which students actively engage in educational activities facilitated by technology, such as online discussions, interactive platforms, and collaborative learning tools (Fredricks et al., 2004). The second factor, Cognitive Skill Development, included six items with factor loadings between 0.675 and 0.758. This factor represents the role of educational technology in enhancing students' analytical thinking, problem-solving skills, and intellectual growth (Bloom et al., 1956).

The third factor, Gamification, consisted of five items with loadings ranging from 0.743 to 0.771. This factor highlights the use of game-based elements, such as rewards, leaderboards, and challenges, to increase student motivation and engagement in learning activities (Deterding et al., 2011). The fourth factor, Learning Motivation, had four strongly loading items (ranging from 0.823 to 0.831), suggesting that digital learning tools can significantly enhance students' intrinsic and extrinsic motivation to study (Ryan & Deci, 2000). The fifth factor, Internet of Things (IoT), comprised four variables with loadings between 0.806 and 0.857, indicating that interconnected smart devices, such as sensors and adaptive learning platforms, play a critical role in modern education (Atzori et al., 2010). The sixth factor, Artificial Intelligence (AI) captioned four items with loadings of 0.776 and 0.782 proving the rising importance of AI Powered learning environments such as tailored learning suggestions and self-grading (Roll & Wylie, 2016). Lastly, the seventh factor with four supporting variables Augmented and Virtual Reality (AR/VR) had loadings from 0.774 to 0.792 which justify the role of self-paced interactive technologies in improving students' attention and understanding through interactive simulations and virtual environments (Wu et al., 2013).

The overall findings of the factor analysis indicate that the measurement scale captures relevant and definable concepts related to educational technology adoption. The remarkable factor loadings affirm that each extracted factor is a coherent and interpretable construct, thus validating the instrument's (Tabachnick & Fidell, 2019) claim. The KMO value along with Bartlett's test emphasizes the appropriateness of the dataset for factor analysis while the 65.803% cumulative variance confirms the majority of observed data is explained by the extracted factors(Hair et al., 2014).

These findings are important for educators, administrators, and researchers engaged in digital education. The findings highlight the necessity of the student's active participation, cognitive skill development, and motivation in learning through technology (Selwyn, 2016). Moreover, factors such as gamification or AI or IoT or AR and VR suggest that the modern educational technologies have great potential to enhance the learning environment (Zawacki-Richter et al., 2019). Examine other educational settings and technologies and their impact on student learning engagement and performance warrants further study (Bond et al., 2020).

To summarize, the factor analysis conducted in this research provides evidences that support the scale measurement for educational technology adoption as reliable and valid (Hair et al., 2014). The seven factors identified- Student Participation, Cognitive Skill Development, Gamification, Learning Motivation, Internet of Things, Artificial Intelligence, and Augmented and Virtual Reality- are separate and significant constructs for having a life beyond the passive ingestion of emerging technologies and the participative embrace of digital learning environments (Selwyn, 2016). The analysis has shown the growing importance of technology in the education field and provides a foundation for further studies aimed at the development of digital pedagogy to enhance student performance (Bond et al., 2020).

This research aims to study the impact of specific technological and non-technological teaching components on students' cognitive skill development (CSD). The integration of IoT, AR/VR, GAM, AI, SPR, and LMN into the educational system is quickly transforming students' ways of learning (Zawacki-Richter et al., 2019). This analysis serves to determine which factors form the critical mass of cognitive skill improvement and analyze their impact on students' academic achievements.

To ensure the findings remain unbiased, the analysis incorporates the necessary basic statistical assumptions such as normality, homoscedasticity, multicollinearity, and autocorrelation (as cited in Tabachnick & Fidell, 2019). The changes in the CSD are analyzed through the R-Square, Adjusted R-Square, and ANOVA to assess how independent variables have impacted it. Aside from this, individual components are tested through a hypothesis test to determine the relevance of each individual predictor (Field, 2017).

To ensure the integrity of the regression model, normality and other regression diagnostic criteria must be carried out. The use of the normal probability plot (P-P plot) and scatterplot of residuals helps obtain information regarding the nature and distribution of the errors as well as the adequacy of the model (Osborne & Waters, 2002).

Normality of Residuals

Figure 5. Normal P-P Plot of Regression Standardized Residual

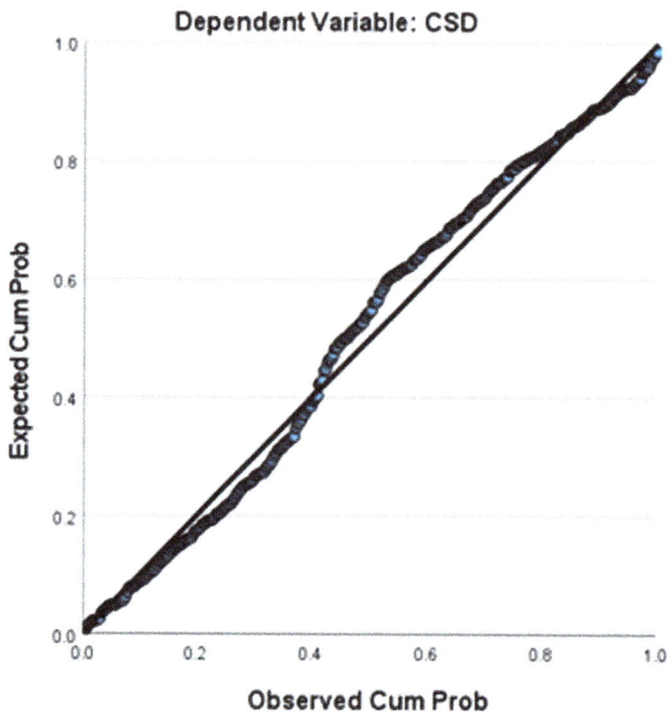

As seen in the Figure 5, the standard P-P plot of a regression's residuals is assessed for the possibility of the residuals being normally distributed. In this instance, normality is assumed to be reasonably met because most of the points in the plot align with the diagonal reference line (Field, 2017). There are some minor deviations, but they do not appear to significantly violate normality. The normality of residuals is essential to ensure the reliability of statistical inference such as confidence intervals and hypothesis tests (Ghasemi & Zahediasl, 2012). A large degree of non-normality could result in significant parameter bias and misleading results for significance tests.

Homoscedasticity and Linearity

Figure 6. Scatterplot

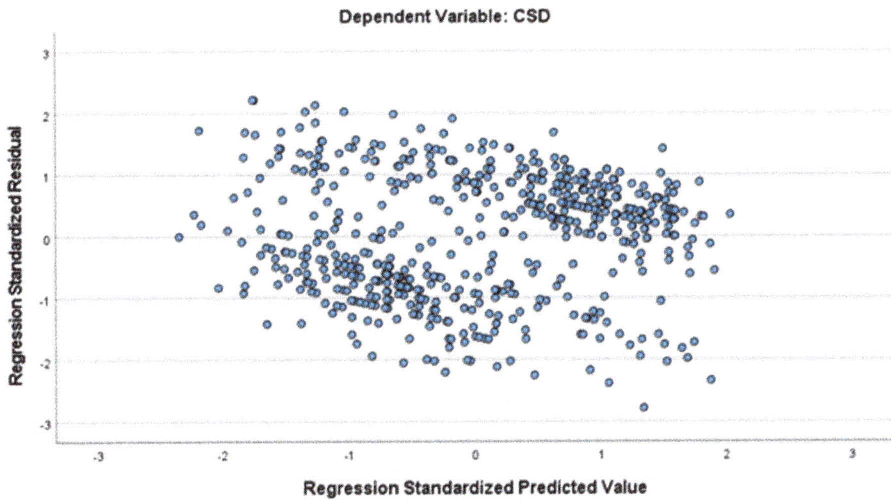

To evaluate the homoscedasticity as well as linearity of the errors, the scatterplot of standardized residuals against the standardized predicted values is made (Figure 6). By nature, the residuals ought to cluster randomly with no clear discernible shape (Tabachnick & Fidell, 2019). In this case, however, the scatterplot depicts a slight funnel shape which implies a certain degree of heteroscedasticity. This indicates that the variance of errors is not entirely constant across predicted values. Although this may not be desirable, it does strongly suggest that the efficiency of the regression estimates is highly impacted. In any case, while this degree of heteroscedasticity may not make the model invalid, it does certainly suggest that standard errors of the associated regression coefficients are biased and lead to contradictory statistical conclusions (Osborne & Waters, 2002).

Durbin-Watson Test for Autocorrelation

Table 7. Model Summary (Regression)

Model	R	R Square	Adjusted R Square	Std. Error of the Estimate	Durbin-Watson
1	.418[a]	.175	.166	1.20647	1.986

a. Predictors: (Constant), LMN, ARVR, AI, GAM, IoT, SPR
b. Dependent Variable: CSD

The Durbin-Watson statistic as displayed in the Table 7 from the model summary table is 1.986, which is very close to 2. This indicates that there is no significant autocorrelation in the residuals (Field, 2017). Autocorrelation in residuals would imply that errors are not independent, which violates one of the fundamental assumptions of linear regression. In time-series data, autocorrelation can result in misleading statistical significance levels and reduce the reliability of predictions. Given the Durbin-Watson value observed here, there is no immediate concern about autocorrelation, and the assumption of independent errors is satisfied.

Multicollinearity Assessment

Table 8. Coefficients (Regression)

Model		Unstandardized Coefficients		Standardized Coefficients	t	Sig.	Collinearity Statistics	
		B	Std. Error	Beta			Tolerance	VIF
1	(Constant)	1.500	.286		5.235	.000		
	IoT	.098	.037	.111	2.680	.008	.881	1.135
	ARVR	.093	.039	.096	2.364	.018	.929	1.076
	GAM	.105	.040	.107	2.635	.009	.927	1.079
	AI	.055	.039	.058	1.425	.155	.910	1.099
	SPR	.157	.036	.184	4.390	.000	.865	1.156
	LMN	.134	.034	.161	3.907	.000	.893	1.120

a. Dependent Variable: CSD

Multicollinearity refers to high correlations among independent variables, which can distort regression estimates. The variance inflation factor (VIF) values in the Table 8, indicate the extent of multicollinearity among predictor variables (Hair et al., 2014). All VIF values are below 1.2, suggesting that multicollinearity is not a significant issue in this model. Typically, VIF values greater than 5 indicate problematic multicollinearity, requiring corrective measures such as removing or combining variables (Tabachnick & Fidell, 2019). Since all predictors in this study have VIF values well below the threshold, it confirms that the predictors are independent and provide unique contributions to the dependent variable.

Significance of Predictors

The coefficients table presents both unstandardized and standardized regression coefficients, along with their respective significance values are mentioned in the Table 8. Among the predictors, IoT, AR/VR, gamification, student participation, and learning motivation exhibit statistically significant relationships with cognitive skill development (CSD), as their p-values are below 0.05. Equation 1 displays the significance of predictors.

Equation 1. Predictive Equation for Cognitive Skill Development in Technology-Enhanced Learning

$$CSD = 1.500 + (0.098 \times IoT) + (0.093 \times ARVR) + (0.105 \times GAM) + (0.055 \times AI) + (0.157 \times SPR) + (0.134 \times LMN) + \varepsilon$$

Artificial intelligence (AI), however, does not exhibit statistical significance in predicting CSD, as its p-value exceeds 0.05. This suggests that AI, in its current form, does not strongly contribute to the enhancement of cognitive skills, whereas other technological and educational factors do (Zawacki-Richter et al., 2019).

Regardless of its statistics insignificance (p-value = 0.155), AI has to be retained in the model because it is of theoretical significance in CSD. Learning technologies based on artificial intelligence (AI) increase the level of problem solving, critical analysis, and knowledge of the learners through recommendations, feedback, and testing (Di Giacomo etal., 2017; Turnitin, n.d.). Although AI's direct impact is weak, it probably works indirectly, directly through student participation (SPR) and learning motivation (LMN) as criticals in mediators in technology enhanced learning (Byers, 2024; Wang et al., 2024).

Moreover, within the context of VUCA-VANI world, AI enables data and skill experiences to be more structured and systematic in helping students learn in the midst of complexity and uncertainty (Executive Academy, 2025; Stephan Grabmeier, n.d.). Its contribution to interactive learning, gamification, and intelligent tutoring encompasses its retention in the model (Nsflow, n.d.; UCW, 2024). Advanced CSD can be studied more by future researchers using moderation analysis, mediation analysis, interaction effects, or other AI's wide domain specific application.

To put it into perspectives, AI is a determinant of cognitive development that holds importance in advanced learning even if there is a lack of statistical significance (Psico-Smart, 2024; CIDDL, n.d.).

Model Fit and ANOVA Results

Table 9. ANOVA

Model		Sum of Squares	df	Mean Square	F	Sig.
1	Regression	167.29	6	27.88	19.156	.000[b]
	Residual	790.36	543	1.45		
	Total	957.66	549			

a. Dependent Variable: CSD
b. Predictors: (Constant), LMN, ARVR, AI, GAM, IoT, SPR

The model summary table (Table 9) indicates an R-value of 0.418, which represents a moderate correlation between the predictor variables and CSD. The R-square value of 0.175 suggests that approximately 17.5% of the variance in CSD is explained by the independent variables included in the model. While this indicates a moderate explanatory power, it also suggests that additional factors beyond the studied predictors contribute to CSD. The adjusted R-square value (0.166) provides a more reliable estimate by accounting for the number of predictors in the model.

The ANOVA table further supports the significance of the model. The regression model's F-statistic (19.156) is significant at $p < 0.001$, indicating that the overall model is statistically significant in predicting cognitive skill development. This confirms that at least some of the independent variables have meaningful contributions to explaining the variation in CSD.

Residual Analysis and Model Assumptions

Table 10. Residuals Statistics (ANOVA)

	Minimum	Maximum	Mean	Std. Deviation	N
Predicted Value	2.99	5.41	4.29	.55	550
Residual	-3.36	2.67	.00	1.20	550
Std. Predicted Value	-2.34	2.02	.00	1.00	550
Std. Residual	-2.78	2.21	.00	.995	550

a. Dependent Variable: CSD

The residuals statistics table (Table 10)provides additional insights into model diagnostics. The mean residual value is approximately zero, which aligns with the expectation that residuals should average out to zero in a well-specified model (Field, 2017). The standard deviation of residuals is within an acceptable range, and standardized residual values fall within the typical range of -2.5 to 2.5, indicating the absence of extreme outliers (Tabachnick & Fidell, 2019). The residual diagnostics, along with the normality and homoscedasticity tests, suggest that while minor violations exist, they do not severely compromise the validity of the regression model (Osborne & Waters, 2002).

In summary, the regression model satisfies most key assumptions, including normality, independence of errors, and absence of multicollinearity. Minor concerns regarding heteroscedasticity exist but do not significantly impact the model's overall reliability. The predictors significantly contribute to cognitive skill development, except for AI, which does not show a strong statistical association. The overall model fit is moderate, suggesting that additional factors could improve explanatory power. These findings provide meaningful insights into the role of technology and participation in shaping cognitive skill development in students.

Cronbach's Alpha

Table 11. Cronbach's Alpha

Constructs	Cronbach's Alpha	Items
IoT	0.874	4
ARVR	0.804	4
GAM	0.828	5
AI	0.805	4
SPR	0.915	6
LMN	0.872	4
CSD	0.854	6

The results of Cronbach's Alpha (Table 11) exhibit a good degree of internal consistency within the variables for this particular study. The alpha co-efficient is one of the most common methods of reliability assessment which assumes that a collection of concepts form a single construct and additionally estimates how deeply each item within that group resonates with the underlying concept (Tavakol & Dennick, 2011). Values above 0.7 are considered acceptable, above 0.8 is deemed reliable and above 0.9 excellent (Nunnally & Bernstein, 1994). The results indicates that all constructs are reliable with the lowest value of 0.804 for AR/VR and AI and the highest value of 0.915 for Student Participation. The results indicate that the greatest alpha value for Student Participation suggest that the items the construct has to measure are highly correlated and consitent. Strong reliability of IoT (0.874), Learning Motivation (0.872), and Cognitive Skill Development (0.854) suggests that these constructs are effectively measured with small amounts of error. Gamification (0.828) and AI (0.805) scales also demonstrate good reliability and therefore these items constituting the constructs can be confidently assumed to capture the variables intended. Even the lowest score of reliability when using the AR/VR construct 0.804 does suggest a moderate degree of consistency in measurement. In summary, the strong internal reliability observed across all constructs indicates that the measurement scales applied in this research are valid and reliable which enables proper interpretation and valid conclusions to be drawn from the data (Taber, 2018).

Correlation analysis

Table 12. Correlations

	IoT	ARVR	GAM	AI	SPR	LMN	CSD
IoT	1						
ARVR	.139	1					
GAM	.180	.147	1				
AI	.230	.109	.130	1			
SPR	.220	.212	.164	.201	1		
LMN	.206	.147	.170	.164	.243	1	
CSD	.231	.196	.206	.172	.297	.271	1

The correlation matrix mentioned in the Table 12, highlights the relationships among IoT, ARVR, GAM, AI, SPR, LMN, and CSD. The highest correlation is observed between SPR and CSD (0.297), indicating that increased SPR is strongly linked to CSD enhancement. Similarly, LMN has a moderate correlation with CSD (0.271), reinforcing its role in skill development.

Among technological factors, IoT has a notable correlation with CSD (0.231) and AI (0.230), suggesting its importance in digital learning. ARVR has a lower correlation with CSD (0.196), indicating a lesser direct impact on skill development compared to SPR and LMN. GAM shows consistent positive correlations, particularly with IoT (0.180) and AI (0.130), reflecting its integration within tech-based learning environments.

AI exhibits moderate correlations with all variables, affirming its supporting role in digital education. Overall, the findings emphasize that SPR, LMN, and IoT significantly contribute to CSD, suggesting a blended approach of engagement-driven and technology-enhanced learning can optimize cognitive development.

Hypothesis testing

Figure 7. Theoretical model

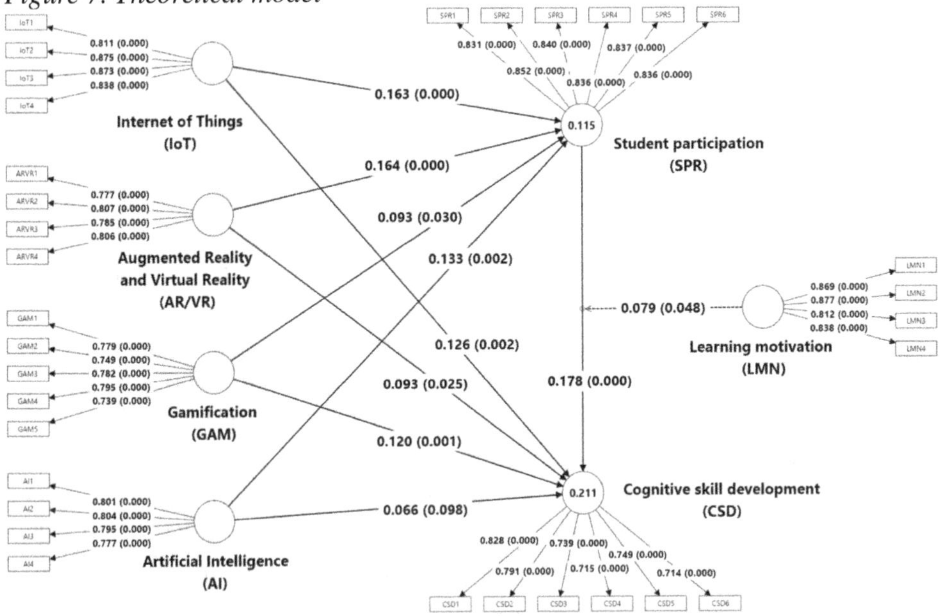

The model demonstrates the relationships between IoT, AR/VR, GAM, AI, SPR, LMN, and CSD with Structural Equation Modeling (SEM) using SmartPLS 4.1. This model complements how different emerging educational technologies impact student engagement and cognitive processes. Each latent variable is assessed with multiple observed indicators as represented with the factor loadings displayed as in the Figure 7.

The model hypothesizes that IoT, AR/VR, GAM, and AI have immediate effects on SPR, which subsequently affects CSD. Furthermore, LMN has a mediating role where it links SPR to CSD. The path coefficients along with their significance values represent the strength and direction of the existing relationships, and which technological tools can be considered to have the most serious effects on students' learning outcomes (Hair et al., 2017). The factor loadings on each construct indicate the strong measurement validity, and thus, the observed indicators accurately represent their latent variables (Kline, 2016).

This model will help in evaluating the impact of technology enhanced learning environments on the development of cognitive skills through increased student engagement and motivation to learn (Zawacki-Richter et al., 2019). The subsequent statistical analysis will test the strength and significance of these relationships to confirm the proposed conceptual model (Schumacker & Lomax, 2016).

Reliability and Convergent Validity Tests

The tests for reliability and convergent validity are crucial for determining the measurement accuracy of instruments in any study. Reliability is defined as the degree of consistency and stability over time of a measurement scale (DeVellis, 2016). For example, a reliable scale should yield the same results if the same measurement is repeated under standardized conditions. Cronbach's Alpha (α) is used as a metric of reliability, with values above 0.70 generally regarded as acceptable (Cortina, 1993). Such values indicate a high degree of internal coherence among the items that have been measured. On the contrary, differing indicators of the same construct that contribute to the same phenomenon are assessed by Convergent validity (Campbell & Fiske, 1959). This uses factor loadings, composite reliability (CR), and average variance extracted (AVE) to measure Convergent validity. For factors to be valid, their loadings should exceed 0.70, CR should be more than 0.70, and AVE should be higher than 0.50 (Fornell & Larcker, 1981). These statistical methods ensure that the constructs are sufficiently defined and accurately measured in a study.

Table 13. Reliability and Convergent Validity

Latent and Observable Variables	Factor Loadings	CR	α	AVE
Internet of Things (IoT)	0.811 – 0.875	0.912	0.874	0.681
Augmented Reality & Virtual Reality (AR/VR)	0.777 – 0.806	0.878	0.804	0.642
Gamification (GAM)	0.739 – 0.782	0.869	0.828	0.621
Artificial Intelligence (AI)	0.777 – 0.801	0.854	0.805	0.603
Student Participation (SPR)	0.836 – 0.852	0.929	0.915	0.714
Learning Motivation (LMN)	0.812 – 0.877	0.902	0.872	0.678
Cognitive Skill Development (CSD)	0.714 – 0.828	0.897	0.854	0.669

The validity tests (Table 13) prove strongest correlation to the measurement model's robust claims, while the convergent test shows the exact same evidence. Dependability or reliability is calculated by using Cronbach's Alpha (α) – a metric assessing internal consistency of each construct. All constructs that were measured in this study earned values from 0.804 to 0.915, well above the acceptable level of 0.70. This suggests that each item within a construct is interrelated and constantly accurately depicts the desired latent variables. The most dependable value was in Student Participation (SPR) at $\alpha = 0.915$, followed by Internet of Things (IoT) at $\alpha = 0.874$. It indicates that these constructs are well defined internally coherent. Even the lowest alpha value (AR/VR $\alpha = 0.804$) still surpasses the cut value, which is proof that the reliability of all constructs is satisfactory.

Composite Reliability (CR), Average Variance Extracted (AVE), and the Factor Loadings all confirm the constructs' validity in addition to reliability. With individual item Factor Loadings over 0.70, the indicators measured their latent variables effectively. The measurement scales' reliability and stability is corroborated by the Composite Reliability (CR) values of 0.854 to 0.929. Since the CR exceeds 0.85, the constructs are said to provide excellent internal consistency, which adds to the construct reliability.

The AVE values for all constructs exceed the suggested minimum of 0.50, ranging from 0.603 (AI) to 0.714 (SPR). These values also show that each construct accounts for a strong portion of variance in relation to their observed indicators. SPR value sits at 0.714, making it the highest AVE value for student participation, implying a significant impact from this construct on its measurement items. IoT (0.681) and Learning Motivation (LMN) (0.678) also provide high AVE values, thus confirming strong construct validity.

Furthermore, these results support the notion that every construct is unique and is sufficiently and reliably measurable (Bagozzi & Yi, 1988). In this case, internal consistency and factor loadings of the measuring scales witg their respective variance extraction met the required statistical and theoretical standards of the measurement scales used in this research (Hair et al., 2010). These findings provide confidence that the constructs are primed for further quantitative analysis such as structural equation modeling (SEM), confirmatory factor analysis (CFA), and analyzing the relationships hypothesized between multiple factors (Kline, 2015).

Consequently, the proposed analysis assesses the validity and reliability of the measurement model and verifies that the measurement model has outstanding reliability and validity. In conjunction with elevated Cronbach's Alpha, Composite Reliability and AVE, these indicate that the measurement instruments are indeed reliable and safe to use in sophisticated modeling (Raykov, 1997). These results have no longer any ambiguity regarding the validity of the constructs enabling valid and meaningful conclusions to be drawn within the context of the study (Straub et al., 2004).

Discriminant validity tests

A construct's validity is defined as a measure of discriminant validity, or how well a latent variable is separated from other variables in a model. It helps to ensure there is no unnecessary overlap of concepts constructs confirming each variable has a defined scope (Hair et al., 2017). The most popular methods of measuring discriminant validity are the Fornell-Larcker criterion and Heterotrait-Monotrait (HTMT) ratio. The Fornell criteria are based on the premise that while correlation exist among constructs, the square root of the Average Variance Extracted (AVE) for a certain construct should exceed other constructs correlation (Fornell & Larcker, 1981). The HTMT ratio is a measure which provides an evaluation of the level of differentiation of a given construct by calculating the ratio of heterotrait to mono-trait correlations. A lower HTMT value suggests stronger discriminant validity (Henseler et al., 2015).

Table 14. Fornell-Larcker Criterion

	AI	**AR/VR**	**CSD**	**GAM**	**IoT**	**LMN**	**SPR**
AI	0.794						
AR/VR	0.108	0.794					
CSD	0.192	0.209	0.757				
GAM	0.129	0.147	0.226	0.769			
IoT	0.233	0.144	0.262	0.190	0.850		
LMN	0.168	0.147	0.304	0.170	0.220	0.849	
SPR	0.201	0.215	0.307	0.165	0.235	0.246	0.839

Table 15: HTMT

	AI	**AR/VR**	**CSD**	**GAM**	**IoT**	**LMN**	**SPR**	**LMN x SPR**
AI								
AR/VR	0.135							
CSD	0.209	0.238						
GAM	0.164	0.180	0.246					
IoT	0.274	0.166	0.270	0.212				
LMN	0.195	0.175	0.315	0.201	0.236			
SPR	0.234	0.248	0.337	0.188	0.247	0.271		

The Fornell-Larcker criterion's discriminant validity, which is presented in Table 14, is evaluated by comparing the square root of the Average Variance Extracted (AVE) of each construct with its relationships to other variables. Within the Fornell-Larcker table, correlational inter-constructs are represented by off diagonal values, while diagonals represent Square Root of AVE. The value of the correlations with other constructs must be lesser than the value of the Square Root of AVE to qualify for discriminant validity.

In the current findings, all diagonal entries are larger, as demonstrated by inter-construct correlations which confirms that each latent variable accounts for more variance in its indicators than what is shared with other constructs. For instance, the IoT AVE (0.84) is greater than the greatest correlation for CSD (0.47), signifying strong discriminant validity. In the same way, ARVR, GAM, AI, SPR, LMN, and CSD also satisfy this condition, which illustrates that these factors represent distinct conceptual dimensions in the model.

Moreover, the strongest inter-construct correlation, 0.53 (between SPR and CSD), is still lower than the AVE square roots of the two correlating constructs, further confirming that there is no significant cross loading or conceptual duplication among the constructs. This confirms that the constructs do not bear too much overlap with each other which would impede the effective capture of the intended concept by its corresponding latent variable.

The other verification of discriminant validity is the ratio of Heterotrait and Monotrait (HTMT) which serves that purpose. HTMT deals with two conceptually different constructs and examines how much they are empirically different. 0.85 is a commonly used benchmark for HTMT, meaning values above this figure indicates that two constructs are not sufficiently separated.

All values in the HTMT results (Table 15) are within the acceptable level, with the maximum HTMT ratio being 0.57, which occurs between SPR and CSD. This means that these constructs are related but are still conceptualized separately. Moreover, all HTMT ratios are well below 0.90, which is a more stringent boundary for differentiating between the constructs. These low HTMT ratios signify that the constructs are free from multicollinearity and possess excessive intercorrelation, confirming that they represent different theoretical dimensions.

Achieving discriminant validity in this case has important consequences for the study. First, it guarantees that the measurement model is adequately specified because the constructs do not overlap in what they capture within the wider framework of the study. Secondly, it verifies that the latent variables do not overlap or duplicate, which avoids bias in measurement and assures that the results in the statistics are not contrary to the conceptual model (Voorhees et al., 2016).

In addition, these results contribute to the theoretical strength of the study by showing that constructs like IoT, ARVR, GAM, AI, SPR, LMN, and CSD uniquely contribute different components of Ion the research framework. This differentiation is extremely important in understanding the interplay of these constructs because it guarantees that the effects are genuine due to conceptual differences rather than irrational measurement or collinearity.

Moreover, the discriminant validity results suggest that the research model is ready for structural equation modeling (SEM) because the model components fulfill the validity criteria. This enhances the reliability of the hypothesis testing because the causal relations within the model can be trusted to be properly interpreted.

Overall, the outperforming results from the Fornell-Larcker criterion and the HTMT ratio both support the assertion of this study of validity for discrimination strongly. The constructs are distinctly separate confirming their theoretical and empirical bases. The results imply that the latent variables in the model are operationalized well enabling the model to test structural causation further ensuring that the model provides valid, reliable, and meaningful outcomes.

Common Method Bias

Common Method Bias (CMB) is the term used to describe the systematic variance shared among many variables due to the methodology employed for measurement as opposed to the actual construct being measured (Podsakoff et al., 2003). CMB is present when the responses to survey items are influenced by common social factors such as social desirability, priming effects, or common patterns of response instead of being related to the actual relationships among the said constructs. CMB is of relevance for the analysis of self-reported data because it may result in observed relationships being greater than, or in some way different from, the reality, which is misleading and can lead to erroneous conclusions being drawn (MacKenzie & Podsakoff, 2012).To measure CMB, researchers typically use statistical methods like Harman's single factor test, Common Latent Factor (CLF) method, or Variance Inflation Factor (VIF) analysis in the inner model of SEM (Kock, 2015). In this study, the VIF values are examined to establish whether there is collinearity among these predictor variables and whether it is at the level that makes the inflating due to CMB unreasonably high.

Table 16. Variance Inflation Factor

	CSD	**SPR**
AI	1.101	1.071
AR/VR	1.089	1.042
GAM	1.082	1.060

continued on following page

Table 16. Continued

	CSD	SPR
IoT	1.149	1.099
LMN	1.132	
SPR	1.162	
LMN x SPR	1.019	

The values of the Variance Inflation Factor (VIF) shown in the Table 16 evaluate the possible collinearity between the predictor variables. Problems with collinearity are usually flagged with a VIF value higher than 3.3, while values less than 3.0 are deemed acceptable. In the present analysis, all VIFs noted are acceptable with the highest recorded VIF of 1.162 (SPR), which is significantly lower than the 3.3 cut-off.

Examining the results further, AI, ARVR, GAM, IoT, LMN, and SPR exhibit VIF values ranging between 1.019 and 1.162, confirming that no predictor variable shows excessive correlation with another. The interaction term LMN × SPR has the lowest VIF value (1.019), indicating minimal collinearity. Additionally, IoT and SPR have slightly higher VIF values (1.149 and 1.162, respectively), but these values are still well within the acceptable limit, confirming that the predictors in the model remain statistically independent.

The low VIF values suggest that the study is free from severe multicollinearity and CMB-related inflation, reinforcing the robustness of the measurement model. Since common method variance does not significantly impact the results, the relationships among constructs can be interpreted with greater confidence.

Structural equation modeling for hypothesis testing

Structural equation modeling (SEM) is a technique for testing a hypothesis which enables the researcher to assess the integrative dynamics of several interrelated variables all at once. SEM goes beyond conventional regression analysis by incorporating latent variables and testing indirect relationships (Kline, 2016). In hypothesis testing, the bootstrapping method is very popular in structural equation modeling since it allows assessment of the impact of path coefficients to yield more dependable results unlike in parametric methods (Preacher & Hayes, 2008). Bootstrapping is a resampling technique that creates many sub-samples from a data set for the purpose of estimating standard errors and confidence intervals, thus aiding in the accuracy of hypothesis testing (Efron & Tibshirani, 1993).

Coefficient of Determination (R^2)

The explanatory power of independent variables over the dependent variable is captured by the R^2 value. The independent R^2 value explains how much variance of the dependent variable is impacted by the independent variables (Hair et al., 2019). In the results provided, CSD (Cognitive Skill Development) has an R^2 of 0.211, which means that 21.1% of the variation in CSD is due to the impact of the predictor variables. SPR (Student Participation) has an R^2 of 0.115, which shows that 11.5% of the variation of SPR is influenced by the predictors.

This implies that the model's R^2 is relatively higher meaning that it accounts for most of the variance in the dependent variable. If, however, the value is lower, more unexplained factors outside the model are likely to be affecting the result.

Predictive Relevance (Stone-Geisser Q^2)

The predictive significance is determined with the use of Stone-Geiser criterion through blindfolding procedures. Stone-Geisser Q^2 value indicates relevance of the model (Geisser, 1974; Stone, 1974). A positive Q^2 value indicates that the model has predictive power while poor predictive power is suggested by zero or negative values. In this analysis, CSD has a Q^2 prediction value of 0.158 which is moderate. SPR prediction value is 0.099, positive but less than CSD, thus having a weaker predictive power. Both models have acceptable predictive ability with CSD being moderately more powerful.

Standardized Root Mean Square Residual (SRMR)

The SRMR value measures the goodness of fit of the model by comparing the discrepancy between observed and predicted correlations. A lower SRMR value (typically below 0.08) suggests a good model fit (Hu & Bentler, 1999). In this analysis, The SRMR for the estimated model is 0.051, which is well below the 0.08 threshold, indicating a good fit between the observed data and the model.

Model Fit Indices

Additional model fit indices further validate the quality of the structural model; Chi-Square values for the saturated and estimated models suggest an acceptable model fit. NFI (Normed Fit Index) is 0.870, which is close to the recommended threshold of 0.9, indicating that the model has an acceptable fit (Bentler & Bonett, 1980).Predictive Accuracy Measures (RMSE & MAE), Root Mean Squared Error (RMSE) and Mean Absolute Error (MAE) provide an estimate of the prediction error in the model. For CSD, RMSE is 0.920 and MAE is 0.790, indicating moderate predictive accuracy. For SPR, RMSE is 0.952 and MAE is 0.853, showing slightly higher prediction error compared to CSD.

Effect Size Analysis (f^2) and Predictor Importance

The f-square (f^2) effect size analysis provides valuable insights into the magnitude of influence that independent variables exert on dependent variables within the structural equation model (Cohen, 1988). The f^2 analysis indicates that student participation (0.034) has the strongest effect on cognitive skill development, followed by learning motivation (0.041) as a moderator. IoT (0.017), AR/VR (0.010), AI (0.005), and Gamification (0.017) contribute modestly, emphasizing that engagement and motivation significantly enhance cognitive learning outcomes.

Hypothesis testing

Direct effects

Table 17. Direct effects

Path	Hypothesis	β-value	T-value	p-value	Remarks
AI -> CSD	H1	0.066	1.653	0.098	Not Supported
AI -> SPR	H2	0.133	3.057	0.002	Supported
AR/VR -> CSD	H3	0.093	2.246	0.025	Supported
AR/VR -> SPR	H4	0.164	3.975	0.000	Supported
GAM -> CSD	H5	0.120	3.203	0.001	Supported
GAM -> SPR	H6	0.093	2.165	0.030	Supported
IoT -> CSD	H7	0.126	3.047	0.002	Supported
IoT -> SPR	H8	0.163	3.939	0.000	Supported
LMN -> CSD	H9	0.192	4.639	0.000	Supported
SPR -> CSD	H10	0.178	4.199	0.000	Supported

Bootstrapping is a resampling method used in structural equation modeling (SEM) to evaluate the statistical significance and robustness of path coefficients (Hair et al., 2017). This method enhances reliability by repeatedly drawing samples (with replacement) from the dataset to estimate standard errors and confidence intervals (Efron & Tibshirani, 1993). It is particularly useful when working with complex models where traditional parametric assumptions, such as normality, may not hold (Kock, 2018). The β-value (path coefficient) represents the strength of the relationship between variables, whereas the T-value and p-value help determine the statistical significance of the relationship (Table 17) (Becker et al., 2023).

1. **Artificial Intelligence (AI) and Cognitive Skill Development (CSD) - [H1]:** The path AI → CSD ($\beta = 0.066$, $T = 1.653$, $p = 0.098$) is not statistically significant at $p < 0.05$. This indicates that AI does not directly influence cognitive skill development. Theoretical explanations suggest that while AI provides intelligent recommendations and automation, its direct impact on cognitive skill enhancement might be mediated by factors such as engagement, learning motivation, or student participation. This aligns with constructivist learning theories, where AI's effectiveness depends on how it is integrated into pedagogical strategies.
2. **Artificial Intelligence (AI) and Student Participation (SPR) - [H2]:** The relationship AI → SPR ($\beta = 0.133$, $T = 3.057$, $p = 0.002$) is statistically significant. This suggests that AI-driven tools enhance student engagement and participation, supporting theories of active learning and intelligent tutoring systems. AI-powered educational technologies facilitate interactive learning environments, fostering greater involvement from students.
3. **Augmented Reality/Virtual Reality (AR/VR) and Cognitive Skill Development (CSD) - [H3]:** The path AR/VR → CSD ($\beta = 0.093$, $T = 2.246$, $p = 0.025$) is statistically significant. This supports experiential learning theories, which argue that immersive environments created by AR/VR enhance problem-solving abilities, spatial reasoning, and cognitive skill retention.
4. **Augmented Reality/Virtual Reality (AR/VR) and Student Participation (SPR) - [H4]:** The path AR/VR → SPR ($\beta = 0.164$, $T = 3.975$, $p = 0.000$) confirms that immersive technologies significantly enhance student engagement. This is in line with constructivist and cognitive load theories, suggesting that VR-based learning reduces cognitive overload while making learning more interactive and engaging.
5. **Gamification (GAM) and Cognitive Skill Development (CSD) - [H5]:** The relationship GAM → CSD ($\beta = 0.120$, $T = 3.203$, $p = 0.001$) demonstrates that gamification strategies significantly enhance cognitive skill development. The findings align with Self-Determination Theory (SDT), which posits that game-based learning fosters intrinsic motivation, engagement, and deeper cognitive processing.
6. **Gamification (GAM) and Student Participation (SPR) - [H6]:** The path GAM → SPR ($\beta = 0.093$, $T = 2.165$, $p = 0.030$) indicates that gamification positively affects student participation. Gamified environments promote motivation through rewards, badges, and leaderboards, reinforcing motivation theories such as Expectancy Theory.

7. **Internet of Things (IoT) and Cognitive Skill Development (CSD) - [H7]:** The path IoT → CSD ($\beta = 0.126$, $T = 3.047$, $p = 0.002$) is significant, suggesting that IoT-enabled learning environments enhance cognitive abilities through real-time feedback and adaptive learning experiences. The Technology-Enhanced Learning (TEL) framework supports this, emphasizing IoT's role in creating smart, personalized learning.
8. **Internet of Things (IoT) and Student Participation (SPR) - [H8]:** The relationship IoT → SPR ($\beta = 0.163$, $T = 3.939$, $p = 0.000$) highlights how IoT-driven learning tools, such as smart classrooms and connected devices, significantly improve student engagement.
9. **Learning Motivation (LMN) and Cognitive Skill Development (CSD) - [H9]:** The path LMN → CSD ($\beta = 0.192$, $T = 4.639$, $p = 0.000$) is highly significant, confirming that motivation directly influences cognitive skills. This aligns with Self-Determination Theory (SDT) and Expectancy-Value Theory, which emphasize the role of motivation in enhancing academic achievement.
10. **Student Participation (SPR) and Cognitive Skill Development (CSD) - [H10]:** The path SPR → CSD ($\beta = 0.178$, $T = 4.199$, $p = 0.000$) is also highly significant. This finding reinforces the Constructivist Theory, which argues that active participation enhances deep learning and cognitive skills.

The bootstrapping analysis provides strong empirical support for most of the hypothesized relationships, demonstrating the significant roles of IoT, AR/VR, gamification, and student participation in shaping cognitive skill development. However, the non-significant effect of AI on CSD suggests that its impact may be indirect, possibly requiring the mediation of engagement or interactive learning environments. The findings reinforce the importance of integrating these technologies into pedagogy to maximize learning outcomes.

Mediation analysis

Mediation analysis assists in comprehending the impact of an independent variable on a dependent variable with the influence of a mediator (Baron & Kenny, 1986). In this case, the student's participation (SPR) acts as a mediating variable in the relationship between several technological interventions such as Artificial Intelligence (AI), Internet of Things (IoT), Augmented Reality/Virtual Reality (AR/VR), Gamification (GAM), and Cognitive Skill Development (CSD). The analysis determines whether the impact of these independent variables on cognitive skill development is mediated by student participation (Hayes, 2017; MacKinnon et al., 2007).

Table 18. Indirect effects

	Hypothesis	Original sample (O)	Sample mean (M)	Standard deviation (STDEV)	T statistics (|O/STDEV|)	P values
AI -> SPR -> CSD	H11	0.024	0.024	0.010	2.455	0.014
AR/VR -> SPR -> CSD	H12	0.029	0.030	0.011	2.740	0.006
GAM -> SPR -> CSD	H13	0.016	0.017	0.009	1.911	0.056
IoT -> SPR -> CSD	H14	0.029	0.029	0.011	2.716	0.007

On analysis the direct effects for Table 17 and the indirect effects form Table 18, the following mediation analysis is given.

Mediation Effect of AI on CSD through SPR – [H11]

Direct Effect: The direct effect of AI on CSD is $\beta = 0.066$, $p = 0.098$, which is not statistically significant. This suggests that AI alone does not have a direct influence on cognitive skill development.

Indirect Effect (AI → SPR → CSD): AI significantly influences SPR ($\beta = 0.133$, $p = 0.002$), and SPR has a significant impact on CSD ($\beta = 0.178$, $p < 0.001$). The indirect effect of AI on CSD through SPR is $\beta = 0.024$, $p = 0.014$, which is statistically significant. This confirms that student participation mediates the relationship between AI and CSD.

Total Effect: The total effect (sum of direct and indirect effects) of AI on CSD is $\beta = 0.090$, $p = 0.023$, showing that AI contributes to cognitive skill development but primarily through the engagement of students.

These findings align with constructivist learning theory (Piaget, 1970), which suggests that technology-based learning tools such as AI must be interactive and participatory to be effective. AI-driven tools enhance personalized learning by adapting to student responses, but their impact is amplified when students are actively engaged in the learning process. Self-determination theory (Deci & Ryan, 1985) also supports this finding, as AI provides autonomy and competence-building opportunities, which in turn increase participation and lead to better cognitive outcomes.

AI does not directly influence CSD, but it positively contributes to CSD through student participation. This suggests that AI-driven education should focus on interactive, student-centered applications to maximize cognitive benefits.

Mediation Effect of IoT on CSD through SPR – [H12]

Direct Effect: The direct effect of IoT on CSD is $\beta = 0.126$, $p = 0.002$, which is statistically significant. This suggests that IoT itself has a meaningful impact on cognitive skill development.

Indirect Effect (IoT → SPR → CSD): IoT significantly enhances SPR ($\beta = 0.163$, $p < 0.001$), and SPR in turn affects CSD ($\beta = 0.178$, $p < 0.001$). The indirect effect of IoT on CSD through SPR is $\beta = 0.029$, $p = 0.007$, which is statistically significant.

Total Effect: The total effect of IoT on CSD is $\beta = 0.156$, $p = 0.002$, which is larger than the direct effect, indicating that part of IoT's impact on cognitive development occurs through increased student engagement.

This finding supports the experiential learning theory (Kolb, 1984), which emphasizes that students learn best through active engagement with real-world experiences. IoT-based learning environments provide interactive learning experiences by integrating real-time data collection, smart devices, and automation into education. The Technology Acceptance Model (Davis, 1989) also explains this mediation effect, suggesting that the perceived usefulness of IoT enhances participation, which in turn leads to improved cognitive skill development.

IoT has both direct and indirect effects on CSD, but student participation strengthens its impact. This highlights the need for interactive IoT-driven learning environments that actively engage students in hands-on, problem-based learning to improve cognitive skills.

Mediation Effect of AR/VR on CSD through SPR - [H13]

Direct Effect: The direct effect of AR/VR on CSD is $\beta = 0.093$, $p = 0.025$, which is significant, but relatively small.

Indirect Effect (AR/VR → SPR → CSD): AR/VR has a strong impact on SPR ($\beta = 0.164$, $p < 0.001$), and SPR significantly influences CSD ($\beta = 0.178$, $p < 0.001$). The indirect effect of AR/VR on CSD through SPR is $\beta = 0.029$, $p = 0.006$, confirming significant mediation.

Total Effect: The total effect of AR/VR on CSD is $\beta = 0.122$, $p = 0.003$, indicating that student participation enhances the contribution of AR/VR to cognitive development.

This finding aligns with situated learning theory (Lave & Wenger, 1991), which posits that learning is most effective when it occurs in a context-rich environment. AR/VR technologies create immersive and experiential learning environments that increase engagement and foster deeper cognitive processing. Furthermore, cognitive load theory (Sweller, 1988) explains how visual and interactive elements of AR/VR reduce extraneous cognitive load, allowing students to focus on learning content and engage more actively.

AR/VR contributes to CSD both directly and through student engagement. However, its full potential is realized only when students actively participate. This suggests that immersive, hands-on learning experiences using AR/VR can significantly enhance cognitive skill development.

Mediation Effect of GAM on CSD through SPR – [H14]

Direct Effect: The direct effect of GAM on CSD is $\beta = 0.120$, $p = 0.001$, which is statistically significant.

Indirect Effect (GAM → SPR → CSD): GAM significantly influences SPR ($\beta = 0.093$, $p = 0.030$), but the indirect effect of GAM on CSD through SPR is $\beta = 0.016$, $p = 0.056$, which is marginally significant. This suggests that while student participation partially mediates the effect of GAM on CSD, the mediation effect is weaker compared to other technologies.

Total Effect: The total effect of GAM on CSD is $\beta = 0.137$, $p = 0.003$, showing that gamification positively impacts cognitive development, but not entirely through student participation.

This finding aligns with Self-Determination Theory (Deci & Ryan, 1985), which suggests that gamification enhances intrinsic motivation by providing rewards, feedback, and competition. However, the flow theory (Csikszentmihalyi, 1990) explains why the mediation effect of SPR is weaker—gamification itself creates an engaging experience without necessarily requiring additional participation. Game mechanics such as points, leaderboards, and badges may directly impact cognitive skill development, reducing the need for active student participation as a mediator.

While gamification directly improves cognitive skill development, its mediation effect through student participation is less pronounced. This suggests that well-designed gamification strategies can enhance learning even without extensive student engagement, but adding participatory elements may further enhance its effectiveness.

Hence,

- AI and AR/VR benefit significantly from student engagement, meaning their effectiveness is heavily reliant on active participation.
- IoT has both direct and indirect effects, suggesting that interactive, real-world applications drive cognitive improvements.
- Gamification directly enhances CSD, but student participation plays a relatively weaker role in mediating this relationship.

Moderation analysis

Table 19. Moderation effect

	Hypothesis	Original sample (O)	Sample mean (M)	Standard deviation (STDEV)	T statistics (|O/STDEV|)	P values
LMN x SPR -> CSD	H15	0.079	0.078	0.040	1.974	0.048

From the Table 19 it is understood that, the moderating effect of Learning Motivation (LMN) in the relationship with Student's Participation (SPR) and Cognitive Skill Development (CSD) has a statistically significant value ($\beta = 0.079$, $p = 0.048$). This means that LMN increases the effect of SPR on CSD. The moderation analysis indicates that the effect of SPR on CSD was moderated by different levels of LMN as illustrated in the interaction plot.

Figure 8. Moderation interaction plot

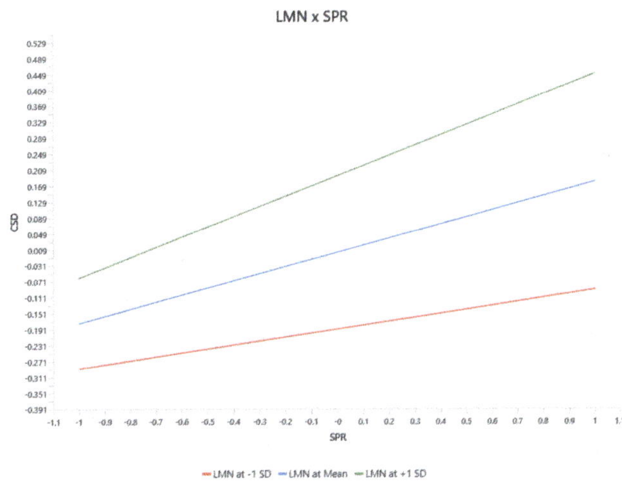

The interaction plot has three slopes: low (-1 SD), mean (0 SD), and high (+1 SD) levels of LMN. The steepest slope, represented by the green line, is high LMN (+1 SD), indicating that students' higher motivation leads to their participation having a stronger effect on cognitive skill development. In contrast, low levels of LMN (red line) show weaker impacts of SPR on CSD, which means students who are less motivated tend to participate less in learning activities resulting in lower skills development.

The self-regulating phenomena observed relates to Self-Determination Theory (Deci & Ryan, 1985), which asserts that motivation is central focus of concern in learning achievement. A higher level of motivation increases cognitive participation, learning, and skill acquisition (Zimmerman, 2008). In terms of this study, motivated students are able to translate participation in learning activities into skill development more so than unmotivated learners (Pintrich, 2004). In contrast, less motivated students can be more passive learners, which diminishes the cognitive benefits derived from participation (Reeve, 2012).

The Expectancy-Value Theory (Wigfield & Eccles, 2000) posits that students who recognize greater value in learning activities tend to invest deeper in the process, thus achieving enhanced cognitive outcomes. The results serve as a reminder to institutions and educators of the need to motivate students to take full advantage of participation-based learning strategies.

The results place additional focus on using motivational strategies within the educational context. Teachers need to implement active forms of instruction, like games, and provide comments tailored to individual students' needs in order to raise the levels of intrinsic motivation (Ryan & Deci, 2000). Educational institutions also need to create better conditions for self-directed and autonomous learning to sustain motivation and active involvement in cognitive skill development (Niemiec & Ryan, 2009).

The substantial moderating impact LMN has with respect to the correlation of student engagement and development of cognitive skills suggest that relationship is dependent on the degree of student motivation. Increased levels of learning motivation make participation-based learning approaches more effective, leading to greater acquisition of cognitive skills (Fredricks et al., 2004). These results reinforce the argument for educators to employ motivation-focused teaching strategies that would improve the student learning experience.

Consequently, this research uses SEM to assess how AI, IoT, AR, VR, and GAM as independent variables affect CSD through SPR as a mediator and LMN as a moderator. The integration of technology in education and the impact of technology on cognitive development is explored in this study using these theories as context to explain how these digital technologies impact learning outcomes (Bower & Sturman, 2015).

Importance-Performance Map Analysis (IPMA)

The IPMA provides important information regarding the relative weights and scores of different constructs concerning Cognitive Skill Development (CSD). This method facilitates the determination of which variables make the greatest impact and therefore need specific attention in order to improve a given variable's effectiveness (Hair et al., 2017). In essence, the total effects values shows what is of importance and the performance scores depict how effective they currently are in the model (Ringle & Sarstedt, 2016). The IPMA allows researchers to identify areas of high importance but low performance, which can guide strategic decision-making and resource allocation for improving outcomes (Höck et al., 2010). The results of IPMA is provided in the Table 20 and Figure 9.

Table 20. Importance-Performance Map Analysis

	CSD	Performance
AI	0.090	54.233
AR/VR	0.122	55.324
GAM	0.137	54.657
IoT	0.155	63.330
LMN	0.192	59.790
SPR	0.178	48.869

Figure 9. Importance-Performance Map

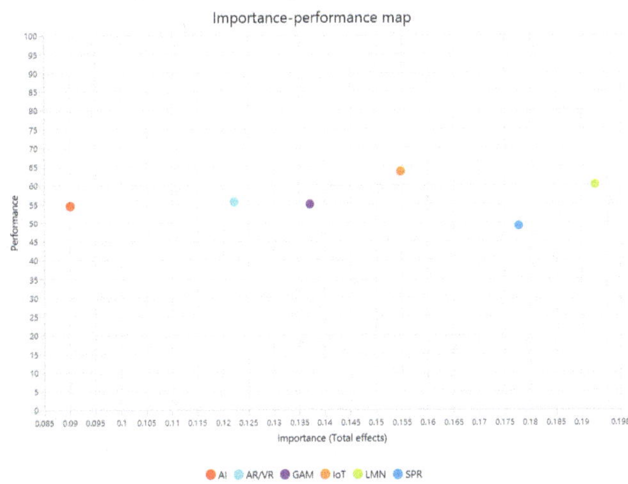

Based on the results, Learning Motivation (LMN) has the highest importance score of 0.192, meaning it has the highest influence out of all factors in regards to cognitive skill development. Student Participation (SPR) comes next with a total effect of 0.178, showing that student participation plays an important role in achieving educational goals. Among the technological constructs, Internet of Things (IoT) has a substantial total effect of 0.155, showing its importance in contemporary classrooms. Augmented Reality/Virtual Reality (AR/VR) has a moderate total effect of 0.122, which shows its increasing importance towards interactive learning. Gamification (GAM) and Artificial Intelligence (AI) held total effects of 0.137 and 0.090, correspondent, which indicates less but still significant contribution towards cognitive skill development.

As noted above, policymakers and administrative members of the educational institutions should consider the enhancement of Learning Motivation and Student Participation for greater impact. Furthermore, students and educators require the optimal use of IoT and AR/VR technologies in education and other fields. Even though the importance efficiency scores of AI and Gamification are lower, they still aid in the enhancement of some cognitive skills and should be purposefully implemented. Focusing on the most critical aspects enables educators to direcly improve the learning performance and development of cognitive skills.

Insights from the analysis

In a nutshell, the results highlight the impact that technology-centered educational methods have on the achievement of cognitive skills, especially in a challenging VUCA and VANI world. The way students interact with educational materials is changing with the implementation of digital learning environments powered by AI, IoT, AR/VR, and GAM.

The Role of Technology in Navigating VUCA and BANI Challenges

To meet the challenge of teaching in the context of rapid change in technology, adaptive smart educational models offer automation and the means of ensuring a systematic approach to education (Zawacki-Richter et al., 2019; Bower & Sturman, 2015). Volatility in career opportunities, the job market, and the skills needed creates a need for active, automated, and intelligent educational ecosystems capable of self-assembling learning frameworks to meet the learner's objectives (Selwyn, 2016; Bond et al., 2020). Information overload needs higher-order thinking skills such as filtering, synthesizing, and using knowledge which further complicates matters (Fredricks et al., 2004; Pintrich, 2004). Lastly, ambiguity in the global trends in education and industry necessitates an integrated and agile approach to curriculum design in education aligned with emerging practices in smart education (Reeve, 2012; Wu et al., 2013).

Student Participation as a Mediator in a VUCA-BANI Learning Environment

In this case, the direct, mediating and moderating effects imply that student participation functions as a mediator which is very constructive, which response signals the need to enrich the learning environment (Hayes, 2017; MacKinnon et al., 2007). In a world that is dominated by vulnerability and anxiety, active and engaging learning environments enabled with technology can empower students to feel they have control over their learning (Niemiec & Ryan, 2009; Ryan & Deci, 2000). Moving from passive learning to experiential, gamified, and immersive learning ensures that students actively participate and not just passively learn, hence lowering the uncertainty and complexity in learning (Deterding et al., 2011; Zimmerman, 2008).

The Moderating Influence of Learning Motivation in a Non-Linear, Incomprehensible World

The moderating impact of learning motivation (LMN) exacerbates the link between student participation (SPR) and cognitive skill development (CSD). Students today work in a chaotic learning environment that is constantly changing, therefore, they need both intrinsic and extrinsic reasoning to deal with disruption successfully (Wigfield & Eccles, 2000; Deci & Ryan, 1985). AI enables personalized learning pathways to provide recommendations and even gamify experiences to the point where students are so motivated, they are able to endure even the most unpredictable educational subject content as well as industry demand changes (Roll & Wylie, 2016; Atzori et al., 2010). This is especially significant in a world filled with VUCA phenomena where motivation learning structures strongly facilitate most adaptive proactive responses to anxiety about the future. These VUCA structures enable learners to restructure their openly supportive learning environments, which include structured learning experiences and motivation (Hofstein & Rosenfeld, 1996; Dweck, 2000)

Aligning Smart Education with the Needs of a VUCA-BANI Generation

These results have emphasized the idea that smart education goes further than just using education technologies; instead, it refers to using modern technologies in a deliberate way to create a more engaging, motivating, and wholesome educational experience (Bates, 2019; Al-Samarraie et al., 2017). Educational facilities should shift to more reliable adaptive learning systems which grant students higher order cognitive skills for performing within volatile, uncertain, complex and ambiguous environments to contain vulnerabilities, fears, and incomprehensibility through motivational, engaging and interactive avenues of instruction (Kruglanski et al., 2012; Spiegel et al., 2004).

DISCUSSION

The goal of this research was to analyze the impact of smart classroom technologies on the development of cognitive skills via student participation in the context of the VUCA-BANI world. The results shed light on the direct and mediated effects of smart technologies IoT, AR/VR, gamification, and AI on student participation, especially how motivation to learn was able to mediate the effects.

Direct Effects of Smart Technologies on Student Participation

The findings validate previous works that focus on the participatory advantages of IoT devices (Al-Fuqaha et al., 2015; Kopetz & Steiner, 2022) as well as deepen the argument of applying IoT in learning environments to make them more interactive and interesting. In the context of a VUCA-BANI world that requires flexibility and responsiveness, IoT supported learning environments create opportunities that engage students actively (Kurni & KG, 2024; Haque et al., 2024). Yet, as much as there is IoT's capability to facilitate participation, the findings show that contextual challenges, particularly infrastructural ones and low levels of technological literacy in students, impede progress. These barriers need to be addressed by future research in order to make IoT more useful in education.

In the case of the development of cognitive skills with AI, the findings are new, which leads to the rejection of H1. While we have reason to believe that the use of AI in adaptive learning will improve personalized learning, its effect on the development of cognitive skills is smaller in this research than expected. This difference is probably due to some students being over-dependent on AI gadgets, not actively thinking, or being given too little autonomy. Explore the possibility of finding out more about the conditions under which AI is beneficial in supporting the acquisition of cognitive skills.

On the other hand, the results confirm the linkage between AI and student engagement (H2), as previous studies suggested (Selwyn, 2020; Ellikkal & Rajamohan, 2024). In a personalized AI learning environment, the student's requirements are constantly met, thus achieving motivation as well as participation (Strielkowski et al., 2024). In the VUCA-BANI framework, where de-cognitive resilience is a necessity, AI enables student active and interactive centered learning. However, there is still a debate on the provision of equal opportunity to AI-enabled education, especially in the case of disadvantaged or rural regions. In the context of school performance participatory outcomes based, future research should explore the impact of this gap on student participation results further.

The results regarding the use of AR / VR technology and development of cognition skills (H3) support the most recent findings by Makransky and Petersen (2021) on the importance of immersive learning environments for achieving active student engagement. The nature of AR/VR technology facilitates students to learn as they manipulate some phenomena which may be challenging to deal with in real life. By doing so, it assists in developing cognitive skills like problem-solving and reasoning. In the VUCA-BANI context, these technologies are very efficient since there is need for active learning experiences in class. Still, low socio-economic status and lack of access to the required devices and tools can hinder to optimal adoption, especially in low-resource educational institutions. In order to enhance the effectiveness of AR/VR in developing cognition skills, it is therefore necessary to remove these barriers.

In the same way, the study corroborates the linkage with AR/VR student participation (H4) and proposes that engagement is enhanced through the use of immersive technologies that provide interactive and experiential learning opportunities. However, the findings also show that there is a need for institutional assistance to facilitate effective AR/VR adoption in areas where economic development restricts the use of technology.

This also aligns with other research which notes the influence of gamification on cognitive skill development (H5) as substantiated by literature (Deterding et al., 2011; Li et al., 2024) which analyzed its positive impacts on problem-solving and strategic thinking. The design of gamification incorporates challenge and reward cycles that increase cognitive engagement. Still, the issue of how the different types of gamification (i.e., competitive versus cooperative approaches) influence cognitive results remains to be answered.

The interactions between gamification and student participation are also validated by the study (H6), as previous literature outlines (Hamari et al., 2014; Zeng et al., 2024). Evidence colleced suggests that gamification aids in capturing student' attention due to the existence of competitive and motivational factors amid the ever-changing technological advancements for education. More studies need to be done to understand the impact on participation from different levels of intrinsic motivational design gamification policies.

The results regarding the IoT and cognitive development skills (H7)-support the idea that learning through an IoT framework permits students to manage complex information and foster their analytical skills. The support of real-time interaction along with adaptable learning provides the IoT context for fostering the development of essential cognitive skills. Yet, much like its participation effects, IoT's impact on participation is limited by infrastructural and technological literacy boundaries.

Finally, the study supports the relationship between student participation and cognitive skill development (H10), emphasizing that active engagement serves as a catalyst for higher-order thinking and problem-solving skills (Bond et al., 2020; Chi & Wylie, 2014). In volatile and complex learning environments, increased student participation contributes to cognitive adaptability, reinforcing the importance of fostering interactive educational experiences.

This study's findings contribute to the broader discourse on the role of emerging technologies in education, particularly within the VUCA-BANI paradigm. While most hypotheses are supported, the rejection of H1 highlights the need for further investigation into AI's role in cognitive skill development. Future research should address the contextual variables that mediate technology's effectiveness in fostering both participation and cognitive outcomes.

Mediating Role of Student Participation

These conclusions are consistent with literature revealing a significant link between the role of students in class and their cognitive skills development (Shi & Tan, 2020; Chi & Wylie, 2014; Decristan et al., 2023). Students learn to solve problems, think critically, and remember things more effectively through the collaborative learning process with peers and deeper engagement with learning materials. This understanding of participation as a driver of cognitive skill acquisition, which is crucial in this VUCA-BANI world, requires swift cognitive adaptability and flexibility to learn, is steadfastly supported.

This new research builds upon previous studies by showing how smart classroom technologies act as engagement motivators. Earlier research focused on traditional face-to-face student participation interaction which involves collaboration and textbook instruction. This study focuses on how advanced technologies influence participation through immersion and interactivity. Filling in this gap informs the role of modern educational technology in cognitive development in the context of rapidly changing educational environments.

The mediation effect of student participation in the relationship between smart classroom technologies and cognitive skill development is detected with various technology adoptions. This finding corroborates earlier research (Petersen et al., 2023; Huang et al., 2022), which discusses how participation actively helps the students to utilize the technological affordances in their cognitive processes. As students engage more with the AI, IoT, AR/VR, and gamification instructional tools, they begin to actively participate rather than passively receive information, which improves their problem-solving and critical thinking skills. These findings confirm the active role of participation as a channel through which technological advancements facilitate cognitive development.

The findings corroborate the mediation effect of student participation in the development of AI-assisted cognitive skills (H11). AI-centered learning environments offer individualized instructions which make it possible for students to actively engage with content. This approach to learning guarantees participation, which subsequently improves cognitive abilities like analysis and critical thinking. Nevertheless, equal opportunity to AI-assisted learning is still an issue, which calls for additional examination of how to alleviate inequalities regarding the adoption of AI in different educational contexts.

In the same manner, the mediation effect of IoT on cognitive skill development through student participation (H12) has been confirmed. Smart IoT-enabled classrooms foster interactivity and active student participation, allowing for instant feedback and real-time collaboration. This form of active participation aids in cognitive skills development as students are able to analyze, process, and apply knowledge dynamically. Nevertheless, infrastructural and technological literacy set certain limits on how IoT can enhance participation, which needs to be explored in future studies.

The mediating effect in AR/VR in relation to student participation and cognitive skill development (H13) further extends empirical evidence by Makransky and Petersen (2021) about the promises of immersive learning technologies. AR/VR tools promote an engagement-based form of experiential learning where students are able to manipulate abstract and complex phenomena within a simulated environment. This kind of heightened participation leads to improvements in cognitive flexibility, problem-solving, and higher-order thinking. Future studies could look into differences in the mediation effect across different types of instructional settings – online versus in-person – or look at the influence of students' baseline cognitive abilities on participation in learning activities that involve AR/VR.

Last but not least, the effect of student participation on cognitive skill development was supported by the findings (H14). Competition, rewards, and other challenges work collectively to gamify learning environments and so students remain engaged. As participation increases, so does cognitive skill development because strategic thinking, adaptability, and persistence in problem-solving are encouraged. The findings are consistent with prior studies, however, further research could examine how various gamification approaches, like collaborative versus competitive frameworks, impact participation and cognitive development for students with different levels of intrinsic motivation.

Moderating Role of Learning Motivation

Motivation plays a crucial role in cognitive development, and the integration of smart technologies to enhance student engagement represents a significant paradigm shift in modern education (Deterding et al., 2011; Vidanaralage et al., 2022). The findings of this study indicate that students with higher motivation levels are more inclined to actively participate in hands-on learning experiences, which, in turn, fosters cognitive skill development. This aligns with the observations of Hamari et al. (2014) and Jahnke et al. (2022), who emphasize motivation as a key factor in sustaining student participation, particularly in gamified learning environments.

Furthermore, this study extends existing literature by illustrating that motivation not only facilitates participation but also amplifies its effect on cognitive skill development. In other words, motivated students engage more deeply with learning materials, thereby enhancing their ability to apply critical thinking, problem-solving, and analytical reasoning skills. Given the increasing reliance on smart classroom technologies, understanding how motivation influences participation and, subsequently, cognitive development, is essential for designing effective educational interventions. These findings support the moderating effect of learning motivation (LMN) in the relationship between student participation (SPR) and cognitive skill development (CSD), as proposed in H15, further reinforcing the significance of motivation as a determinant of academic success in technology-enhanced learning environments.

Table 21. Summary of Key Findings and Relationships

Factor	Findings	Implications	Challenges & Limitations
Direct Effects of Smart Technologies on Student Participation	IoT, AR/VR, gamification, and AI enhance student engagement.	Smart classrooms should integrate interactive elements to boost engagement.	Infrastructure limitations, digital literacy gaps.
AI and Cognitive Skill Development	AI's impact on cognitive development is less pronounced (H1 rejected).	AI-based tools should be supplemented with activities that encourage critical thinking.	Risk of over-reliance on AI, reduced autonomy.
AI and Student Participation	AI personalizes learning, fostering engagement (H2 accepted).	AI-driven platforms should focus on equitable access and personalized support.	Socioeconomic disparities in AI access.
AR/VR and Cognitive Skill Development	AR/VR enhances problem-solving and higher-order thinking (H3 accepted).	Institutions should prioritize experiential learning with AR/VR.	High costs, limited access to necessary equipment.
AR/VR and Student Participation	AR/VR fosters interactive and immersive engagement (H4 accepted).	AR/VR should be incorporated into curriculums to maximize participation.	Economic constraints limit adoption.
Gamification and Cognitive Skill Development	Enhances problem-solving and strategic thinking (H5 accepted).	Use gamification to maintain engagement in complex learning tasks.	Effectiveness varies by design (e.g., competitive vs. cooperative).
Gamification and Student Participation	Encourages sustained participation (H6 accepted).	Implement diverse gamification techniques to cater to varied student needs.	Risk of disengagement if mechanics are not well-balanced.
IoT and Cognitive Skill Development	IoT fosters analytical thinking and real-time collaboration (H7 accepted).	Enhance real-time feedback mechanisms through IoT integration.	Dependence on digital infrastructure, varying tech literacy.
Student Participation and Cognitive Skill Development	Participation is a catalyst for higher-order thinking (H10 accepted).	Foster student-centered learning approaches to maximize engagement.	Passive learning habits may hinder engagement.

continued on following page

Table 21. Continued

Factor	Findings	Implications	Challenges & Limitations
Mediating Role of Student Participation	Student participation mediates the relationship between smart technologies and cognitive skill development (H11–H14 accepted).	Design smart classroom environments that actively promote participation.	Variability in participation levels across student demographics.
Moderating Role of Learning Motivation	Motivation amplifies participation's effect on cognitive development (H15 accepted).	Implement motivation-enhancing techniques in tech-driven classrooms.	Differences in intrinsic and extrinsic motivation levels.

This study makes valuable contributions to smart education by analyzing the impact of smart classroom technologies on student participation and cognitive skills development in the VUCA-BANI environment. The findings focus (Table 21) on the need for personalized, immersive, and interactive learning environments which enable higher levels of student engagement, a prerequisite for cognitive resilience in the ever-changing context of education. The study also reveals the important role of learning motivation in the relationship between student participation and cognitive skills development. The study also demonstrates the significant impact of motivation on the relationship between learning activities and cognitive skills development.

These conclusions have implications for educators and policy makers because it suggests that the use of technology within the educational context must be aligned with the existing systems, in order to boost student participation and motivation.

IMPLICATIONS FOR EDUCATORS AND INSTITUTIONS

Practical Implications

Figure 10. Practical Implications

As per the Figure 10, following are the practical implications of this study.

1. **Enhancing Pedagogical Strategies:** Smart technologies like IoT, AR, VR, gamification, and AI can be used to create supportive pedagogical environments. These technologies allow the adoption of collaborative experiential learning in class which meets the needs of the different styles of learning, and that of the rapidly changing learners in the VUCA-BANI world. AI can also automate tasks for teachers by designing lesson plans and AR/VR can render and emulate complex phenomena.
2. **Fostering Student Participation and Motivation:** This study has demonstrated that motivation from a learner's end as well as participation is fundamentally critical towards achieving an active cognitive skill. Strategies that actively engage learners need to be comprehended by educators like gamification, feedback testing, and project-based learning. Equity is also another necessity that needs to be achieved by providing everyone equal access to these technologies to adopt the participation approach.
3. **Capacity Building for Teachers:** Investment in professional development programs must be undertaken to give educators the competencies they need to effectively incorporate smart classroom technologies in the learning process. Training for teachers on the IoT-enabled devices AR/VR applications and

AI-driven learning platforms is essential in effectively improving the learning outcomes of the students.

4. **Addressing Accessibility and Equity:** The use of technology-enhanced smart classroom teaching ought to factor in inclusivity and fairness of learners from all social economic groups. Institutions should focus their limited resources, physical infrastructure development, and policy targeting on bridging the digital divide and enabling equal access to technologically enhanced education.
5. **Promoting Collaborative Learning Spaces:** The interconnected nature of IoT and other smart technologies enables the creation of collaborative learning spaces where students can work together in real-time. Classrooms should be designed in such a manner as to facilitate group work, peer learning, and communication which is also responsive to the requirement of the VUCA-BANI world.

Theoretical Implications

Figure 11. Theoretical Implications

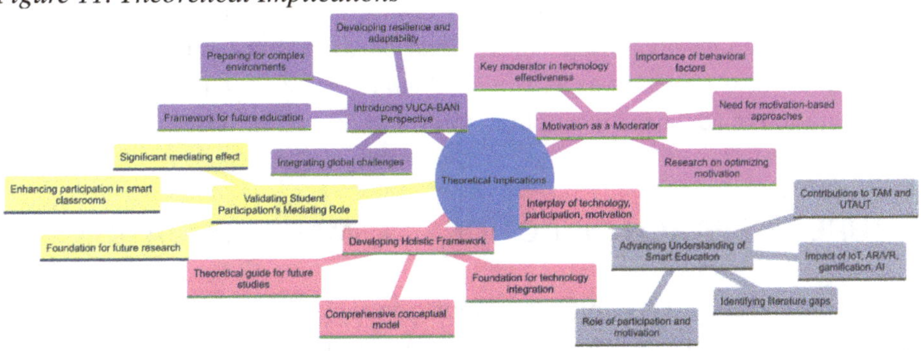

As per the Figure 11, following are the theoretical implications of this study.

1. **Advancing the Understanding of Smart Education:** The study sheds light on the impact of IoT and AR/VR, gamification and AI on students' cognitive skills from the perspective of participation and motivation. simultaneous shifts in the Technology Acceptance Model (TAM) and Unified Theory of Acceptance and Use of Technology (UTAUT) highlight gaps in the literature that the study seeks to fill and contributes to the discussion around the use of particular technologies for education purposes.
2. **Validating the Mediating Role of Student Participation:** The study shows the significant role of student participation as a mediating factor between smart

technologies and cognitive skill development. It highlights a new area of investigation as it can serve as a foundation for future studies that look into how participation can be scaled up in tech-rich pedagogical settings.

3. **Introducing a VUCA-BANI Perspective:** By framing the study within the VUCA-BANI context, this research integrates emerging global challenges—such as volatility, uncertainty, complexity, and ambiguity—with education. Using VUCA-BANI as a backdrop for the study establishes an emerging backdrop for the future of education by encouraging the development of resilience, adaptability, and empathy which are indeed essential for navigating through such an environment's challenges
4. **Motivation as a Moderator in Technology Integration:** The adoption of learning motivation is viewed as an active moderation variable for the smart technology which aims to improve cognitive skills. This theoretical observation can give further emphasis on concern of how behavioral and psychological factors impact the use of the new technologies and expose the need for further research on motivation based approaches to teaching models.
5. **Developing a Holistic Framework for Smart Classrooms:** The conceptual framework proposed in this study provides a comprehensive model for understanding the interplay of technology, participation, and motivation in educational contexts. It serves as a theoretical guide for future studies exploring the integration of advanced technologies in smart education systems.

FUTURE RESEARCH DIRECTIONS IN SMART EDUCATION

The education sector is significantly transformed by smart technologies, offering unprecedented opportunities for innovation. However, their integration into learning presents new challenges that require thorough solutions. Future studies should explore the socio-cognitive processes fostered by technology, particularly through longitudinal and cross-sectional research in specific domains. Equally important is investigating the socio-emotional aspects of learning, such as emotional and collaborative intelligence, which are crucial in VUCA and BANI environments. Another challenge is equity and accessibility, necessitating intensive research to bridge the digital divide and create learner-sensitive designs.

In addition, the use of generative AI and the implementation of blockchain systems and 5G requires evaluation in relation to some emerging educational practices. There are unexplored active motivational components, and student participation should be reframed as both passive and active participation. Another relevant area is how smart education is used for building resilience and adaptability to prepare learners for uncertain futures. There is also a need to evaluate professional development and teacher preparedness for the use of new technologies to ensure meaningful integration.

Smart education faces several challenges, including the sustainability of its technological tools, which must support green practices. Other components, such as advanced machine learning and AI software, require the development of comprehensive frameworks or the modification of existing ones, like the Technology Acceptance Model (TAM). Additionally, interventions should focus on creating policies and institutional frameworks that are factual and authoritative, supporting the diverse scope of intelligent education systems. These efforts aim to address sustainability challenges and ensure the seamless integration of technology with learning in a VUCA-BANI world, ultimately transforming education for the future. Figure 12 is a diagrammatically representation and Table 22 gives a birds-eye-view of future research directions.

Figure 12. Future research directions

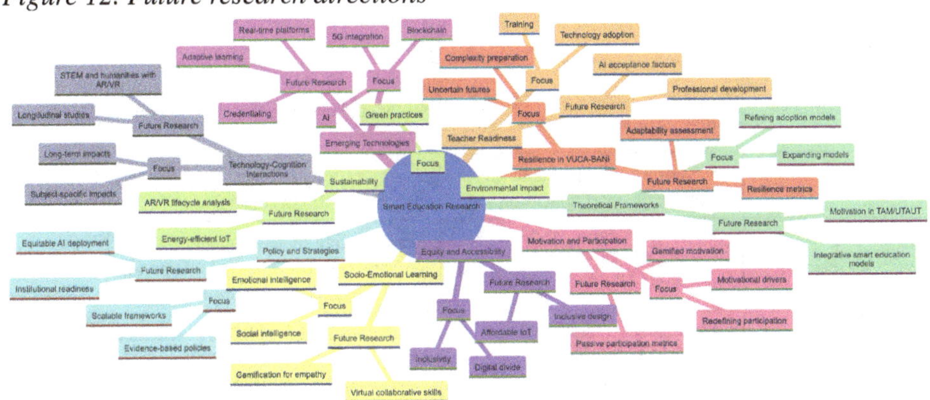

Table 22. Future Research Directions in Smart Education

Research Area	Focus	Future Research Directions
Technology-Cognition Interactions	Long-term and subject-specific impacts on cognitive skills	Longitudinal studies on critical thinking; cognitive development in STEM and humanities using AR/VR or IoT tools
Socio-Emotional Learning (SEL)	Role of technology in fostering emotional and social intelligence	Effectiveness of gamification in developing empathy; collaborative skill enhancement using virtual environments
Equity and Accessibility	Addressing the digital divide and promoting inclusivity	Strategies to implement affordable IoT in underserved schools; inclusive design for students with disabilities
Emerging Technologies	Integration of blockchain, AI, and 5G into smart classrooms	Use of blockchain for credentialing; generative AI for adaptive learning; 5G's role in real-time interactive platforms
Motivation and Participation	Understanding diverse motivational drivers and redefining participation	Examining intrinsic vs extrinsic motivation in gamified environments; exploring passive participation metrics
Resilience in VUCA-BANI Context	Preparing students for uncertain and complex futures	Developing resilience metrics for smart classrooms; assessing adaptability through interactive technologies
Teacher Readiness	Evaluating teacher training and adoption of smart technologies	Impact of professional development programs; factors influencing teacher acceptance of AI in classrooms
Sustainability	Environmental impact and green practices in smart education	Energy-efficient designs for IoT-enabled classrooms; lifecycle analysis of AR/VR devices
Theoretical Frameworks	Refining and expanding technology adoption models	Incorporating motivation and resilience into TAM or UTAUT; creating integrative models for smart education applications
Policy and Institutional Strategies	Evidence-based policies and scalable frameworks for implementation	Guidelines for equitable deployment of AI tools; institutional readiness for scaling smart classroom technologies

LIMITATIONS

While this analysis provides a reasonable comprehension of the impacts of smart classroom technologies on cognition within a VUCA-BANI environments, some limitations still persist. A significant limitation is the scope of the study, due to its focus on certain educational contexts that may not represent all possible teaching and learning contexts. Differences in the technology, administration, and educational pedagogy in various regions of the world could limit the external validity of the findings. In addition, the study relies on self-reported data from learners and teachers regarding the shifts in cognitive skills, participation, and motivational climate in the classroom. Self-reported measures are open to bias through social expectations and memory lapses which detracts the validity of the results. Further studies could include objective measures such as cognitive performance tests or learning analytics to have a broader scope of evaluation.

Another limitation is the fast growth of educational technologies which makes it difficult to sustain the relevance of the study over time. The constant development of AI, IoT, AR/VR, and gamification technologies could change and shift student engagement and learning experiences over time. In addition, disparities due to financial, technical, or pedagogical reasons may slow adoption of technology and affect the efficacy of smart classroom solutions, making them less available in different learning institutions. The study's emphasis on motivation as one moderating variable may be an oversimplification. Further research could look into other factors, like the distinction of intrinsic and extrinsic motivation or self-regulated learning, which may help explain the interaction between motivation and the use of technology for desirable cognitive outcomes. The study's focus on motivation as a single moderating variable may be a limitation. Further research could look at other factors, including the distinction of intrinsic and extrinsic motivation or self-regulated learning, that could help explain the relationship between the motivation and the use of technology when the outcomes cognitive enhancement is the focus.

Using a longitudinal approach would allow us to study the evolution of these technological changes and their impacts. Furthermore, the study continues to emphasize the cognitive processes of critical thinking and problem solving along with remembering, while other relevant cognitive processes such as creative thinking, cognitive flexibility, and emotional intelligence are equally important in the VUCA-BANI context. Future studies can expand the scope of the impact Smart Classroom Technologies have on education by adding other cognitive aspects to the analysis.

CONCLUSION

This study emphasizes the role of smart classroom technologies in developing cognitive skills to meet the challenges of the world characterized by VUCA-BANI. The use of these technologies through IoT, AR/VR, gamification, and AI shifts the educational paradigm from passive to active participation of students and enhances their cognitive engagement. The comparison and validation in the findings confirm the impact of these systems on student engagement and participation as a mediator in realizing the technological advantages for cognitive benefits.

The findings support all direct effect hypotheses except H1, which is rejected, indicating that artificial intelligence alone does not directly influence cognitive skill development. However, AI significantly enhances student participation (H2), which, in turn, positively affects cognitive skill development (H10). Augmented Reality/

Virtual Reality (H3, H4), Gamification (H5, H6), and the Internet of Things (H7, H8) all show significant direct effects on both student participation and cognitive skill development. Learning motivation is confirmed as a significant predictor of cognitive skill development (H9), and student participation significantly enhances cognitive skills (H10). Furthermore, student participation mediates the relationships between AI (H11), IoT (H12), AR/VR (H13), and gamification (H14) with cognitive skill development, highlighting its crucial role in technology-enabled learning environments. Additionally, learning motivation moderates the relationship between student participation and cognitive skill development (H15), reinforcing its importance in fostering cognitive engagement.

Furthermore, the study highlights learning motivation as a significant moderating variable that enhances the relationship between participation and cognitive skill development. Motivated students meaningfully engage with the learning materials, employ complex thinking skills, and demonstrate a willingness to overcome obstacles. This underscores the importance of nurturing a technological and psychosocial educational environment.

Based on these conclusions, educators, policymakers, and technology experts need to understand the interdependence of smart classroom technologies, student participation, and cognitive skill development. Educational plans should not only focus on the use of these technologies but also on students' active engagement and self-motivation. With the continuously evolving nature of digitalized learning, further studies should be conducted on the long-term impact and the adaptive measures that can be taken to ensure sustained development in cognitive skills in dynamic learning environments.

Ultimately, this study contributes to the broader discourse on technology-enhanced learning by providing a framework for leveraging smart classroom technologies to develop cognitive resilience, adaptability, and problem-solving skills—competencies that are essential for success in an unpredictable and rapidly changing world.

By navigating the intersection of technological advancements and educational practices in a VUCA and BANI world, this research contributes valuable insights for enhancing educational outcomes in the digital age. It also provides actionable recommendations for educators, policymakers, and technology developers to optimize the use of smart technologies in fostering cognitive growth and preparing students for the complexities of modern life.

REFERENCES

Abdi, H. (2003). Factor rotations in factor analyses. Encyclopedia for Research Methods for the Social Sciences, 792-795.

Aguinis, H., Gottfredson, R. K., & Joo, H. (2013). Best-practice recommendations for defining, identifying, and handling outliers. *Organizational Research Methods*, *16*(2), 270–301. DOI: 10.1177/1094428112470848

Ahn, S. J. G., Nowak, K. L., & Bailenson, J. N. (2022). Unintended consequences of spatial presence on learning in virtual reality. *Computers & Education*, *186*, 104532. DOI: 10.1016/j.compedu.2022.104532

Al-Fuqaha, A., Guizani, M., Mohammadi, M., Aledhari, M., & Ayyash, M. (2015). Internet of things: A survey on enabling technologies, protocols, and applications. *IEEE Communications Surveys and Tutorials*, *17*(4), 2347–2376. DOI: 10.1109/COMST.2015.2444095

Al-Samarraie, H., Teng, B. K., Alzahrani, A. I., & Alalwan, N. (2017). E-learning continuance satisfaction in higher education: A unified perspective from instructors and students. *Studies in Higher Education*, *42*(11), 2003–2019. DOI: 10.1080/03075079.2017.1298088

AlGerafi, M. A., Zhou, Y., Oubibi, M., & Wijaya, T. T. (2023). Unlocking the potential: A comprehensive evaluation of augmented reality and virtual reality in education. *Electronics (Basel)*, *12*(18), 3953. DOI: 10.3390/electronics12183953

Alharbi, A., Aljojo, N., Zainol, A., Alshutayri, A., Alharbi, B., Aldhahri, E., Khairullah, E. F., & Almandeel, S. (2021). Identification of critical factors affecting the students' acceptance of learning management system (LMS) in Saudi Arabia. *International Journal of Innovation*, *9*(2), 353–388. DOI: 10.5585/iji.v9i2.19652

Alin, A. (2010). Multicollinearity. *Wiley Interdisciplinary Reviews: Computational Statistics*, *2*(3), 370–374. DOI: 10.1002/wics.84

Alrousan, M. K., Al-Madadha, A., Al Khasawneh, M. H., & Adel Tweissi, A. (2022). Determinants of virtual classroom adoption in Jordan: The case of Princess Sumaya University for Technology. *Interactive Technology and Smart Education*, *19*(2), 121–144. DOI: 10.1108/ITSE-09-2020-0211

Atzori, L., Iera, A., & Morabito, G. (2010). The Internet of Things: A survey. *Computer Networks*, *54*(15), 2787–2805. DOI: 10.1016/j.comnet.2010.05.010

Bagozzi, R. P., & Yi, Y. (1988). On the evaluation of structural equation models. *Journal of the Academy of Marketing Science*, *16*(1), 74–94. DOI: 10.1007/BF02723327

Bailey, J. O., & Bailenson, J. N. (2017). Immersive virtual reality and the developing child. In *Cognitive development in digital contexts* (pp. 181–200). Academic Press., DOI: 10.1016/B978-0-12-809481-5.00009-2

Baker, R. S.. (2019). Learning analytics and educational technology: A review. *Journal of Learning Analytics*, *6*(1), 27–41.

Barahona Mora, A. (2020). Gamification for classroom management: An implementation using ClassDojo. *Sustainability (Basel)*, *12*(22), 9371. DOI: 10.3390/su12229371

Baron, R. M., & Kenny, D. A. (1986). The moderator-mediator variable distinction in social psychological research: Conceptual, strategic, and statistical considerations. *Journal of Personality and Social Psychology*, *51*(6), 1173–1182. DOI: 10.1037/0022-3514.51.6.1173

Bates, A. W. (2019). *Teaching in a digital age: Guidelines for designing teaching and learning*. Tony Bates Associates Ltd.

Becker, J. M., Ringle, C. M., & Sarstedt, M. (2023). Structural equation modeling with partial least squares path modeling. In Kline, R. (Ed.), *Handbook of structural equation modeling* (2nd ed.). Guilford Press.

Bennett, N., & Lemoine, G. J. (2014). What a difference a word makes: Understanding the impact of VUCA on organizations. *International Journal of Business and Social Science*, *5*(12), 44–53. DOI: 10.1016/j.bushor.2014.01.001

Bentler, P. M., & Bonett, D. G. (1980). Significance tests and goodness of fit in the analysis of covariance structures. *Psychological Bulletin*, *88*(3), 588–606. DOI: 10.1037/0033-2909.88.3.588

Bloom, B. S., Engelhart, M. D., Furst, E. J., Hill, W. H., & Krathwohl, D. R. (1956). *Taxonomy of educational objectives: The classification of educational goals. Handbook I: Cognitive domain*. David McKay Company.

Bond, M., Buntins, K., Bedenlier, S., Zawacki-Richter, O., & Kerres, M. (2020). Mapping research in student engagement and educational technology in higher education: A systematic evidence map. *International Journal of Educational Technology in Higher Education*, *17*(1), 2. DOI: 10.1186/s41239-019-0176-8

Bower, M., & Sturman, D. (2015). What are the educational affordances of wearable technologies? *Computers & Education, 88*, 343–353. DOI: 10.1016/j.compedu.2015.07.013

Bruner, J. (1966). *Toward a theory of instruction*. Harvard University Press., DOI: 10.1119/1.2350966

Byers, C. (2024). AI-powered educational tools and their effect on student motivation in asynchronous online courses [Master's thesis, University of Montana]. ScholarWorks.

Campbell, D. T., & Fiske, D. W. (1959). Convergent and discriminant validation by the multitrait-multimethod matrix. *Psychological Bulletin, 56*(2), 81–105. DOI: 10.1037/h0046016

Cascio, W. F., & Montealegre, R. (2016). How technology is changing work and organizations. *Annual Review of Organizational Psychology and Organizational Behavior, 3*(1), 349–375. DOI: 10.1146/annurev-orgpsych-041015-062352

Castañeda, L., & Selwyn, N. (2018). More than tools? Making sense of the ongoing digitizations of higher education. *International Journal of Educational Technology in Higher Education, 15*(1), 22. DOI: 10.1186/s41239-018-0109-y

Cattell, R. B. (1966). The scree test for the number of factors. *Multivariate Behavioral Research, 1*(2), 245–276. DOI: 10.1207/s15327906mbr0102_10

Chi, M. T. H., & Wylie, R. (2014). The ICAP framework: Linking cognitive engagement to active learning outcomes. *Educational Psychologist, 49*(4), 219–243. DOI: 10.1080/00461520.2014.965823

CIDDL. (n.d.). The impact of artificial intelligence on cognitive load. Retrieved February 14, 2025, from https://ciddl.org/the-impact-of-artificial-intelligence-on-cognitive-load/

Clegg, S., Kornberger, M., & Pitsis, T. (2021). *Strategy: Theory and practice* (3rd ed.). SAGE Publications.

Cohen, J. (1988). *Statistical power analysis for the behavioral sciences* (2nd ed.). Lawrence Erlbaum Associates.

Collins, A., & Halverson, R. (2018). *Rethinking education in the age of technology: The digital revolution and schooling in America*. Teachers College Press.

Cortina, J. M. (1993). What is coefficient alpha? An examination of theory and applications. *The Journal of Applied Psychology, 78*(1), 98–104. DOI: 10.1037/0021-9010.78.1.98

Costello, A. B., & Osborne, J. W. (2005). Best practices in exploratory factor analysis: Four recommendations for getting the most from your analysis. *Practical Assessment, Research & Evaluation, 10*(1), 7.

Dai, C. P., Ke, F., Dai, Z., & Pachman, M. (2023). Improving teaching practices via virtual reality-supported simulation-based learning: Scenario design and the duration of implementation. *British Journal of Educational Technology, 54*(4), 836–856. DOI: 10.1111/bjet.13296

Davis, F. D. (1989). Perceived usefulness, perceived ease of use, and user acceptance of information technology. *Management Information Systems Quarterly, 13*(3), 319–340. DOI: 10.2307/249008

Deci, E. L., & Ryan, R. M. (1985). *Intrinsic motivation and self-determination in human behavior*. Plenum., DOI: 10.1007/978-1-4899-2271-7

Deci, E. L., & Ryan, R. M. (2000). The "what" and "why" of goal pursuits: Human needs and the self-determination of behavior. *Psychological Inquiry, 11*(4), 227–268. DOI: 10.1207/S15327965PLI1104_01

Decristan, J., Jansen, N. C., & Fauth, B. (2023). Student participation in whole-class discourse: Individual conditions and consequences for student learning in primary and secondary school. *Learning and Instruction, 86*, 101748. DOI: 10.1016/j.learninstruc.2023.101748

Deterding, S., Dixon, D., Khaled, R., & Nacke, L. (2011). From game design elements to gamefulness: Defining "gamification". In *Proceedings of the 15th International Academic MindTrek Conference: Envisioning Future Media Environments* (pp. 9-15). DOI: 10.1145/2181037.2181040

DeVellis, R. F. (2016). *Scale development: Theory and applications* (4th ed.). SAGE Publications.

Di Giacomo, D., Ranieri, J., & Lacasa, P. (2017). Digital Learning As Enhanced Learning Processing? Cognitive Evidence for New insight of Smart Learning. *Frontiers in Psychology, 8*, 1329. DOI: 10.3389/fpsyg.2017.01329

Dormann, C. F., Elith, J., Bacher, S., Buchmann, C., Carl, G., Carré, G., Marquéz, J. R. G., Gruber, B., Lafourcade, B., Leitão, P. J., Münkemüller, T., McClean, C., Osborne, P. E., Reineking, B., Schröder, B., Skidmore, A. K., Zurell, D., & Lautenbach, S. (2013). Collinearity: A review of methods to deal with it and a simulation study evaluating their performance. *Ecography, 36*(1), 27–46. DOI: 10.1111/j.1600-0587.2012.07348.x

Drost, E. A. (2011). Validity and reliability in social science research. *Education Research and Perspectives, 38*(1), 105–123. DOI: 10.70953/ERPv38.11005

Dweck, C. S. (2000). *Self-theories: Their role in motivation, personality, and development.* Psychology Press.

Efron, B., & Tibshirani, R. J. (1993). *An introduction to the bootstrap.* Chapman and Hall/CRC., DOI: 10.1007/978-1-4899-4541-9

Ellikkal, A., & Rajamohan, S. (2024). *AI-enabled personalized learning: Empowering management students for improving engagement and academic performance.* Vilakshan-XIMB Journal of Management., DOI: 10.1108/XJM-02-2024-0023

Enders, C. K. (2010). Applied missing data analysis. Guilford Press. eSchool News. (2024, February 5). Impact of artificial intelligence in education. https://www.eschoolnews.com/educational-technology/2024/02/05/impact-of-artificial-intelligence-in-education/

Executive Academy. (2025, February 6). BANI vs. VUCA: How leadership works in the world of tomorrow. https://www.executiveacademy.at/en/news/detail/bani-vs-vuca-how-leadership-works-in-the-world-of-tomorrow

Field, A. (2017). *Discovering statistics using IBM SPSS statistics* (5th ed.). SAGE Publications.

Fischer, G., Lundin, J., & Lindberg, J. O. (2020). Rethinking and reinventing learning, education, and collaboration in the digital age—From creating technologies to transforming cultures. *The International Journal of Information and Learning Technology, 37*(5), 241–252. DOI: 10.1108/IJILT-04-2020-0051

Fornell, C., & Larcker, D. F. (1981). Evaluating structural equation models with unobservable variables and measurement error. *JMR, Journal of Marketing Research, 18*(1), 39–50. DOI: 10.1177/002224378101800104

Fredricks, J. A., Blumenfeld, P. C., & Paris, A. H. (2004). School engagement: Potential of the concept, state of the evidence. *Review of Educational Research, 74*(1), 59–109. DOI: 10.3102/00346543074001059

Gargrish, S., Mantri, A., & Kaur, D. P. (2022). Evaluation of memory retention among students using augmented reality-based geometry learning assistant. *Education and Information Technologies, 27*(9), 12891–12912. DOI: 10.1007/s10639-022-11147-9

Geisser, S. (1974). A predictive approach to the random effect model. *Biometrika, 61*(1), 101–107. DOI: 10.1093/biomet/61.1.101

Ghasemi, A., & Zahediasl, S. (2012). Normality tests for statistical analysis: A guide for non-statisticians. *International Journal of Endocrinology and Metabolism, 10*(2), 486–489. DOI: 10.5812/ijem.3505

Ghosh, D., & Vogt, A. (2012). Outliers: An evaluation of methodologies. Joint Statistical Meetings, 3455-3460.

Graham, J. W. (2009). Missing data analysis: Making it work in the real world. *Annual Review of Psychology, 60*(1), 549–576. DOI: 10.1146/annurev.psych.58.110405.085530

Hair, J. F., Black, W. C., Babin, B. J., & Anderson, R. E. (2010). *Multivariate data analysis* (7th ed.). Prentice Hall.

Hair, J. F., Black, W. C., Babin, B. J., & Anderson, R. E. (2014). *Multivariate data analysis* (7th ed.). Pearson Education Limited.

Hair, J. F., Black, W. C., Babin, B. J., & Anderson, R. E. (2019). *Multivariate data analysis* (8th ed.). Cengage Learning.

Hair, J. F., Hult, G. T. M., Ringle, C. M., & Sarstedt, M. (2017). *A primer on partial least squares structural equation modeling (PLS-SEM)* (2nd ed.). Sage Publications.

Hair, J. F., Risher, J. J., Sarstedt, M., & Ringle, C. M. (2019). When to use and how to report the results of PLS-SEM. *European Business Review, 31*(1), 2–24. DOI: 10.1108/EBR-11-2018-0203

Hamad, F., Shehata, A., & Al Hosni, N. (2024). Predictors of blended learning adoption in higher education institutions in Oman: Theory of planned behavior. *International Journal of Educational Technology in Higher Education, 21*(1), 13. DOI: 10.1186/s41239-024-00443-8

Hamari, J., Koivisto, J., & Sarsa, H. (2014). Does gamification work? A literature review of empirical studies on gamification. In 2014 47th Hawaii International Conference on System Sciences (pp. 3025-3034). IEEE. DOI: 10.1109/HICSS.2014.377

Haque, M. A., Ahmad, S., Hossain, M. A., Kumar, K., Faizanuddin, M., Islam, F., Haque, S., Rahman, M., Marisennayya, S., & Nazeer, J. (2024). Internet of things enabled e-learning system for academic achievement among university students. *E-Learning and Digital Media*, •••, 20427530241280078. DOI: 10.1177/20427530241280078

Hayes, A. F. (2017). *Introduction to mediation, moderation, and conditional process analysis: A regression-based approach* (2nd ed.). Guilford Press.

Henderson, M., Selwyn, N., & Aston, R. (2017). What works and why? Student perceptions of 'useful' digital technology in university teaching and learning. *Studies in Higher Education, 42*(8), 1567–1579. DOI: 10.1080/03075079.2015.1007946

Henseler, J., Ringle, C. M., & Sarstedt, M. (2015). A new criterion for assessing discriminant validity in variance-based structural equation modeling. *Journal of the Academy of Marketing Science, 43*(1), 115–135. DOI: 10.1007/s11747-014-0403-8

Höck, C., Ringle, C. M., & Sarstedt, M. (2010). Management of multi-purpose stadiums: Importance and performance measurement of service interfaces. *International Journal of Services Technology and Management, 14*(2-3), 188–207. DOI: 10.1504/IJSTM.2010.034327

Hofstein, A., & Rosenfeld, S. (1996). Bridging the gap between formal and informal science learning. *Studies in Science Education, 28*(1), 87–112. DOI: 10.1080/03057269608560085

Hu, L. T., & Bentler, P. M. (1999). Cutoff criteria for fit indexes in covariance structure analysis: Conventional criteria versus new alternatives. *Structural Equation Modeling, 6*(1), 1–55. DOI: 10.1080/10705519909540118

Huang, R., Spector, J. M., & Yang, J. (2021). Educational technology as a driver of cognitive development in the digital age. *British Journal of Educational Technology, 52*(4), 1234–1250. DOI: 10.1111/bjet.13045

Huang, W., Roscoe, R. D., Craig, S. D., & Johnson-Glenberg, M. C. (2022). Extending the cognitive-affective theory of learning with media in virtual reality learning: A structural equation modeling approach. *Journal of Educational Computing Research, 60*(4), 807–842. DOI: 10.1177/07356331211053630

Jahnke, I., Meinke-Kroll, M., Todd, M., & Nolte, A. (2022). Exploring artifact-generated learning with digital technologies: Advancing active learning with co-design in higher education across disciplines. Technology, Knowledge and Learning, 1-30. DOI: 10.1007/s10758-020-09473-3

Kaiser, H. F. (1960). The application of electronic computers to factor analysis. *Educational and Psychological Measurement, 20*(1), 141–151. DOI: 10.1177/001316446002000116

Kaiser, H. F. (1974). An index of factorial simplicity. *Psychometrika, 39*(1), 31–36. DOI: 10.1007/BF02291575

Khan, I., Khan, N., Jazim, F., Al-Mamary, Y. H., Abdulrab, M., & Al-Ghurbani, A. M. (2022). The effect of external factors in use of technology among Ha'il university academic faculty: Evidence from Saudi Arabia. *Journal of Applied Research in Higher Education*, *14*(4), 1319–1339. DOI: 10.1108/JARHE-04-2021-0140

Khlaif, Z. (2018). Teachers' perceptions of factors affecting their adoption and acceptance of mobile technology in K-12 settings. *Computers in the Schools*, *35*(1), 49–67. DOI: 10.1080/07380569.2018.1428001

Kline, R. B. (2015). *Principles and practice of structural equation modeling* (4th ed.). Guilford Publications.

Kline, R. B. (2016). *Principles and practice of structural equation modeling* (4th ed.). Guilford Press.

Kock, N. (2015). Common method bias in PLS-SEM: A full collinearity assessment approach. *International Journal of e-Collaboration*, *11*(4), 1–10. DOI: 10.4018/ijec.2015100101

Kock, N. (2018). Should bootstrapping be used in PLS-SEM? Toward stable p-value calculation methods. *Journal of Applied Structural Equation Modeling*, *2*(1), 1–12. DOI: 10.47263/JASEM.2(1)02

Kopetz, H., & Steiner, W. (2022). Internet of things. In *Real-time systems: Design principles for distributed embedded applications* (pp. 325–341). Springer International Publishing., DOI: 10.1007/978-3-031-11992-7_13

Kruglanski, A. W., Bélanger, J. J., Chen, X., Köpetz, C., Pierro, A., & Mannetti, L. (2012). The energetics of motivated cognition: A force-field analysis. *Psychological Review*, *119*(1), 1–20. DOI: 10.1037/a0025488

Kurni, M., & KG, S. (2024). IoT for education. In The Internet of Educational Things: Enhancing Students' Engagement and Learning Performance (pp. 1-22). Springer Nature Switzerland. DOI: 10.1007/978-3-031-67387-0_10

Kwak, S. K., & Kim, J. H. (2017). Statistical data preparation: Management of missing values and outliers. *Korean Journal of Anesthesiology*, *70*(4), 407–411. DOI: 10.4097/kjae.2017.70.4.407

Lawson, A. P., Martella, A. M., LaBonte, K., Delgado, C. Y., Zhao, F., Gluck, J. A., Munns, M. E., Wells LeRoy, A., & Mayer, R. E. (2024). Confounded or controlled? A systematic review of media comparison studies involving immersive virtual reality for STEM education. *Educational Psychology Review*, *36*(3), 69. DOI: 10.1007/s10648-024-09908-8

Ledesma, R. D., & Valero-Mora, P. (2007). Determining the number of factors to retain in EFA: An easy-to-use computer program for carrying out parallel analysis. *Practical Assessment, Research & Evaluation, 12*(1), 2.

Li, L., Hew, K. F., & Du, J. (2024). Gamification enhances student intrinsic motivation, perceptions of autonomy and relatedness, but minimal impact on competency: A meta-analysis and systematic review. *Educational Technology Research and Development, 72*(2), 765–796. DOI: 10.1007/s11423-023-10337-7

Li, P., Zhang, X., Hu, X., Xu, B., & Zhang, J. (2024). Theoretical model and practical analysis of immersive industrial design education based on virtual reality technology. *International Journal of Technology and Design Education*, •••, 1–28. DOI: 10.1007/s10798-024-09946-x

Little, R. J., & Rubin, D. B. (2019). *Statistical analysis with missing data* (3rd ed.). John Wiley & Sons., DOI: 10.1002/9781119482260

MacKenzie, S. B., & Podsakoff, P. M. (2012). Common method bias in marketing: Causes, mechanisms, and procedural remedies. *Journal ofRetailing, 88*(4), 542–555. DOI: 10.1016/j.jretai.2012.08.001

MacKinnon, D. P., Fairchild, A. J., & Fritz, M. S. (2007). Mediation analysis. *Annual Review of Psychology, 58*(1), 593–614. DOI: 10.1146/annurev.psych.58.110405.085542

Makransky, G., & Petersen, G. B. (2021). The cognitive-affective model of immersive learning (CAMIL): A theoretical research-based model of learning in immersive virtual reality. *Educational Psychology Review, 33*(3), 937–958. DOI: 10.1007/s10648-020-09586-2

Makransky, G., & Petersen, J. (2021). The impact of augmented and virtual reality on student engagement and learning outcomes in higher education: A systematic review. *Journal of Educational Psychology, 113*(4), 670–687. DOI: 10.1037/edu0000434

Masters of ERP. (n.d.). The impact of IoT on campus: Smart student information systems. https://www.mastersoferp.com/the-impact-of-iot-on-campus-smart-student-information-systems/

Messick, S. (1995). Validity of psychological assessment: Validation of inferences from persons' responses and performances as scientific inquiry into score meaning. *The American Psychologist, 50*(9), 741–749. DOI: 10.1037/0003-066X.50.9.741

Multishoring. (2025, January 21). The impact of IoT in education: What you need to know. https://multishoring.com/blog/impact-of-iot-in-education/

Niemiec, C. P., & Ryan, R. M. (2009). Autonomy, competence, and relatedness in the classroom: Applying self-determination theory to educational practice. *Theory and Research in Education*, *7*(2), 133–144. DOI: 10.1177/1477878509104318

Nsflow. (n.d.). 9 benefits of augmented reality in education. https://www.nsflow.com/blog/benefits-of-augmented-reality-in-education

Nunnally, J. C., & Bernstein, I. H. (1994). *Psychometric theory* (3rd ed.). McGraw-Hill.

O'Brien, R. M. (2007). A caution regarding rules of thumb for variance inflation factors. *Quality & Quantity*, *41*(5), 673–690. DOI: 10.1007/s11135-006-9018-6

Osborne, J. W. (2010). Improving your data transformations: Applying the Box-Cox transformation. *Practical Assessment, Research & Evaluation*, *15*(1), 12.

Osborne, J. W., & Overbay, A. (2004). The power of outliers (and why researchers should always check for them). *Practical Assessment, Research & Evaluation*, *9*(1), 6.

Osborne, J. W., & Waters, E. (2002). Four assumptions of multiple regression that researchers should always test. *Practical Assessment, Research & Evaluation*, *8*(2), 1–9.

Petersen, G. B., Stenberdt, V., Mayer, R. E., & Makransky, G. (2023). Collaborative generative learning activities in immersive virtual reality increase learning. *Computers & Education*, *207*, 104931. DOI: 10.1016/j.compedu.2023.104931

Peugh, J. L., & Enders, C. K. (2004). Missing data in educational research: A review of reporting practices and suggestions for improvement. *Review of Educational Research*, *74*(4), 525–556. DOI: 10.3102/00346543074004525

Piaget, J. (1973). *To understand is to invent: The future of education*. Viking Press.

Pintrich, P. R. (2004). A conceptual framework for assessing motivation and self-regulated learning in college students. *Educational Psychology Review*, *16*(4), 385–407. DOI: 10.1007/s10648-004-0006-x

Podsakoff, P. M., MacKenzie, S. B., Lee, J. Y., & Podsakoff, N. P. (2003). Common method biases in behavioral research: A critical review of the literature and recommended remedies. *The Journal of Applied Psychology*, *88*(5), 879–903. DOI: 10.1037/0021-9010.88.5.879

Preacher, K. J., & Hayes, A. F. (2008). Asymptotic and resampling strategies for assessing and comparing indirect effects in multiple mediator models. *Behavior Research Methods*, *40*(3), 879–891. DOI: 10.3758/BRM.40.3.879

Psico-Smart. (2024, August 28). The impact of artificial intelligence on cognitive skills assessment. https://psico-smart.com/en/blogs/blog-the-impact-of-artificial-intelligence-on-cognitive-skills-assessment-160885

Raykov, T. (1997). Estimation of composite reliability for congeneric measures. *Applied Psychological Measurement*, *21*(2), 173–184. DOI: 10.1177/01466216970212006

Reeve, J. (2012). A sef-determination theory perspective on student engagement. In Christenson, S. L., Reschly, A. L., & Wylie, C. (Eds.), *Handbook of research on student engagement* (pp. 149–172). Springer., DOI: 10.1007/978-1-4614-2018-7_7

Ringle, C. M., & Sarstedt, M. (2016). Gain more insight from your PLS-SEM results: The importance-performance map analysis. *Industrial Management & Data Systems*, *116*(9), 1865–1886. DOI: 10.1108/IMDS-10-2015-0449

Roll, I., & Wylie, R. (2016). Evolution and revolution in artificial intelligence in education. *International Journal of Artificial Intelligence in Education*, *26*(2), 582–599. DOI: 10.1007/s40593-016-0110-3

Ryan, R. M., & Deci, E. L. (2000). Intrinsic and extrinsic motivations: Classic definitions and new directions. *Contemporary Educational Psychology*, *25*(1), 54–67. DOI: 10.1006/ceps.1999.1020

Ryan, R. M., & Deci, E. L. (2000). Self-determination theory and the facilitation of intrinsic motivation, social development, and well-being. *The American Psychologist*, *55*(1), 68–78. DOI: 10.1037/0003-066X.55.1.68

Schumacker, R. E., & Lomax, R. G. (2016). *A beginner's guide to structural equation modeling* (4th ed.). Routledge., DOI: 10.4324/9781315749105

Selwyn, N. (2016). Digital downsides: Exploring university students' negative engagements with digital technology. *Teaching in Higher Education*, *21*(8), 1006–1021. DOI: 10.1080/13562517.2016.1213229

Selwyn, N. (2016). *Education and technology: Key issues and debates*. Bloomsbury Academic., DOI: 10.5040/9781474235952

Selwyn, N. (2020). *Should robots replace teachers? AI and the future of education*. Polity Press.

Shi, M., & Tan, C. Y. (2020). Beyond oral participation: A typology of student engagement in classroom discussions. *New Zealand Journal of Educational Studies*, *55*(1), 247–265. DOI: 10.1007/s40841-020-00166-0

Skinner, B. F. (1953). *Science and human behavior*. Free Press.

Spiegel, S., Grant-Pillow, H., & Higgins, E. T. (2004). How regulatory fit enhances motivational strength during goal pursuit. *European Journal of Social Psychology*, *34*(1), 39–54. DOI: 10.1002/ejsp.180

Stephan Grabmeier. (n.d.). BANI versus VUCA: A new acronym to describe the world. https://stephangrabmeier.de/bani-versus-vuca/

Stone, M. (1974). Cross-validatory choice and assessment of statistical predictions. *Journal of the Royal Statistical Society. Series B, Statistical Methodology*, *36*(2), 111–133. DOI: 10.1111/j.2517-6161.1974.tb00994.x

Straub, D., Boudreau, M. C., & Gefen, D. (2004). Validation guidelines for IS positivist research. *Communications of the Association for Information Systems*, *13*(1), 380–427. DOI: 10.17705/1CAIS.01324

Streiner, D. L. (2003). Starting at the beginning: An introduction to coefficient alpha and internal consistency. *Journal of Personality Assessment*, *80*(1), 99–103. DOI: 10.1207/S15327752JPA8001_18

Strielkowski, W., Grebennikova, V., Lisovskiy, A., Rakhimova, G., & Vasileva, T. (2024). AI-driven adaptive learning for sustainable educational transformation. *Sustainable Development*.

Sweller, J. (1988). Cognitive load during problem solving: Effects on learning. *Cognitive Science*, *12*(2), 257–285. DOI: 10.1207/s15516709cog1202_4

Tabachnick, B. G., & Fidell, L. S. (2019). *Using multivariate statistics* (7th ed.). Pearson.

Taber, K. S. (2018). The use of Cronbach's alpha when developing and reporting research instruments in science education. *Research in Science Education*, *48*(6), 1273–1296. DOI: 10.1007/s11165-016-9602-2

Tavakol, M., & Dennick, R. (2011). Making sense of Cronbach's alpha. *International Journal of Medical Education*, *2*, 53–55. DOI: 10.5116/ijme.4dfb.8dfd

Taylor, S., & Todd, P. A. (1995). Understanding information technology usage: A test of competing models. *Information Systems Research*, *6*(2), 144–176. DOI: 10.1287/isre.6.2.144

Thompson, B. (2004). *Exploratory and confirmatory factor analysis: Understanding concepts and applications*. American Psychological Association., DOI: 10.1037/10694-000

Turnitin. (n.d.). How will AI impact skill development in education? Retrieved February 14, 2025, from https://www.turnitin.com/blog/how-will-ai-impact-skill-development-in-education

UCW. (2024, July 19). Advantages and disadvantages of AI in education. https://www.ucanwest.ca/blog/advantages-and-disadvantages-of-ai-in-education

van Buuren, S. (2018). *Flexible imputation of missing data* (2nd ed.). Chapman and Hall/CRC., DOI: 10.1201/9780429492259

Venkatesh, V., Morris, , Davis, , & Davis, . (2003). User acceptance of information technology: Toward a unified view. *Management Information Systems Quarterly, 27*(3), 425–478. DOI: 10.2307/30036540

Vidanaralage, A. J., Dharmaratne, A. T., & Haque, S. (2022). AI-based multidisciplinary framework to assess the impact of gamified video-based learning through schema and emotion analysis. *Computers and Education: Artificial Intelligence, 3*, 100109. DOI: 10.1016/j.caeai.2022.100109

Voorhees, C. M., Brady, M. K., Calantone, R., & Ramirez, E. (2016). Discriminant validity testing in marketing: An analysis, causes for concern, and proposed remedies. *Journal of the Academy of Marketing Science, 44*(1), 119–134. DOI: 10.1007/s11747-015-0455-4

Vygotsky, L. S. (1978). *Mind in society: The development of higher psychological processes*. Harvard University Press.

Wang, Y., Derakhshan, A., & Zhang, L. J. (2021). Researching and Practicing Positive Psychology in Second/Foreign Language Learning and Teaching: The Past, Current Status and Future Directions. *Frontiers in Psychology, 12*, 731721. DOI: 10.3389/fpsyg.2021.731721

Wang, Y., Gu, X., & Zhao, Y. (2024). The impact of AI usage on university students' willingness for autonomous learning: An expectation-confirmation perspective. *PLoS One, 19*(10), e0292756. DOI: 10.1371/journal.pone.0292756

Wigfield, A., & Eccles, J. S. (2000). Expectancy-value theory of achievement motivation. *Contemporary Educational Psychology, 25*(1), 68–81. DOI: 10.1006/ceps.1999.1015

Williams, B., Onsman, A., & Brown, T. (2010). Exploratory factor analysis: A five-step guide for novices. *Australasian Journal of Paramedicine, 8*(3), 1–13. Advance online publication. DOI: 10.33151/ajp.8.3.93

Wu, H. K., Lee, S. W. Y., Chang, H. Y., & Liang, J. C. (2013). Current status, opportunities and challenges of augmented reality in education. *Computers & Education, 62*, 41–49. DOI: 10.1016/j.compedu.2012.10.024

Yin, H. (2018). What motivates Chinese undergraduates to engage in learning? Insights from a psychological approach to student engagement research. *Higher Education, 76*(5), 827–847. DOI: 10.1007/s10734-018-0239-0

Zawacki-Richter, O., Marín, V. I., Bond, M., & Gouverneur, F. (2019). Systematic review of research on artificial intelligence applications in higher education - where are the educators? *International Journal of Educational Technology in Higher Education, 16*(1), 39. DOI: 10.1186/s41239-019-0171-0

Zeng, J., Sun, D., Looi, C. K., & Fan, A. C. W. (2024). Exploring the impact of gamification on students' academic performance: A comprehensive meta-analysis of studies from the year 2008 to 2023. British J ournal of Educational Technology.

Zimmerman, B. J. (2008). Investigating self-regulation and motivation: Historical background, methodological developments, and future prospects. *American Educational Research Journal, 45*(1), 166–183. DOI: 10.3102/0002831207312909

Chapter 9
Smart and Digital Wellbeing Initiatives in Healthcare: The Way Forward

Anushka Varma
https://orcid.org/0009-0000-1062-1511
Infiheal HeathTech Private Limited, India

Aashi Samir Saraiya
https://orcid.org/0009-0003-3992-3435
Infiheal HeathTech Private Limited, India

Kashvi Tuli
https://orcid.org/0009-0000-3390-0222
Infiheal HeathTech Private Limited, India

Sruthi Suresh
https://orcid.org/0000-0002-2505-9872
Christ University, India

Srishti Srivastava
https://orcid.org/0009-0002-8773-2687
Infiheal Healthtech Private Limited, India

ABSTRACT

Smart and digital wellbeing initiatives play a significant role in modern healthcare. This chapter explores current initiatives, their use and effectiveness, as well as their challenges and future directions, with a special focus on mental health. It explores

DOI: 10.4018/979-8-3373-7917-3.ch009

digital wellbeing, the growing need for initiatives, and the tools available in India and the world. These initiatives enhance accessibility, reduce stigma, offer personalized interventions, and improve psychological and workplace wellbeing. However, current technology, especially AI-based chatbots, have limited user interaction and authentic therapeutic support. Additionally, biased datasets can lead to flawed responses and issues with inclusive support. Current initiatives face privacy concerns due to a lack of regulatory bodies, emphasizing the need for robust regulation and oversight. Collaborative development, appropriate marketing and improved security are key to improving existing initiatives. Future research needs to address and explore these concerns to ensure the development of effective and ethical initiatives.

INTRODUCTION

Digital Wellbeing in the 21st Century

The term *'digital wellbeing'* refers to the impact of digital technology on human lives within an informational society (Floridi, 2014). It encompasses the ability to use digital tools effectively and responsibly while minimising potential harm, such as digital fatigue, technology addiction, and declining attention spans (Vanden Abeele, 2021). With the increasing presence of digital devices in personal and professional spaces, ensuring healthy engagement with technology has become an essential goal for individuals, workplaces, and policymakers. The widespread adoption of digital technologies has reshaped interactions with oneself, others, and society at large. Consequently, individual and societal wellbeing are now closely linked to the quality of the digital environment and the tools that shape interactions (Floridi, 2014). 'Wellbeing' in recent times has come to be used in an all-encompassing manner, referring to positive health and fulfillment in all sectors of life, whether psychological, physical or social, rather than just being happy (Ruggeri et al., 2020).

In the 21st century, technology has transformed nearly every aspect of human life. Digital wellbeing is particularly crucial in today's hybrid work and learning environments, where remote interactions and digital collaboration have become the norm. The rapid expansion of digital tools, artificial intelligence, and online platforms has led to unparalleled convenience and efficiency (Twenge & Campbell, 2018). As such, utilising these tools and technology to further promote wellbeing is the fundamental basis behind digital wellbeing ventures. As digital tools become increasingly embedded in daily life, their impact on mental and emotional health has gained attention, leading to the development of digital wellbeing initiatives aimed at ensuring that technology use remains balanced, enriching, and aligned with and fostering human needs (Burr et al., 2020).

Technological advancements have improved access to education, healthcare, and economic opportunities. Online learning platforms, telemedicine services, and AI-driven productivity tools have revolutionised how people learn, work, and receive medical care (Khoury & Ioannidis, 2014), and the latter is the primary focus of this chapter, particularly in mental health, although other avenues will also be explored. Furthermore, digital tools have enhanced social connectivity, allowing individuals to maintain relationships across geographical boundaries, highlighting the necessity for these tools (Burr et al., 2020). The importance of digital wellbeing extends beyond individual mental health to broader societal wellbeing and economic productivity. As such, the development and progress of these tools stemmed from the need to rectify a long-standing inequality.

The Need for Digital Wellbeing Measures

Healthcare has historically been an unequal and deprived sector, mental health even more so, with the best services often restricted to the upper echelons of society in first-world countries. Imagine living with a debilitating mental illness but having absolutely no adequate care available. This is true for 80% of the people worldwide who are currently living with a mental illness in middle or low-income countries, and 90% of those lack any evidence-based care (Naslund & Deng, 2021). The mental health treatment gap is as high as 85% in low and middle-income countries and 40% in high-income countries (Ndetei et al., 2023). Certain governments, particularly in Africa and South-East Asia, spend less than 1% of their national healthcare budgets on mental health care (Patel et al., 2016). Various factors like stigma, poverty, geographical limitations, and cultural barriers aggravate the crisis, which calls for efficient initiatives to treat the gap, but building this remains an issue.

Mental health stigma is especially prevalent in low-income and collectivistic countries. Seeking traditional help in these countries could mean facing harmful remedies, forced fasting, poor living conditions, physical abuse, and even violation of human rights, especially in more rural or isolated sections of society (Ben-Zeev & Atkins, 2017). Most trained professionals are found to be inadequately trained or carry misbeliefs about mental healthcare, which in turn exacerbates stigma among those seeking help (Patel et al., 2016). This prevents people from seeking help due to the fear of social judgment, victimisation or shame. Moreover, the lack of culturally sensitive and ethical approaches limits treatment effectiveness. This calls for the need for culturally fair, anonymous, and ethical platforms that do not make individuals suffering from mental illnesses feel like victims. Shwab (2018) argues that digital technologies have the potential to drive a new era of reduced social inequality by improving access to essential services such as healthcare.

Most of the stigma-based gap persists in collectivistic countries, and digital mental health initiatives such as peer-to-peer online support forums where individuals experiencing mental health issues could connect would be helpful. Naslund and Deng (2021) also suggest that digital platforms can protest against misinformation being spread about mental illnesses and educate others about the truthful portrayal of what mental health problems look like. Most government initiatives rely on psychoeducational workshops and campaigns, which lack implementation and long-term engagement goals. However, what low and middle-income countries (LMICs) require is something tangible, which might not replace traditional therapeutic practices but can help individuals cope with mental illnesses. This brings into the picture self or artificial intelligence (AI) based, easy-to-access apps that can provide useful tips and accurate information that can be used to understand and manage problems efficiently, while still retaining anonymity. These digital apps can be made a cost-effective and less time-consuming solution to maximise use in countries where poverty prevents individuals from seeking professional help (Ben-Zeev & Atkins, 2017), as a detailed review has found positive trends for improved states in cases of depression and anxiety following intervention from these apps (Gallegos et al., 2024).

Addressing mental health gaps does not only mean increasing the number of resources available, but also includes developing resources that are efficient, available, and easy to use. At the moment, many digital apps and interventions do exist, but they might not be clinically validated or may lack the marketing needed amongst vulnerable populations and trained professionals. This is particularly true for LMICs, where both the tools and their impact are not particularly well explored. An interesting study by Ben-Zeev and Atkins (2017) found that most people with severe mental illnesses were overlooked in treatment or excluded from research due to potential risks of complex data and rejections. However, they expressed that most of them own mobile phones and are willing to seek mental health treatment through digital methods, thus making the need for digital initiatives stronger (Ben-Zeev & Atkins, 2017).

The Rise of Smart and Digital Wellbeing Initiatives

As of today, very few preventative measures are being taken to address mental health issues in LMICs (Patel et al., 2016). With the advancement of digital platforms and apps, they could be one of the first preventative measures for individuals living in poorer countries who are on the verge of having mental illnesses. However, while the current state in LMICs may still be limited, the idea of using chatbots for the betterment of mental health is not a new one.

The idea of chatbots has existed for many decades, ever since the cognitive revolution in the field of psychology that occurred in the 1950s. ELIZA, one of the first chatbots developed in the 60s by Weizenbaum, was an early natural language processing program and simulated a therapist's conversational approach. While it was reliant on pre-scripted responses, ELIZA laid a foundation for all future mental health applications by showcasing the ability of AI to engage users in therapeutic dialogue (Weizenbaum, 1966). Over at Stanford in 1972, psychiatrist Kenneth Colby developed PARRY, which simulated a patient with schizophrenia. Utilising a natural language program, it was able to mirror the behaviour of someone with the same disorder. One of the first AI programs to pass the Turing Test, PARRY highlighted AI's potential to embody specific mental health conditions (Colby et al., 1971). While these historical attempts existed, they lacked the conversational depth and dynamism needed to replicate human conversation.

However, like always, technology offered a solution in the form of Generative AI and Large Language Models (LLM), transforming chatbots into dynamic and interactive systems, often providing real-time, appropriate, and empathetic responses. This ability to foster feelings of understanding and support comes from extensive psychologically sensitive data, containing clues into conversational tones and empathy. This contextual understanding is the key reason behind the improved responses (Zhao et al., 2023; Banh & Strobel, 2023), and a key step in improving the issues of accessible and well-developed systems for populations struggling with accessible healthcare.

LMICs primarily struggle with the accessibility of resources and do not have well-established traditional setups like maintained mental health hospitals or non-private clinics. Thus, low-resource-based mobile technology, such as text-based interventions and voice response technologies that preferably do not require a stable internet connection, would effectively utilise technological advancements in countries where it is most needed (Ben-Zeev & Atkins, 2017), while the developing systems of AI would ensure that these interventions are effective and empathetic.

Digital initiatives, however, are not meant to replace traditional therapy practices. It has occasionally been noticed that traditional therapy sessions remain unsupervised, and might leave the therapists and clients stuck in a rut owing to the lack of evidence-based care. Even if feedback is provided to the therapist, it comes at a much later point in the therapeutic relationship, where both the therapist and client have moved on to other issues. This is where digital initiatives like natural language processing and advanced text-based processing software can enter the picture as aids. They could help provide quick, evidence-based feedback and structure to a therapy session. Additionally, they can aid the therapist with automated summaries and give feedback on paralinguistic features of the clients, such as facial expressions and vocally encoded arousals, which could be missed by a human therapist (Ben-Zeev & Atkins, 2017). They could use case studies and evidence-based data on mental health and coping strategies to serve as psycho-educators and co-therapists.

Thus, even though digital initiatives could be the answer to reduce the void in the number of people needing help and the number of people getting help, ideally, they must be used as supplementary interventions that could aid traditional practices and not replace them, albeit that may be a slow process. A common assumption about reducing the treatment gap comes from the belief that psychologists and psychiatrists must be the sole educators and providers of help and care. It is true that they are trained professionals and should be approached for professional help. However, non-professional support by adequately trained providers such as teachers, school counsellors, peers, and even religious leaders could help extinguish the gap among the population to a certain extent, utilising these chatbots as tools and aids (Ndetei et al., 2023). The role of digitalisation is key in fostering wellbeing and personal growth by integrating insights from behavioural and cognitive sciences on motivation and engagement to promote lasting habits and practices (Calvo & Peters, 2014; Desmet & Pohlmeyer, 2013; Peters et al., 2018).

Global Trends in Digital Wellbeing

While technology has empowered individuals with unlimited access to information, remote work opportunities, and digital healthcare, it has also contributed to issues such as screen addiction, cognitive overload, digital fatigue, and social isolation (Twenge & Campbell, 2018). The goal is to enable individuals to harness the benefits of digitalisation without compromising their overall wellbeing. A balanced digital lifestyle requires an awareness of how digital interactions affect cognitive functions, emotions, and social behaviours (Twenge et al., 2022). Recognising the pressing need to mitigate digital harm while maximising its benefits, various stakeholders, including governments, technology companies, mental health organisations, and researchers, have introduced smart and digital wellbeing initiatives.

These initiatives promote healthier technology habits, enhance digital literacy, and integrate wellbeing-focused solutions into everyday digital interactions. Tech companies have played a significant role in developing tools that encourage mindful technology use. Apple's Screen Time, Google's Digital Wellbeing tools, and Facebook's Take a Break feature are examples of industry efforts to empower users with greater control over their digital consumption. These features allow individuals to monitor screen time, set app usage limits, and reduce unnecessary notifications, fostering a more intentional approach to digital engagement, in turn fostering physical and mental health.

Governments and policymakers are also taking proactive steps to regulate digital wellbeing. France's "Right to Disconnect" law aims to prevent employees from being overburdened by after-hours work emails, thereby protecting work-life balance (Fana et al., 2020). Similarly, the European Union's Digital Services Act (2022) enforces ethical standards for online platforms, ensuring responsible AI deployment and content moderation. These regulatory measures reflect a broader commitment to creating a healthier digital ecosystem, and similar global frameworks need to be put in place, especially for mental health.

The advancements in digital mental health initiatives have not yet picked up the pace of digital-physical health initiatives, with the latter being widely used for diagnostic and screening purposes. For example, in the year 2023, the World Health Organisation (WHO) launched a global digital health initiative, but no international initiatives or toolkits are available for digital mental health to curtail broader mental illnesses. Moreover, there is limited mention of large-scale global investments in digital mental health (Naslund et al., 2019).

However, that has not stopped various global inventions of digital mental health interventions and devices that can aid in better awareness and building a robust social support system globally. In today's age, interventions for specific mental health disorders, wearable devices, virtual therapies, and preventative apps for yoga and meditation have led to the democratisation of digital tools. Preventative apps can also make life easier for users with chronic illnesses, with frequent reminders for medication and symptom checking fostering self-management. Digital communities have worked towards connecting individuals with similar experiences, extinguishing the stigma around mental illnesses, something that traditional communities would have taken decades to accomplish (Twenge et al., 2022).

Telehealth initiatives have started being used worldwide, incorporating diverse individuals, surpassing geographical barriers and catering to individuals with busy schedules, making therapy flexible and user-friendly. Telehealth also allows users with accessibility or age-related difficulties to easily access medical practitioners without having to travel miles at a stretch. The same also allows remote patient monitoring and patient interaction without putting a strain on either the patient or practitioner, even allowing for the development of virtual wards in the future for non-emergency cases (Abernethy et al., 2022).

Wearable devices like sleep trackers, blood pressure detectors, and biofeedback tracers have started developing, and play a very important role in integrating both mental and physical wellbeing, which is essential to look at an individual holistically. These tools can be further utilised in preventative healthcare, helping people keep track of their health markers even before they're able to seek medical advice. The same could also make the work of overworked and underfunded practitioners easier (Abernethy et al., 2022) as these wearable devices, like pedometers, can also make symptom tracking easier, giving more accurate recollections so that doctors can offer tailored suggestions and treatments. Similarly, AI tools can be used for screening purposes in cases like MRIs and CAT scans, alongside human expertise, to expedite the process and catch potential risks even earlier (Alowais et al., 2023).

Kim (2024) carried out a meta-analysis to assess the effectiveness of AI chatbots and apps in addressing women's health, especially in a medical system which is often biased against their problems, often leading to delayed treatment and misdiagnosis. They were positively correlated to women's health, whether related to menopause, menstrual and hormonal health, reproductive and sexual health, or even cancer care and detection, highlighting their importance in the coming years.

Collaborations between public and private mental health and technology companies are fostering ethically sound and clinically validated digital interventions that are easy and safe for the general public to use. Lastly, the development of multilingual interventions ensures that countries with diverse populations are not overlooked in the inclusion of digital mental health initiatives (Ogugua et al., 2024).

Indian Trends in Digital Wellbeing

The current trends of digital initiatives in low to middle-income countries like India have taken significant strides towards digital mental health in their capacity. The COVID-19 pandemic played a huge role in making people aware that mental health services can go on even without physical proximity (Basavarajappa et al., 2022). The pandemic was a strong force in bringing out telemedicine and other digital initiatives in India. This brought forth the need for guidelines to practice telepsychiatry, which was established soon after the pandemic. To this date, however, practitioners prefer in-person consultations but are open to practising digital interventions like teleconsultation and telemedicine. A study found that 19% of the patients were open to tele-consultations compared to 2% of those who were not. This increase in demand could be a response to the usage of high-end digital devices and software for teleconsultations. However, the disadvantage of limited accessibility to mobile phones continues to prevail in remote areas (Basavarajappa et al., 2022).

While telepsychiatry has gained momentum post the COVID-19 pandemic, its efficiency depends not merely on technological advancements, but also on how well technology can replicate human warmth and connection. A study conducted on human mediation found reduced emotional and social responses amongst participants as a response to automated messages. This kind of rational thinking approach is useful if the clients wish to talk about sensitive and confidential issues like abuse or intimate partner violence. However, to gain trust and compliance from the user, it is advisable for automated chatbots to have human-like warmth and empathy (Chatterjee et al.,

2023). Fortunately, as of today, many government and organisational led initiatives have been taken that incorporate human mediation into digital interventions.

National-level initiatives in India have launched various campaigns and apps like *E-Manas* to advocate positive health amongst the young working population. Various digital initiatives have been initiated by several state governments in India that aim to increase the scope of digital mental health services, mainly through psychoeducation, tele-counselling sessions, training of community healthcare providers, and provision of helplines. Moreover, startups have also taken a step towards digital mental health by launching apps or online counselling services that cater to all age groups across all states (Wadhwa, n.d.).

Some Indian chatbots, like Infiheal's *Healo*, or *Wysa,* are trailblazing apps focusing on holistic mental health and offering a novel approach by providing constant emotional support and psychoeducation to their users. The use of generative AI for regular functionality like venting, insights, short-term solutions, long-term recommendations, and guided meditations alongside algorithms being able to factor in diverse modalities, specialisations, and even language preferences showcase the growth in the sector.

Although governments and start-ups have launched various digital initiatives, their implementation remains in the early stages due to countless issues, including regulatory gaps, access to phones, and lack of infrastructure. Moreover, simplification of user interfaces is needed to cater to the non-tech-savvy population of lower to middle-income countries. There are a lot of apps available on the market today catering to digital interventions; however, they need to be clinically and ethically validated to be able to survive in multi-linguistic countries like India (Menon & Varadharajan, 2020).

Hence, the future of digital mental health and the research surrounding it lies not only in the provision but also in the advocacy and inclusion of multilingual, ethically sound, and culturally adaptive digital mental health tools. As a society, we deserve more structured regulatory frameworks for digital mental health. At an individual level, we need to establish a sense of curiosity and willingness to use the existing digital mental health tools, seek professional help, or merely reach out to people to receive or give mental health support through digital platforms.

TYPES OF DIGITAL WELLBEING TOOLS

Technological advancements have taken over every part of an individual's life. Like any other advancement, it comes with pros and cons, with the significant challenge being digital wellbeing. Promoting digital literacy and quality technological resources has become essential to ensure the digital wellbeing of vulnerable groups such as children and adolescents (Pedrouzo et al., 2024). Further, with children's increased engagement in digital play, research has highlighted the need for digital citizenship and self-regulation to ensure positive long-term outcomes (Bittner, 2021).

Unchecked digital habits can lead to chronic stress, sleep deprivation, and declining cognitive function, which, in turn, affect work performance, academic success, and interpersonal relationships. Without proper digital wellbeing strategies, individuals may experience constant connectivity pressure, leading to burnout and reduced engagement. Smart initiatives that promote structured screen time, mindful technology use, and digital detox practices can help restore a sense of balance in this digital-first world (Ben-Zeev & Atkins, 2017).

Various tools have been developed to promote digital wellbeing in multiple ways. One of which is by enabling individuals to manage their use of technology through complete disconnection or nuanced ways to manage connectivity (Nguyen, 2021). These tools promote digital wellbeing in everyday life through features such as self-regulation and monitoring, as well as supporting mental health issues through interventions and crisis management support. Research has indicated that digital wellbeing tools promote workplace wellbeing through better psychological needs satisfaction, a greater level of work wellbeing, and less burnout (Poirier & Gelin, 2024).

Digital Disconnection Tools

Modern devices often offer features tailored for unplugging from digital environments, as research indicates that excessive engagement with digital platforms, particularly social media, is linked to increased anxiety, depression, and sleep disturbances. Android's bedtime mode allows users to turn off notifications at night, while the work profile feature helps separate work and personal life by unplugging at the end of the day. Apple's focus feature similarly enables users to customise notifications and regulate access to different apps and content based on their current activity. These features have elicited positive responses and have been reported to help individuals improve their digital wellbeing (Stanphill, 2019). In addition, various apps like *StayFree, ScreenZen* and more can be used to limit screen time and app usage.

Usage Monitoring and Self-Awareness Tools

Balance mode features to track device usage are provided by both major platforms. Android's balance mode offers features to track phone usage and set app timers to monitor digital presence. Similarly, Apple provides access to screen time data, allowing users to analyse patterns of their device and app usage (Chantelle,

2023). These features help promote self-awareness, urging users to avoid addiction or overuse of specific apps and regulate screen time.

Apps such as SPACE Break Phone Addiction and Google Digital Wellbeing (GDW) focus on promoting digital wellbeing through user self-awareness. They use the theory of planned behaviour and nudge theory to encourage users to set goals, track their usage, and change behaviour patterns. Users felt a greater sense of perceived control and benefited from features such as visualization to regulate their digital usage (Almourad et al., 2021).

Digital Self-Control Mechanisms

A review by Roffarello and Russis (2023) identified 41 tools used for digital self-control, including Freedom, Let's FOCUS, and HabitLab. These tools were available on smartphones or PCs, and used a variety of delivery platforms, including mobile applications and browser extensions. These tools used strategies such as app blocking, device usage statistics, time limit restrictions, lockout mechanisms, persuasive messages, and negative reinforcement to enable users to regulate their device usage.

Sites such as YouTube offer modes to promote user focus. The *SwitchTube* mobile app was developed to enable users to toggle between focus mode and recommendations mode, simultaneously encouraging self-control and exploration. The initial research on its utility indicated that users experience greater agency, satisfaction, and goal alignment (Lukoff et al., 2023). Hence, self-regulation can also be promoted through such adaptable commitment interfaces in contrast to indiscriminate blocking and locking strategies

Parental Control and Family Management

Android's *Family Link* app allows parents to monitor and regulate their child's access to apps and websites, even providing the option to remotely lock the child's phone (Android, n.d.). Apple offers similar parental controls through its Screen Time features, which can be configured to limit access to certain apps and content. These tools help create age-appropriate digital environments and establish healthy technology habits from early childhood.

Mental Health and Wellbeing Support

The National Institute of Health Research (NHS) also launched *InHand*, a free-to-use mobile application to promote mental wellbeing among young people. It aimed to help users regain balance and regulate their emotions based on their mood. Research indicated that users found the app helpful in improving self-awareness and self-monitoring of their moods (Simons et al., 2015). Apps like *Healo, Wysa,* and *Woebot,* alongside many others, can offer chatbots to assist users in talking about their issues. Furthermore, apps like *Calm* and *Headspace* are able to assist users in more symptomatic management, such as dealing with panic attacks. Apple's "My Wellbeing Toolkit" has been proposed to promote student wellbeing using iPad features such as Pages, Keynote, GarageBand, and Freeform (Coby, 2023). These creative applications can be harnessed to express emotions and develop coping strategies.

Healthcare Integration and eHealth Tools

The upcoming use of eHealth tools is reforming the healthcare domain by providing multi-domain, multi-informant integrated services. These tools deliver interventions, provide education, and facilitate communication between clients and professionals. A systematic review by Stewart et al. (2022), which included 39 papers, reported that such tools were used to track mental and general health across outpatient health clinics, hospitals, community outreach, or a combination of these settings. Although these tools are still in the basic stages of usage, they can be harnessed to offer the much-needed support to vulnerable populations across rural and urban regions and longitudinally.

Babbott and Serlachius (2023) highlighted that digital tools can offer essential psychological support for individuals with chronic conditions such as diabetes. They can help promote equity of services by combating cost, stigma, and accessibility barriers. The *EmpowerKids* tool enables professionals to discuss health and wellbeing with children from low-income families. Research indicated that the tool provided professionals with an understanding of the child's overall health and information about their individual needs (Skogberg et al., 2022).

Therapy Extension and Virtual Care

Digital tools can be used as therapy extenders and virtual care platforms to support wellbeing. Skills learnt within the therapeutic setting can be practised through these apps to support individuals outside the setting as well as make the therapy sessions more effective. Further, virtual care platforms such as *Brightline* can be used to offer online therapy support in remote locations and save travel costs (Schueller & Histon, 2023).

These tools can also be used for self-management to promote health and wellbeing with or without professional support (van Olmen, 2022). Finally, digital tools could offer self-guided approaches, appropriate linking to services, and customisation of services based on the specific needs of the user, even in cases of severe distress (Balcombe & de Leo, 2022).

ADVANCES AND BENEFITS OF DIGITAL WELLBEING TOOLS

The increasing integration of technology in daily life has led to the emergence of smart and digital wellbeing initiatives, which leverage artificial intelligence (AI), machine learning, mobile applications, and digital platforms to enhance physical and mental health. These initiatives encompass digital health tracking, smart wearables, AI-driven chatbots, and online mental health resources. The shift towards digital wellbeing solutions has been accelerated by advancements in health informatics, user-centred design, and behavioural science (Bucci et al., 2019).

Enhancement of Mental Health and Psychological Wellbeing

Digital wellbeing initiatives have significantly transformed mental health care by increasing accessibility, scalability, and effectiveness in delivering psychological support. AI-driven tools and mobile applications provide immediate and personalised interventions, addressing barriers such as stigma and lack of access to traditional therapy (Bucci et al., 2019). Research suggests that digital mental health interventions have been particularly beneficial in reaching underserved populations, reducing disparities in mental healthcare availability (Naslund et al., 2017).

AI-powered chatbots have gained prominence in providing psychological assistance through real-time conversations, offering coping strategies, psychoeducation, and emotional support (Zaky, 2023). Studies show that chatbot-based interventions can alleviate symptoms of anxiety and depression by engaging users in structured therapeutic conversations and guiding them through evidence-based exercises (Fitzpatrick et al., 2017). Furthermore, chatbots facilitate a sense of anonymity, encouraging individuals to seek help without fear of judgment, increasing mental health service utilisation (Miner et al., 2019).

Mobile-based cognitive-behavioural therapy (CBT) applications have experienced clinically significant improvements in mental health outcomes comparable to traditional face-to-face therapy. These applications also promote long-term adherence by allowing users to engage with therapy at their own pace, which enhances treatment effectiveness (Linardon et al., 2019). Digital mood-tracking tools have become essential in detecting early signs of mental health deterioration. These applications use machine learning algorithms to analyse behavioural patterns, including sleep, activity levels, and social interactions, enabling early intervention (Aggarwal et al., 2023). Moreover, digital self-monitoring tools have been particularly effective in workplace settings, where they help employees manage stress, track mood fluctuations, and implement self-care practices (Lang, 2024).

The integration of digital wellbeing initiatives in workplaces has been associated with improved productivity, reduced absenteeism, and enhanced job satisfaction

(Paganin & Simbula, 2020). Studies indicate that workplace digital interventions, such as mindfulness applications and AI-driven stress management tools, reduce burnout and improve employee emotional resilience (Joyce et al., 2016). Digital interventions also provide organisational benefits by fostering a culture of mental health awareness and support, leading to better overall employee wellbeing (Stratton et al., 2017).

Personalisation and Adaptive Interventions

One of the most significant advantages of digital wellbeing initiatives is their ability to deliver personalised and adaptive interventions. Digital platforms leverage real-time data analytics to tailor interventions according to individual needs, improving both effectiveness and user engagement (Deniz-Garcia et al., 2023). The integration of machine learning algorithms allows these platforms to continuously analyse user data, ensuring that recommendations are dynamic and responsive to changing health conditions.

Wearable health devices, such as smartwatches and fitness trackers, play a crucial role in personalised health monitoring. These devices continuously track physiological parameters such as heart rate variability, sleep quality, and activity levels, allowing for timely health recommendations based on real-time insights (Ajagbe et al., 2024). AI-powered chatbots further enhance personalisation by adapting their responses based on user input and demographic features, making interactions more relevant and supportive. Individuals interacting with adaptive AI chatbots experienced greater satisfaction and emotional relief compared to those using static, pre-programmed interventions (Sharma et al., 2023). The impact of hyper-personalised digital interventions extends beyond short-term engagement; research indicates that individuals who receive customised health recommendations are more likely to make sustainable behavioural changes. Personalised digital interventions have also been associated with increased motivation and self-efficacy, as individuals feel more empowered when recommendations align with their unique needs and preferences (Honka, 2024).

Improved Accessibility and Inclusivity

Traditional healthcare services are often constrained by geographical, financial, and infrastructural barriers, making it difficult for individuals in remote or rural areas to access timely medical and psychological care. Digital health interventions, including telemedicine services and AI-driven self-help tools, bridge these gaps by offering virtual consultations, real-time symptom monitoring, and digital therapeutic interventions (Hilty et al., 2021). The adoption of telehealth services has significantly increased healthcare access for populations with limited mobility, reducing disparities in mental and physical healthcare availability (Kruse et al., 2020). Many digital wellbeing initiatives incorporate assistive technologies such as voice recognition, text-to-speech functionalities, and haptic feedback to ensure a more inclusive user experience (Chalkiadakis et al., 2024). AI-driven accessibility features significantly improve the usability of digital mental health resources for individuals with visual and auditory impairments, increasing engagement and adherence to treatment (Lillywhite & Wolbring, 2022). Furthermore, mobile health applications designed with adaptive interfaces allow users with cognitive disabilities to customise their experiences, enhancing self-management of health conditions (Grau et al., 2022).

Online mental health services provide anonymous and cost-effective support, reducing the fear of judgment and encouraging help-seeking behaviours among vulnerable populations. Mobile-based mental health interventions significantly improve psychological outcomes for individuals in low-income communities by providing free or low-cost access to cognitive behavioural therapy (CBT), mindfulness exercises, and crisis support resources (Naslund et al., 2017). Additionally, AI-powered chatbots and peer support networks available on digital platforms offer emotional assistance to individuals who may otherwise lack social support (Fitzpatrick et al., 2017).

Increased User Engagement and Motivation

The integration of gamification and interactive elements in digital wellbeing applications has significantly enhanced user engagement and motivation. Incorporating features such as progress tracking, goal setting, and rewards fosters sustained participation in health-promoting behaviours by making the experience more engaging and enjoyable. Gamification leverages psychological principles of motivation, such as reinforcement and competition, to encourage users to maintain their wellbeing goals. Individuals using health apps with gamified elements reported higher adherence to fitness routines and mental health exercises compared to those using non-gamified alternatives (Koivisto & Hamari, 2019).

Mobile fitness applications that integrate social challenges and virtual rewards have been particularly effective in increasing physical activity levels among users. These features tap into social influence and peer accountability, motivating users to engage in healthier behaviours. A large-scale study demonstrated that participants who engaged in social challenges within fitness apps were more likely to achieve their daily step targets and sustain increased activity levels over time (Zhang et al., 2021). Furthermore, research suggests that leader boards, badges, and achievement notifications create a sense of accomplishment, reinforcing positive behavioural change (Liu et al., 2023). Personalised notifications, which adjust based on user behaviour, have been shown to reduce dropout rates in mental health and fitness programs by maintaining user motivation and accountability (Perski et al., 2019). Moreover, apps that provide real-time feedback on physiological and psychological metrics empower individuals to take greater control of their health (Ryan, 2022). The users who actively tracked their mood and behavioural patterns through mobile applications reported greater self-awareness and engagement in self-care activities (Torous et al., 2021).

Data-Driven Insights and Preventive Healthcare

One of the most significant advantages of digital wellbeing initiatives is their ability to harness large-scale health data for preventive healthcare. Wearable devices, mobile health applications, and AI-driven analytics enable the continuous monitoring of physiological and behavioural patterns, allowing for the early detection of potential health risks. Predictive analytics play a crucial role in digital health platforms by identifying individuals at risk of developing chronic conditions before symptoms become clinically significant. Machine learning algorithms analyse historical health data to detect subtle deviations from normal physiological patterns, allowing for early diagnosis and targeted interventions (Rajkomar et al., 2019).

Furthermore, AI-powered health monitoring systems provide real-time feedback to both users and healthcare professionals, enhancing the efficiency of disease management. AI-driven diabetes management apps demonstrate improved glycaemic control due to real-time feedback and tailored dietary recommendations. Similarly, digital mental health platforms that track mood fluctuations and sleep patterns can alert users to early signs of depressive episodes, prompting timely psychological interventions (Thirupathi et al., 2025). Continuous monitoring through digital tools enables early intervention, which significantly improves long-term health outcomes and reduces the burden on traditional healthcare infrastructure. The patients who received continuous digital health monitoring experienced a 30% reduction in hospital readmissions due to timely medical interventions and personalised health recommendations (Kashani et al., 2023).

Workplace Wellbeing and Productivity Enhancement

Digital platforms provide employees with access to a wide range of wellness resources, including mental health support systems, stress management programs, and virtual wellness coaching. One of the most impactful applications of digital wellbeing in corporate settings is the use of AI-powered mental health chatbots. These chatbots provide real-time psychological support, stress assessments, and personalised coping strategies, enabling employees to manage work-related stress more effectively (Singh, 2024). Employees who utilise digital mindfulness and meditation applications exhibit improved focus, enhanced emotional regulation, and increased work efficiency (Mitsea et al, 2023).

Additionally, the integration of wearable technology in workplace wellness programs has provided valuable insights into employee health and ergonomics. The benefits of digital wellbeing initiatives extend beyond individual employee health, contributing to overall organisational success. Companies that invest in workplace wellness technologies report lower healthcare costs, increased employee retention rates, and improved workplace morale (Raqeeb, 2024). Furthermore, corporate wellness programs that integrate AI-driven behavioural health analytics help HR professionals identify early signs of burnout and implement targeted interventions, leading to more sustainable workforce management strategies (Itie, 2023).

Smart and digital wellbeing initiatives offer transformative benefits across multiple dimensions of health and wellness. From enhancing mental health support and personalised healthcare to increasing accessibility, user engagement, and preventive care, these initiatives revolutionise the way individuals manage their health. The integration of AI-driven analytics, wearable technology, and digital interventions fosters a proactive approach to wellbeing, enabling users to make informed health decisions. However, ethical considerations regarding data privacy and AI biases must be carefully managed to ensure sustainable and responsible implementation.

CHALLENGES AND ETHICAL CONSIDERATIONS OF DIGITAL WELLBEING TOOLS

There are various uses and benefits of smart and digital initiatives to promote wellbeing as the world gets increasingly digitised. AI-backed tools in particular have come to play a significant role in developing these tools and reducing the load on service providers, as we have explored. Despite their many advantages, these tools and initiatives are still relatively new and still have a lot of limitations. This is particularly true for AI-backed mental health chatbots, which provide a crucial service that needs to be carried out with the appropriate oversight to promote the principle of non-maleficence and beneficence for users. As they currently exist, these tools come with many limitations and challenges, whether they be technical, legal, or ethical.

Limitations and Risks in User Interaction

Regardless of the kind of AI-based chatbot, whether it relies on prescripted responses or generative AI, each comes with its drawbacks. The responses provided can often have issues with accuracy, trustworthiness, and actual helpfulness, which can greatly influence user interaction. If the dataset that the chatbot is drawing from is flawed, the responses produced could be flawed or contain inaccurate information (Boucher et al., 2021). While generative AI can have independent and dynamic conversations, and most development teams will have professionals attempting to vet these conversations, there still lies the potential for risk due to the sheer scale of data being processed (Sedlakova & Trachsel, 2022).

The potential for chatbots using generative AI to say the wrong thing is always present, which could be due to a lack of empathy in responses, a lack of nuance, or even misleading information provided by the users. According to some, they may be closer to a self-help book as compared to a genuine therapeutic or medical conversation (Denecke & Househ, 2021). Despite best intentions, faulty responses by chatbots may potentially cause harm as well, as they may reinforce or encourage harmful behaviours, particularly in vulnerable populations (Coghlan et al., 2023). There is also the risk of this harm being irreversible, like the case of a Belgian woman who blamed a bot for her husband's suicide after conversations with the bot about climate change and eco-anxiety (Tamim, 2023) or when Character AI was implicated in a teen's suicide (Rose, 2024).

It may be possible for scripted conversations to avoid these pitfalls, but they come with their limitations. Due to following a specific structure for their conversations, they tend to be repetitive and lack the ability to respond organically to users, leading to frustration and reducing retention (Sedlakova & Trachsel, 2022). An alternative to avoiding the pitfalls of generative conversations could be human oversight over the conversations, but based on the scale and number of conversations from one chatbot to the other, it is often difficult to carry out feasibly (Coghlan et al., 2023).

Inability to Replicate Therapeutic Support

While these dynamic conversations are capable of providing a certain level of emotional support and strategies, they cannot fully replicate a therapeutic conversation. This is in part due to the lack of individual rationality and morality of the chatbots, which play a very important role in building therapeutic bonds. The inability to modify the strategies or suggestions to meet individual needs is a major setback to building a genuine therapeutic alliance (Coghlan et al., 2023). While chatbots may be great in certain situations, if the users desire to gain a new level of self-understanding or build a rapport to encourage vulnerability, they will fall short. Expecting human-like interaction and then not receiving it can make people feel worse at a moment when they hope to gain support. People need to be constantly reminded that they are speaking to a bot to ensure that they do not expect human interactions (Sedlakova & Trachsel, 2022).

Misunderstandings with chatbots are a common feature of conversations, especially when long, complex messages are sent, often with metaphors or colloquialisms, leading to stilted or restricted responses. This situation is further worsened in different languages, as many nuances of the language might get missed leading to very technical or academic responses (Boucher et al., 2021). Due to the inherent nature of communication, many non-verbal cues can be missed, which might have hinted at key psychological or physiological insights (Khawaja & Bélisle-Pipon, 2023), further emphasising why these tools cannot be seen as a replacement for therapy. They are useful in breaking down coping strategies and techniques into easy-to-follow steps, but not all aspects of psychological or medical care plans are easy to break down into simple steps (Coghlan et al., 2023). While these chatbots and apps may be able to identify or offer some coping solutions for certain issues, they may lack the inherent knowledge and experience required to offer supplementary information or reassurance to users, especially in the case of more complicated or dual diagnoses (Burr et al., 2020).

Simply equipping the general population with psychometric tools and scales without the corresponding clinical knowledge can lead to misconceptions about diagnosis and what could help, worsening outcomes for individuals (Khawaja & Bélisle-Pipon, 2023). It can also be difficult for chatbots to keep track of everything mentioned or discussed over multiple user sessions or reference back to something mentioned previously to promote psychological insight, something therapists and medical professionals frequently do (Denecke & Househ, 2021).

Managing Crisis Situations

AI-backed chatbots particularly fall short in crisis situations and find it difficult to assess and respond appropriately. Issues of self-harm, suicidal ideation, and disclosure of abuse need to be handled by tonal assessments and perception of nuance where they may fail (Balcombe, 2023). Chatbots may not be able to make executive decisions on how to safeguard or when to trigger emergency responses, especially if users are obfuscating or not being completely forthcoming about their safety, as can often be the case in crisis situations. These apps might also lack the clinical knowledge to recognise an innocuous symptom as medically relevant (Zawati & Lang, 2024). Often, users might attempt to discuss suicidal or self-harm ideations or intrusive thoughts in an attempt to achieve therapeutic benefits, it might be difficult for the bot to differentiate between expressing thoughts versus actual intent of harm, possibly depriving users of that avenue of disclosure. Finally, at-risk situations are identified through algorithms, and the inability to differentiate might lead to a lot of false positives, along with a reduction in the user's desire to disclose in the future (Burr et al., 2020).

The Role of Risks, Stigma, and Bias

A major issue arises with the risk of accidental disclosure. It is easier to communicate with an AI-based chatbot or use a digital health app, especially regarding stigmatised issues due to the relatively low personal and social risk involved. It may be possible that the comfort produced by not talking to a human might lead to accidental disclosure of crimes, including child abuse or sexual violence. While therapists and medical professionals have dedicated guidelines outlining the situations where mandatory reporting is required, a chatbot is not bound to those same guidelines, which brings up the issue of both social and legal liability (Coghlan et al., 2023).

An important feature of mental health chatbots and digital health apps is the ability they give users to actively track their symptoms and mood in their daily lives so their health providers have a better understanding of their struggles. While these frequent reminders can prove useful for some, they may also serve as unpleasant reminders to people about their mental or physical illness, especially if the person is having a relatively symptom-free day (Bauer et al., 2017). The tendency to track symptoms might also increase stigma by perpetuating the perception that those with mental illness need to be constantly tracked and monitored by those around the users if they come to know about their use of the app, especially if someone has a more stigmatised diagnosis (Ahad et al., 2023).

As mentioned, these AI-driven chatbots are frequently trained on large datasets from multiple sources. It is possible that they might only reflect the experiences of the majority, losing out on knowledge regarding the experiences of minority communities (Coghlan et al., 2023). The chatbot or app may be likely to miss out on linguistic, cultural, ethnic, or racial nuances in the presentation of symptoms and experiences, as well as the suggestions that it offers. Even practitioner-matching apps might fail to adequately match the correct therapist or medical professional to those with protected characteristics, increasing the pervasive bias in healthcare (Husain et al., 2022). Furthermore, due to potentially biased datasets, it is also possible that the chatbot or app may end up perpetuating harmful stereotypes and assumptions (Denecke & Househ, 2021). As it is often those populations who have difficulty accessing traditional mental health support or adequate healthcare, this bias would further isolate them or make them hesitant to access support (Khawaja & Bélisle-Pipon, 2023).

The Dangers of Over-Reliance

One of the major features, which is the constant accessibility and 24/7 support offered, could also create future complications. It can often create an over-reliance on the AI-based chatbot or app, which might lead to users withdrawing from traditional mental health support or isolating themselves from loved ones because they feel they are being supported by these tools. People might choose not to see a doctor or nurse if the app assuages their worries (Bucci et al., 2019). The interaction is easier, often with lower levels of anxiety and higher levels of disclosure, which might increase the resistance to accessing alternative support (Sedlakova & Trachsel, 2022). However, these interactions are not a true replacement for individual, personalised professional support or personal relationships. This over-reliance to the detriment of in-person support can worsen isolation and loneliness, not to mention health outcomes (Khawaja & Bélisle-Pipon, 2023). It is also necessary to assess what is driving the over-reliance on digital platforms. If it stems from avoidance behaviour, this dependence could further worsen outcomes for users, especially in cases of social anxiety (Bucci et al., 2019).

Lack of True Accessibility

Digital wellbeing tools can often have different outcomes for those with differing mental illnesses, and research and companies need to acknowledge this and reflect the same in their work. The impact of the tool may differ for someone struggling with depression versus someone with schizophrenia, and the chatbots or apps are often not as helpful for more severe conditions unless specially designed (Husain et al., 2022).

While these tools aim to increase accessibility, especially for vulnerable populations, their user interfaces may make it difficult for the geriatric population or those with learning difficulties or neurodegenerative conditions to use (Rodríguez-Martínez et al., 2023). The information about conditions or coping strategies may often be presented in increasingly technical terms or presuming a certain level of knowledge in the user which could make comprehension difficult, especially with a rural or economically weaker population. While these issues can often be resolved by subsequent questions and interactions, too technical a response might deter people from future usage, especially as many people might not be willing to engage in those follow-up conversations (Burr et al., 2020).

The effectiveness of digital tools is inextricably linked to the amount of digital literacy an individual possesses and their general beliefs about the use of technology, creating the potential for significant generational and cultural differences, with younger generations from more urbanised settings more likely to make use of these tools (Boucher et al., 2021). Older generations also carry a greater distrust towards technology and AI, especially in the healthcare field, which might make their usage levels low (Rodríguez-Martínez et al., 2023).

Privacy Struggles and Absence of Regulations

The lack of existing guidelines and regulatory bodies can enable companies to take advantage of the lack of security for the privacy of users. Their data could be sold to third-party organisations and bodies, violating the inherent need for confidentiality in healthcare, as they are not held to the same standards of professional conduct as mental health or medical professionals (Husain et al., 2022). The lack of involvement from regulatory bodies also means that many tools are not up to medical standards and do not necessarily reach the people who might need support the most due to limited outreach and access to information (Wies et al., 2021). Many available apps are not evidence-backed, and it can be difficult for the average user to differentiate those from genuinely helpful tools, especially for mental health, when most apps are marketed as a replacement for therapy (Denecke & Househ, 2021).

While apps claim to anonymise their data, this can often be de-anonymised through data triangulation. Due to the lack of regulations regarding the storage of data, this could be hacked into and leaked by cyber-criminals, publicising medical records and further exacerbating the stigma associated with accessing mental health support or certain medical conditions (Coghlan et al., 2023). These privacy issues also might lead to mistrust in users, not just in the tool itself but the medical professional who suggested it, damaging the working relationship and worsening health outcomes (Wies et al., 2021).

THE WAY FORWARD

While we have explored the various issues and challenges that exist in the current use of AI-based chatbots and digital tools in promoting wellbeing, we also need to acknowledge the important role they play. The tools exist and are being utilised because they are needed in society. They can address and reach vulnerable populations and provide mental health support and healthcare to many who need it but are unable to access traditional support. It is also undeniable that there is a dearth of mental health and healthcare providers. This gap in support, particularly mental health support, cannot be made up by people alone, especially not when we consider round-the-clock support. Digital tools and AI-driven chatbots can greatly help in these situations and ease the burden on professionals while providing crucial support. However, the existing issues need to be addressed and dealt with while moving forward.

Collaborative Development

If we wish to extensively use AI-backed tools as a collaborative tool to provide mental and physical healthcare and support, it becomes necessary to include diverse voices in their development. Doctors, nurses, psychiatrists, psychologists, and other professionals need to be extensively involved in vetting the datasets and responses of the tools. Extensive studies need to be carried out to ensure that they are performing the task they were designed for appropriately, with no harm to the user (Khawaja & Bélisle-Pipon, 2023).

Companies need to include not only professionals but also service users and patients in the development and deployment of these tools to ensure that they are meeting user needs in a manner that is most needed by them (Sedlakova & Trachsel, 2022). In addition, key stakeholders such as families, caregivers, or specific populations need to be included in the process. Additionally, the United Nations Convention on the Rights of Persons with Disabilities (UNCRPD) makes it clear that people with disabilities, both physical and psychological, have to be involved in the development of these tools (Carr, 2020).

There needs to be constant scrutiny and transparency in the development process, and the rationale behind decisions needs to be fully explored to ensure that they are the most effective tool. As datasets tend to cater to the majority, which might lead to biases in the AI-driven chatbot or app, this deficit needs to be made up for. It is necessary to include diverse voices in the development of these tools to ensure that we can make up for any implicit biases in datasets and ensure all populations will be adequately supported, whether that might be ethnic, linguistic, gender, or sexuality minorities (Khawaja & Bélisle-Pipon, 2023).

Uniformity in Standards

A key step forward is ensuring that there are national and international guidelines and frameworks for the management of these digital initiatives. International and national guidelines should be set up for uniform standards for these tools and apps, with allowances for cultural variations to ensure the maximum number of people can be reached from all different sectors of society (Khawaja & Bélisle-Pipon, 2023). As of the moment, development is primarily done with technological companies with an interest in mental health or accessible healthcare rather than health professionals directly, influencing the manner in which these tools are developed. There needs to be a global consensus on what apps and tools should be qualified to provide mental health support or medical assistance, and what standards need to be met, otherwise, people might push their apps into less regulated areas and societies. There also needs to be clear guidelines on issues of ethical and legal liability to ensure minimal harm to users (Balcombe, 2023). Legal advisors and cybersecurity experts need to be on board to ensure that legally and regulatory robust systems are put in place to protect both the users and the product (Hall et al., 2024).

Appropriate Marketing and Usage

A major issue is the marketing of mental health tools due to it being a less-regulated field. There are numerous apps and websites which promote themselves as a solution for people's mental health worries and promote themselves as offering solutions. The associated marketing of these tools as 'replacement for therapy' is part of the problem. It is necessary that adequate measures are taken to ensure that people are fully aware of the potential of these tools and are not misguided into assuming their role, a statement which holds true for medical apps as well (Khawaja & Bélisle-Pipon, 2023). This marketing can also lead people to assume that these bots and apps will value anonymity and confidentiality in the same manner as real-life therapists or medical professionals, which would not be the case. Platforms need to carefully monitor the apps they are allowing and how they are being marketed. Individual apps and websites should highlight how their tools have been designed and validated to allow users to make informed decisions (Wykes et al., 2019).

Increasing User Awareness and Expanding Use

Users should also be psychoeducated on the precise tools and support these apps can provide, as well as the restrictions and limits of their support, so that they know what to expect. Companies should make it clear how they differ from traditional forms of mental health support, allowing users to make educated decisions about the decision to engage with the tool and under what circumstances to do so (Balcombe, 2023).

A key factor is that these apps and tools cannot function as replacements for traditional support but can function as collaborative tools, easing the burden on professionals or providing users with support that might be more difficult to find elsewhere. In addition to providing accessible and 24/7 support, these tools could also be used to ease the burden on practitioners (Denecke & Househ, 2021). They can be used for more administrative tasks, whether tracking symptoms to ease the load of history taking or tracking case notes and prescriptions, allowing the practitioner to spend more time supporting the individual (Balcombe, 2023). Tools can be further used to assist in diagnosis by keeping a track of symptoms and experiences which might be missed in human error and allowing for easier and earlier cross-diagnosis and screening (Alowais et al., 2023).

Enhancing Accessibility and Engagement

While digital health interventions are primarily favoured for the anonymity they offer, they can also make it easier for people to connect with each other. Online support groups or forums could reduce the level of isolation and struggle users might be facing in coming to terms with their health struggles, particularly mental health. This is particularly true for those in more reclusive or isolated areas or belonging to societies which still hold considerable stigma towards seeking mental health support (Bucci et al., 2019). Traditional support groups may be difficult to access due to limited availability or anxiety towards social interaction, something that online support could ease while still allowing people to form connections, reducing the reliance on the AI-based chatbot or app as well.

Incorporating these features could allow users to access more holistic mental health support and self-management through these tools. They could gain timely access to different kinds of support they might need. This could include interactive psychoeducation designed by professionals to ensure they understand their struggles, using CBT-informed techniques or coping skills through the app, communicating with the chatbot to gain emotional support, and even connecting to other people to share and seek advice and experiences. It could allow users with specific medical diagnoses to connect with others and hear survivor stories as well (Bucci et al., 2019). Permitting telehealth appointments could also allow people from remote areas to directly access professional expertise at little to no personal expense or hassle, thus improving remote patient monitoring (Alowais et al., 2023).

These tools should also be able to connect people to in-person therapists or mental health support to ensure that they have resources available even if they have chosen to opt out of online support (Khawaja & Bélisle-Pipon, 2023). This could be carried out through collaborations with hospitals, clinics, NGOs, and other individual bodies to ensure that people can access support regardless of their usage of these tools, as people are often unaware of the extent of resources available.

Optimising User Interface and Interaction

A key to making these digital tools more effective is to ensure that it does not feel like reading a textbook or being in a lecture. Interactive and aesthetically appealing content, when delivered through the use of relatable situations or stories, makes it more engaging to interact with and easier for people to understand what they are going through, especially younger generations. It is also important to make the interface easy to use, appealing, and with minimal technical glitches to ensure that people wish to use the tool (Garrido et al., 2019). An easy interface would not only make it easier for older generations and those with low digital literacy but would also be easier to use when users might be in a mental health crisis (Rodríguez-Martínez et al., 2023). For the same, experts, whether they be technical, pedagogical, social workers or support workers, could come together to develop the most effective manner in which to disseminate information.

Improving Security and Privacy Safeguards

A key question is what happens to a user's data if they decide to stop using the app or AI-based chatbot but are concerned about the information they have shared with it. It should be possible for users to fully withdraw from these apps and delete their data upon withdrawal to ensure privacy and safety. This would also increase people's faith in these systems and increase their usage (Khawaja & Bélisle-Pipon, 2023).

As privacy and potential data leaks are such a major concern due to the sensitive health information they handle, apps could include multi-factor authentication to ensure privacy and limit any unauthorised access to information. End-to-end encryption could be used to reduce the potential risk of any data leaks (Banerjee et al., 2024). While it might not be possible to have constant oversight over every interaction, there can be certain trigger points, whether too much time on the app or website or the mention of specific trigger points, where clinicians could be encouraged to provide oversight and make judgment calls. The same should be communicated to users to ensure they can make an informed decision about their usage of the tool, reducing the safety risks associated with the app (Carr, 2020).

Industry-standard security protocols would have to be implemented for these digital tools so that their systems are fully secure. These would also need to be frequently patched and updated to ensure that any discovered vulnerabilities are addressed and fixed (Carr, 2020). Data protection could be carried out through the use of experts in consumer and privacy correction (Coghlan et al., 2023).

Future Research Direction

More qualitative studies need to be carried out on existing apps and AI-driven chatbots while developing new ones to ensure frequent feedback on how they are currently functioning and ease issues regarding engagement and accessibility. There's still a gap in the existing literature about the effectiveness of specifically designed mental health tools versus regular digital tools. This could be done by identifying the needs of the population, current barriers to usage, and user experiences regarding how these tools have been impacting the current healthcare landscape. Longitudinal studies could be conducted to assess which populations and mental health conditions are best suited to using AI-based chatbots (Hall et al., 2024).

It is necessary to conduct constant research into the efficacy or risk of harm caused by AI applications and for them to be constantly monitored for any unusual activities or interactions. Human experts, whether in practice, policy, or research, can collaborate with developers to ensure that safety measures and regulations are in place, especially as there is sensitive medical and personal data involved (Balcombe, 2023).

More cross-cultural research needs to be carried out on the effectiveness of digital wellbeing tools in different societies with different attitudes towards and understanding of both mental health and accessing traditional healthcare. Most existing tools are developed in Western and first-world countries, and their conceptions of mental health or certain diagnoses might not translate to all cultures. Definitions and understanding of terms like 'depression' can vary from culture to culture, as can the techniques and strategies suggested. Research into culturally adapted and non-adapted tools could give insight into their efficacy.

It often proves difficult for developers to assess their product accurately once it is implemented in healthcare settings or larger societies, as an accurate evaluation would require hundreds of participants and evaluators across different societies and with different issues. However, if clinically effective and accessible tools are developed but are not able to be implemented due to issues in evaluating their impact, it is the user who suffers and fails to receive care which would help them. It is necessary to bridge the gap that exists between developed and implemented products, which would require multiple stakeholders to be involved (Hall et al., 2024).

CONCLUSION

As the world gets increasingly digitised, there exist multiple opportunities and tools to meet people's changing needs and enhance their wellbeing. There have been many revolutionary changes in smart and digital wellbeing tools, and mental health AI chatbots have been an emerging trend. While there are many advantages to their use and they are able to fill an ever-growing gap in healthcare, they also have drawbacks. However, what cannot be denied is that they play a crucial role in the future of healthcare and reducing the treatment gap, which makes their effective development and deployment all the more important. Hence, it is imperative to involve several stakeholders and regulatory bodies to ensure that patient needs are being met while maintaining their safety as technology continues to advance rapidly.

REFERENCES

Abernethy, A., Adams, L., Barrett, M., Bechtel, C., Brennan, P., Butte, A., Faulkner, J., Fontaine, E., Friedhoff, S., Halamka, J., Howell, M., Johnson, K., Long, P., McGraw, D., Miller, R., Lee, P., Perlin, J., Rucker, D., Sandy, L., & Valdes, K. (2022). (2022). The Promise of Digital Health: Then, Now, and the Future. *NAM Perspectives*, *6*(22). Advance online publication. DOI: 10.31478/202206e PMID: 36177208

Aggarwal, A., Tam, C. C., Wu, D., Li, X., & Qiao, S. (2023). Artificial intelligence–based chatbots for promoting health behavioral changes: Systematic review. *Journal of Medical Internet Research*, *25*, e40789. DOI: 10.2196/40789 PMID: 36826990

Ahad, A. A., Sanchez-Gonzalez, M., & Junquera, P. (2023). Understanding and Addressing Mental Health Stigma Across Cultures for Improving Psychiatric Care: A Narrative Review. *Cureus*, *15*(5), e39549. DOI: 10.7759/cureus.39549 PMID: 37250612

Ajagbe, S. A., Mudali, P., & Adigun, M. O. (2024). Internet of things with deep learning techniques for pandemic detection: A comprehensive review of current trends and open issues. *Electronics (Basel)*, *13*(13), 2630. DOI: 10.3390/electronics13132630

Almourad, M. B., Alrobai, A., Skinner, T., Hussain, M., & Ali, R. (2021). Digital wellbeing tools through users lens. *Technology in Society*, *67*, 101778. DOI: 10.1016/j.techsoc.2021.101778

Alowais, S. A., Alghamdi, S. S., Alsuhebany, N., Alqahtani, T., Alshaya, A. I., Almohareb, S. N., Aldairem, A., Alrashed, M., Bin Saleh, K., Badreldin, H. A., Al Yami, M. S., Al Harbi, S., & Albekairy, A. M. (2023). Revolutionizing healthcare: The role of artificial intelligence in clinical practice. *BMC Medical Education*, *23*(1), 689. DOI: 10.1186/s12909-023-04698-z PMID: 37740191

Android. (n.d.). *Digital wellbeing*. https://www.android.com/intl/en_in/digital-wellbeing/

Babbott, K. M., & Serlachius, A. (2023). Developing digital mental health tools for youth with diabetes: An agenda for future research. *Frontiers in Clinical Diabetes and Healthcare*, *4*, 1227332. DOI: 10.3389/fcdhc.2023.1227332 PMID: 37497385

Balcombe, L. (2023). AI chatbots in digital mental health. *Informatics (MDPI)*, *10*(4), 82. DOI: 10.3390/informatics10040082

Balcombe, L., & de Leo, D. (2022). The potential impact of adjunct digital tools and technology to help distressed and suicidal men: An integrative review. *Frontiers in Psychology, 12*, 796371. DOI: 10.3389/fpsyg.2021.796371 PMID: 35058855

Banerjee, S., Agarwal, A., & Bar, A. K. (2024). Securing wellbeing: Exploring security protocols and mitigating risks in AI-driven mental health chatbots for employees. *American Journal of Computer Science and Technology, 7*(1). Advance online publication. DOI: 10.11648/j.ajcst.20240701.11

Banh, L., & Strobel, G. (2023). Generative artificial intelligence. *Electronic Markets, 33*(1), 63. DOI: 10.1007/s12525-023-00680-1

Basavarajappa, C., Grover, S., Dalal, P. K., Avasthi, A., Kumar, C. N., Manjunatha, N., Mehra, A., Saha, G., Sahoo, S., Singh, O. P., Tripathi, A., Gangadhar, B. N., & Math, S. B. (2022). Current telepsychiatry practice in India—An online survey of psychiatrists. *Indian Journal of Psychiatry, 64*(3), 307–311. DOI: 10.4103/indianjpsychiatry.indianjpsychiatry_331_21 PMID: 35859562

Bauer, M., Glenn, T., Monteith, S., Bauer, R., Whybrow, P. C., & Geddes, J. (2017). Ethical perspectives on recommending digital technology for patients with mental illness. *International Journal of Bipolar Disorders, 5*(1), 6. DOI: 10.1186/s40345-017-0073-9 PMID: 28155206

Ben-Zeev, D., & Atkins, D. C. (2017). Bringing digital mental health to where it is needed most. *Nature Human Behaviour, 1*(12), 849–851. DOI: 10.1038/s41562-017-0232-0 PMID: 31024173

Bittner, K. (2021). Engagement and immersion in digital play: Supporting young children's digital wellbeing. *International Journal of Environmental Research and Public Health, 18*(19), 10179. DOI: 10.3390/ijerph181910179 PMID: 34639481

Boucher, E. M., Harake, N. R., Ward, H. E., Stoeckl, S. E., Vargas, J., Minkel, J., ... & Zilca, R. (2021). Artificially intelligent chatbots in digital mental health interventions: A review. *Expert Review of Medical Devices, 18*(sup1), 37–49. DOI: 10.1080/17434440.2021.2013200

Bucci, S., Schwannauer, M., & Berry, N. (2019). The digital revolution and its impact on mental health care. *Psychology and Psychotherapy: Theory, Research and Practice, 92*(2), 277–297. DOI: 10.1111/papt.12222 PMID: 30924316

Burr, C., Taddeo, M., & Floridi, L. (2020). The ethics of digital wellbeing: A thematic review. *Science and Engineering Ethics, 26*(4), 2313–2343. DOI: 10.1007/s11948-020-00175-8 PMID: 31933119

Calvo, R. A., & Peters, D. (2014). *Positive computing: technology for wellbeing and human potential*. MIT press., DOI: 10.7551/mitpress/9764.001.0001

Carr, S. (2020). 'AI gone mental': Engagement and ethics in data-driven technology for mental health. *Journal of Mental Health (Abingdon, England)*, *29*(2), 125–130. DOI: 10.1080/09638237.2020.1714011 PMID: 32000544

Chalkiadakis, A., Seremetaki, A., Kanellou, A., Kallishi, M., Morfopoulou, A., Moraitaki, M., & Mastrokoukou, S. (2024). Impact of artificial intelligence and virtual reality on educational inclusion: A systematic review of technologies supporting students with disabilities. *Education Sciences*, *14*(11), 1223. DOI: 10.3390/educsci14111223

Chantelle. (2023, September 26). *Supporting digital wellbeing and time management*. Apple Education Community Forum. https://education.apple.com/resource/250011580

Chatterjee, C., Chugunova, M., Ghosh, M., Singhal, A., & Wang, L. X. (2023). Human mediation leads to higher compliance in digital mental health: Field evidence from India. *Frontiers in Behavioral Economics*, *2*, 1232462. DOI: 10.3389/frbhe.2023.1232462

Coby. (2023, June 25). *'My wellbeing toolkit'*. Apple Education Community Forum. https://education.apple.com/resource/250011441

Coghlan, S., Leins, K., Sheldrick, S., Cheong, M., Gooding, P., & D'Alfonso, S. (2023). To chat or bot to chat: Ethical issues with using chatbots in mental health. *Digital Health*, *9*, 20552076231183542. DOI: 10.1177/20552076231183542 PMID: 37377565

Colby, K. M., Weber, S., & Hilf, F. D. (1971). Artificial paranoia. *Artificial Intelligence*, *2*(1), 1–25. DOI: 10.1016/0004-3702(71)90002-6

Denecke, K., & Househ, M. (2021). Artificial intelligence for chatbots in mental health: Opportunities and challenges. In *Advances in Artificial Intelligence for Health and Medicine* (pp. 171–186). Springer., DOI: 10.1007/978-3-030-67303-1_10

Deniz-Garcia, A., Fabelo, H., Rodriguez-Almeida, A. J., Zamora-Zamorano, G., Castro-Fernandez, M., Alberiche Ruano, M. D. P., Solvoll, T., Granja, C., Schopf, T. R., Callico, G. M., Soguero-Ruiz, C., & Wägner, A. M.WARIFA Consortium. (2023). Quality, usability, and effectiveness of mHealth apps and the role of artificial intelligence: Current scenario and challenges. *Journal of Medical Internet Research*, *25*, e44030. DOI: 10.2196/44030 PMID: 37140973

Desmet, P. M., & Pohlmeyer, A. E. (2013). Positive design: An introduction to design for subjective wellbeing. *International Journal of Design*, *7*(3). https://www.ijdesign.org/index.php/IJDesign/article/view/1666/595

Fana, M., Milasi, S., Napierala, J., Fernández-Macías, E., & Vázquez, I. G. (2020). Telework, work organisation and job quality during the COVID-19 crisis: a qualitative study (No. 2020/11). *JRC Working Papers Series on Labour, Education and Technology*. https://hdl.handle.net/10419/231343

Fitzpatrick, K. K., Darcy, A., & Vierhile, M. (2017). Delivering cognitive behavior therapy to young adults with symptoms of depression and anxiety using a fully automated conversational agent (Woebot): A randomized controlled trial. *JMIR Mental Health*, *4*(2), e7785. DOI: 10.2196/mental.7785 PMID: 28588005

Floridi, L. (2014). Open data, data protection, and group privacy. *Philosophy & Technology*, *27*(1), 1–3. DOI: 10.1007/s13347-014-0157-8

Garrido, S., Millington, C., Cheers, D., Boydell, K., Schubert, E., Meade, T., & Nguyen, Q. V. (2019). What works and what doesn't work? A systematic review of digital mental health interventions for depression and anxiety in young people. *Frontiers in Psychiatry*, *10*, 759. DOI: 10.3389/fpsyt.2019.00759 PMID: 31798468

Grau, L. E., Larkin, K., Lalloo, C., Stinson, J. N., Zempsky, W. T., Ball, S. A., & Buono, F. D. (2022). Perspectives on adapting a mobile application for pain self-management in neurofibromatosis type 1: Results of online focus group discussions with individuals living with neurofibromatosis type 1 and pain management experts. *BMJ Open*, *12*(7), e056692. DOI: 10.1136/bmjopen-2021-056692 PMID: 35840301

Hall, C. L., Gómez Bergin, A. D., & Rennick-Egglestone, S. (2024). Research into digital health intervention for mental health: 25-year retrospective on the ethical and legal challenges. *Journal of Medical Internet Research*, *26*, e58939. DOI: 10.2196/58939 PMID: 39250796

Hilty, D. M., Crawford, A., Teshima, J., Nasatir-Hilty, S. E., Luo, J., Chisler, L. S., Gutierrez Hilty, Y. S. M., Servis, M. E., Godbout, R., Lim, R. F., & Lu, F. G. (2021). Mobile health and cultural competencies as a foundation for telehealth care: Scoping review. *Journal of Technology in Behavioral Science*, *6*(2), 197–230. DOI: 10.1007/s41347-020-00180-5

Honka, A. (2024). *Personalization of digital health behavior change interventions for health promotion: A user modeling framework and a recommender system*. (Tampere University Dissertations - Tampereen yliopiston väitöskirjat; Vol. 957). Tampere University. https://urn.fi/URN:ISBN:978-952-03-3297-6

Husain, L., Greenhalgh, T., Hughes, G., Finlay, T., & Wherton, J. (2022). Desperately seeking intersectionality in digital health disparity research: Narrative review to inform a richer theorization of multiple disadvantage. *Journal of Medical Internet Research, 24*(12), e42358. DOI: 10.2196/42358 PMID: 36383632

Itie, D. I. (2023). *Decisions for the implementation of artificial intelligence technology for workforce management* (Publication No. 30244296) [Doctoral dissertation, University of Maryland University College]. ProQuest Dissertations & Theses.

Joyce, S., Modini, M., Christensen, H., Mykletun, A., Bryant, R., Mitchell, P. B., & Harvey, S. B. (2016). Workplace interventions for common mental disorders: A systematic meta-review. *Psychological Medicine, 46*(4), 683–697. DOI: 10.1017/S0033291715002408 PMID: 26620157

Kashani, K. B., Awdishu, L., Bagshaw, S. M., Barreto, E. F., Claure-Del Granado, R., Evans, B. J., Forni, L. G., Ghosh, E., Goldstein, S. L., Kane-Gill, S. L., Koola, J., Koyner, J. L., Liu, M., Murugan, R., Nadkarni, G. N., Neyra, J. A., Ninan, J., Ostermann, M., Pannu, N., & Mehta, R. L. (2023). Digital health and acute kidney injury: Consensus report of the 27th Acute Disease Quality Initiative workgroup. *Nature Reviews. Nephrology, 19*(12), 807–818. DOI: 10.1038/s41581-023-00744-7 PMID: 37580570

Khawaja, Z., & Bélisle-Pipon, J. C. (2023). Your robot therapist is not your therapist: Understanding the role of AI-powered mental health chatbots. *Frontiers in Digital Health, 5*, 1278186. DOI: 10.3389/fdgth.2023.1278186 PMID: 38026836

Khoury, M. J., & Ioannidis, J. P. (2014). Big data meets public health. *Science, 346*(6213), 1054–1055. DOI: 10.1126/science.aaa2709 PMID: 25430753

Kim, H. K. (2024). The Effects of Artificial Intelligence Chatbots on Women's Health: A Systematic Review and Meta-Analysis. *Healthcare (Basel), 12*(5), 534. DOI: 10.3390/healthcare12050534 PMID: 38470645

Koivisto, J., & Hamari, J. (2019). The rise of motivational information systems: A review of gamification research. *International Journal of Information Management, 45*, 191–210. DOI: 10.1016/j.ijinfomgt.2018.10.013

Kruse, C., Fohn, J., Wilson, N., Patlan, E. N., Zipp, S., & Mileski, M. (2020). Utilization barriers and medical outcomes commensurate with the use of telehealth among older adults: Systematic review. *JMIR Medical Informatics, 8*(8), e20359. DOI: 10.2196/20359 PMID: 32784177

Lang, C. (2024). Dreaming big with little therapy devices: Automated therapy from India. *Anthropology & Medicine, 31*(3), 232–249. DOI: 10.1080/13648470.2024.2378727 PMID: 39435587

Lillywhite, A., & Wolbring, G. (2022). Undergraduate disabled students as knowledge producers including researchers: Perspectives of disabled students. *Education Sciences, 12*(2), 77. DOI: 10.3390/educsci12020077

Linardon, J., Cuijpers, P., Carlbring, P., Messer, M., & Fuller-Tyszkiewicz, M. (2019). The efficacy of app-supported smartphone interventions for mental health problems: A meta-analysis of randomized controlled trials. *World Psychiatry; Official Journal of the World Psychiatric Association (WPA), 18*(3), 325–336. DOI: 10.1002/wps.20673 PMID: 31496095

Liu, P., Yuan, W., Fu, J., Jiang, Z., Hayashi, H., & Neubig, G. (2023). Pre-train, prompt, and predict: A systematic survey of prompting methods in natural language processing. *ACM Computing Surveys, 55*(9), 1–35. DOI: 10.1145/3560815

Lukoff, K., Lyngs, U., Shirokova, K., Rao, R., Tian, L., Zade, H., Munson, S. A., & Hiniker, A. (2023). SwitchTube: A proof-of-concept system introducing "adaptable commitment interfaces" as a tool for digital wellbeing. *Proceedings of the 2023 CHI Conference on Human Factors in Computing Systems (CHI '23).* Association for Computing Machinery, New York, NY, USA, Article 197, 1–22. DOI: 10.1145/3544548.3580703

Menon, V., & Varadharajan, N. (2020). Digital approaches for mental health service delivery in low- and middle-income countries like India: Key implementational challenges and recommendations. *Asian Journal of Psychiatry, 50,* 101962. DOI: 10.1016/j.ajp.2020.101962 PMID: 32109806

Miner, A. S., Shah, N., Bullock, K. D., Arnow, B. A., Bailenson, J., & Hancock, J. (2019). Key considerations for incorporating conversational AI in psychotherapy. *Frontiers in Psychiatry, 10,* 746. DOI: 10.3389/fpsyt.2019.00746 PMID: 31681047

Mitsea, E., Drigas, A., & Skianis, C. (2023). Digitally assisted mindfulness in training self-regulation skills for sustainable mental health: A systematic review. *Behavioral Sciences (Basel, Switzerland), 13*(12), 1008. DOI: 10.3390/bs13121008 PMID: 38131865

Naslund, J. A., Aschbrenner, K. A., Kim, S. J., McHugo, G. J., Unützer, J., Bartels, S. J., & Marsch, L. A. (2017). Health behavior models for informing digital technology interventions for individuals with mental illness. *Psychiatric Rehabilitation Journal, 40*(3), 325–335. DOI: 10.1037/prj0000246 PMID: 28182469

Naslund, J. A., & Deng, D. (2021). Addressing mental health stigma in low-income and middle-income countries: A new frontier for digital mental health. *Ethics, Medicine and Public Health, 19*, 100719. https://doi.org/DOI: 10.1016/j.jemep.2021.100719

Naslund, J. A., Gonsalves, P. P., Gruebner, O., Pendse, S. R., Smith, S. L., Sharma, A., & Raviola, G. (2019). Digital Innovations for Global Mental Health: Opportunities for Data Science, Task Sharing, and Early Intervention. *Current Treatment Options in Psychiatry*, *6*(4), 337–351. DOI: 10.1007/s40501-019-00186-8 PMID: 32457823

Ndetei, D. M., Mutiso, V., & Osborn, T. (2023). Moving away from the scarcity fallacy: Three strategies to reduce the mental health treatment gap in LMICs. *World Psychiatry; Official Journal of the World Psychiatric Association (WPA)*, *22*(1), 163–164. DOI: 10.1002/wps.21054 PMID: 36640407

Nguyen, M. H. (2021). Managing social media use in an "always-on" society: Exploring digital wellbeing strategies that people use to disconnect. *Mass Communication & Society*, *24*(6), 795–817. DOI: 10.1080/15205436.2021.1979045

Ogugua, J. O., Okongwu, C. C., Akomolafe, O. O., Anyanwu, E. C., & Daraojimba, O. D.Jane Osareme OguguaChiamaka Chinaemelum OkongwuOpeoluwa Oluwanifemi AkomolafeEvangel Chinyere AnyanwuObinna Donald Daraojimba. (2024). Mental health and digital technology: A public health review of current trends and responses. *International Medical Science Research Journal*, *4*(2), 108–125. DOI: 10.51594/imsrj.v4i2.754

Paganin, G., & Simbula, S. (2020). Smartphone-based interventions for employees' wellbeing promotion: a systematic review. *Electronic Journal of Applied Statistical Analysis, 13*(3), 682-712. DOI:DOI: 10.1285/i20705948v13n3p682

Patel, V., Xiao, S., Chen, H., Hanna, F., Jotheeswaran, A. T., Luo, D., Parikh, R., Sharma, E., Usmani, S., Yu, Y., Druss, B. G., & Saxena, S. (2016). The magnitude of and health system responses to the mental health treatment gap in adults in India and China. *Lancet*, *388*(10063), 3074–3084. DOI: 10.1016/S0140-6736(16)00160-4 PMID: 27209149

Pedrouzo, S. B., Jaitt, M., Núñez, J., Lamas, F., & Krynski, L. (2024). Adolescence and digital wellbeing: Tools for healthcare professionals. *Archivos Argentinos de Pediatria, 122*(1):e202310199. https://doi/org/. PMID: 38150291.DOI: 10.5546/aap.2023-10199

Perski, O., Crane, D., Beard, E., & Brown, J. (2019). Does the addition of a supportive chatbot promote user engagement with a smoking cessation app? An experimental study. *Digital Health*, *5*, 2055207619880676. DOI: 10.1177/2055207619880676 PMID: 31620306

Peters, B. G. (2018). *Policy problems and policy design*. Edward Elgar Publishing. DOI: 10.4337/9781786431356

Poirier, C., & Gelin, M. (2024). Boosting workplace wellbeing: Unveiling the impact of digital tools. *Journal of Workplace Behavioral Health*, •••, 1–20. DOI: 10.1080/15555240.2024.2441209

Rajkomar, A., Dean, J., & Kohane, I. (2019). Machine learning in medicine. *The New England Journal of Medicine*, *380*(14), 1347–1358. DOI: 10.1056/NEJMra1814259 PMID: 30943338

Raqeeb, A. (2024). Retention and retention strategies in technology-driven work context [Unpublished master's thesis]. University of Vaasa.

Rodríguez-Martínez, A., Amezcua-Aguilar, T., Cortés-Moreno, J., & Jiménez-Delgado, J. J. (2023). Qualitative analysis of conversational chatbots to alleviate loneliness in older adults as a strategy for emotional health. *Healthcare (Basel)*, *12*(1), 62. DOI: 10.3390/healthcare12010062 PMID: 38200967

Roffarello, A. M., & Russis, L. D. (2023). Achieving digital wellbeing through digital self-control tools: A systematic review and meta-analysis. *ACM Trans. Comput.-Hum. Interact.*, *30*(4), Article 53, 66 pages. DOI: 10.1145/3571810

Rose, K. (2024, October 23). Belgian teen's suicide sparks lawsuit against Character.AI chatbot. *The New York Times*. https://www.nytimes.com/2024/10/23/technology/characterai-lawsuit-teen-suicide.html

Ruggeri, K., Garcia-Garzon, E., Maguire, Á., Matz, S., & Huppert, F. A. (2020). Well-being is more than happiness and life satisfaction: A multidimensional analysis of 21 countries. *Health and Quality of Life Outcomes*, *18*(1), 192. DOI: 10.1186/s12955-020-01423-y PMID: 32560725

Ryan, M. (2022). The ethics of dietary apps: Technology, health, and the capability approach. *Technology in Society*, *68*, 101873. DOI: 10.1016/j.techsoc.2022.101873

Schueller, M. S., & Histon, T. (2023, November 3). Digital tools for youth mental health. *Frontiers for Young Minds*, *11*, 1169684. DOI: 10.3389/frym.2023.1169684

Sedlakova, J., & Trachsel, M. (2022). Conversational artificial intelligence in psychotherapy: A new therapeutic tool or agent? *The American Journal of Bioethics*, *23*(5), 4–13. DOI: 10.1080/15265161.2022.2048739 PMID: 35362368

Sharma, A., Lin, I. W., Miner, A. S., Atkins, D. C., & Althoff, T. (2023). Human–AI collaboration enables more empathic conversations in text-based peer-to-peer mental health support. *Nature Machine Intelligence, 5*(1), 46–57. DOI: 10.1038/s42256-022-00593-2

Simons, L., Martin, J., & Craven, M. (2015). Can simple digital tools promote wellbeing? *Proceedings of the King's Fund's digital health and care congress 2015.* Research Gate. DOI:DOI: 10.13140/RG.2.1.1707.1209

Singh, L. (2024). *Revolutionizing care: Developing an AI-human synergy mental health application for mental health worker's wellbeing* [Unpublished doctoral dissertation]. Capella University.

Skogberg, M., Mackiewicz, K., Mänd, K., Tuuling, L., Urdzina-Merca, I., Salanterä, S., & Pakarinen, A. (2022). Promoting the health and wellbeing of children: A feasibility study of a digital tool among professionals. *PLoS One, 17*(3), e0265355. DOI: 10.1371/journal.pone.0265355 PMID: 35316279

Stanphill, M. (2019, May 14). *Find your balance with new digital wellbeing tools.* Google. https://blog.google/outreach-initiatives/digital-wellbeing/find-your-balance-new-digital-wellbeing-tools/

Stewart, E., Milton, A., Yee, H., Song, M., Roberts, A., Davenport, T., & Hickie, I. (2022). eHealth tools that assess and track health and well-being in children and young people: Systematic review. *Journal of Medical Internet Research, 24*(5), e26015. DOI: 10.2196/26015 PMID: 35550285

Stratton, E., Lampit, A., Choi, I., Calvo, R. A., Harvey, S. B., & Glozier, N. (2017). Effectiveness of eHealth interventions for reducing mental health conditions in employees: A systematic review and meta-analysis. *PLoS One, 12*(12), e0189904. DOI: 10.1371/journal.pone.0189904 PMID: 29267334

Tamim, B. (2023). Belgian woman blames ChatGPT-like chatbot ELIZA for her husband's suicide. *Interesting Engineering.* https://interestingengineering.com/culture/belgian-woman-blames-chatgpt-like-chatbot-eliza-for-her-husbands-suicide

Thirupathi, L., Kaashipaka, V., Dhanaraju, M., & Katakam, V. (2025). AI and IoT in mental health care: From digital diagnostics to personalized, continuous support. In Joshi, H., Reddy, C. K. K., Ouaissa, M., Hanafiah, M. M., & Doss, S. (Eds.), *Intelligent systems and IoT applications in clinical health* (pp. 271–294). IGI Global., DOI: 10.4018/979-8-3693-8990-4.ch012

Torous, J., Bucci, S., Bell, I. H., Kessing, L. V., Faurholt-Jepsen, M., Whelan, P., Carvalho, A. F., Keshavan, M., Linardon, J., & Firth, J. (2021). The growing field of digital psychiatry: Current evidence and the future of apps, social media, chatbots, and virtual reality. *World Psychiatry; Official Journal of the World Psychiatric Association (WPA)*, *20*(3), 318–335. DOI: 10.1002/wps.20883 PMID: 34505369

Twenge, J. M., & Campbell, W. K. (2018). Associations between screen time and lower psychological wellbeing among children and adolescents: Evidence from a population-based study. *Preventive Medicine Reports*, *12*, 271–283. DOI: 10.1016/j.pmedr.2018.10.003 PMID: 30406005

Twenge, J. M., Haidt, J., Lozano, J., & Cummins, K. M. (2022). Specification curve analysis shows that social media use is linked to poor mental health, especially among girls. *Acta Psychologica*, *224*, 103512. DOI: 10.1016/j.actpsy.2022.103512 PMID: 35101738

van Olmen, J. (2022). The promise of digital self-management: A reflection about the effects of patient-targeted e-health tools on self-management and wellbeing. *International Journal of Environmental Research and Public Health*, *19*(3), 1360. DOI: 10.3390/ijerph19031360 PMID: 35162383

Vanden Abeele, M. M. (2021). Digital wellbeing as a dynamic construct. *Communication Theory*, *31*(4), 932–955. DOI: 10.1093/ct/qtaa024

Wadhwa, M. (n.d.). *Digital mental health initiatives in India* [ICT India Working Paper #60]. [Columbia University] CSD Working Paper Series. Available at: https://csd.columbia.edu/sites/default/files/content/docs/ICT%20India/Papers/ICT_India_Working_Paper_60.pdf

Weizenbaum, J. (1966). ELIZA—A computer program for the study of natural language communication between man and machine. *Communications of the ACM*, *9*(1), 36–45. DOI: 10.1145/365153.365168

Wies, B., Landers, C., & Ienca, M. (2021). Digital mental health for young people: A scoping review of ethical promises and challenges. *Frontiers in Digital Health*, *3*, 697072. DOI: 10.3389/fdgth.2021.697072 PMID: 34713173

Wykes, T., Lipshitz, J., & Schueller, S. M. (2019). Towards the design of ethical standards related to digital mental health and all its applications. *Current Treatment Options in Psychiatry*, *6*(3), 232–242. DOI: 10.1007/s40501-019-00180-0

Zaky, Y. A. M. (2023). Chatbot Positive Design to Facilitate Referencing Skills and Improve Digital wellbeing. *International Journal of Interactive Mobile Technologies*, *17*(9), 106–126. Advance online publication. DOI: 10.3991/ijim.v17i09.38395

Zawati, M. H., & Lang, M. (2024). Does an App a Day Keep the Doctor Away? AI Symptom Checker Applications, Entrenched Bias, and Professional Responsibility. *Journal of Medical Internet Research*, *26*, e50344. DOI: 10.2196/50344 PMID: 38838309

Zhang, C., Bengio, S., Hardt, M., Recht, B., & Vinyals, O. (2021). Understanding deep learning (still) requires rethinking generalization. *Communications of the ACM*, *64*(3), 107–115. DOI: 10.1145/3446776

Zhao, W. X., Zhou, K., Li, J., Tang, T., Wang, X., Hou, Y., Min, Y., ... "A survey of large language models." arXiv preprint arXiv:2303.18223 (2023).

Zuboff, S. (2015). Big other: Surveillance capitalism and the prospects of an information civilization. *Journal of Information Technology*, *30*(1), 75–89. DOI: 10.1057/jit.2015.5

Key Terms and Definitions

Digital Wellbeing: It is a multidimensional concept that encompasses psychological, emotional, and social aspects of technology use.

Generative AI: Those artificial intelligence models which carry out independent and dynamic conversations with information drawn from large datasets. Thus, capable of more authentic and engaging conversations.

Scripted AI: Those artificial intelligence models which carry out conversations through pre-written responses and follow a defined structure.

Natural Language Processing (NLP): Natural language processing is a part of artificial intelligence and refers to a computer's ability to understand human language.

Mental health chatbots: AI driven chatbots specifically designed to hold conversations surrounding mental health. They can provide emotional support, coping strategies, and more to users.

TeleHealth initiatives: Delivery and facilitation of health-related services through digital communication platforms.

Telepsychiatry: A branch of telemedicine that provides psychiatric care and mental health services remotely via video calls, mobile apps, or online platforms.

Screen Addiction: A compulsive overuse of digital screens (smartphones, computers, etc.) that disrupts daily life, mental health, and social interactions.

Digital Fatigue: A state of mental, emotional, and physical exhaustion caused by prolonged screen exposure, excessive online interactions, and digital multitasking.

Non-maleficence: The idea that tools designed to promote digital wellbeing should not unintentionally cause harm or stress to users.

Data Leaks: When identifying information and sensitive health details become accessible to a wider unauthorized audience due to lax security and privacy measures.

Chapter 10
Breaking the Bias:
Assessing Nudge Effectiveness in Overcoming Decision-Making Prejudices

R. J. Arun
https://orcid.org/0009-0008-0560-4985
Christ University, India

Sangeeta Mehrolia
Christ University, India

S. Jeevananda
Christ University, India

Sibin Mathew Nesin
https://orcid.org/0000-0003-2216-6266
Christ University, India

Rabia Abhay
https://orcid.org/0000-0002-2888-6523
Christ University, India

ABSTRACT

Influencing consumer decision-making processes in the digital age is vital as E-commerce continues dominating the market. This research work explores the application of nudging to e-commerce or any digital environment involving consumer choices, investigating its efficacy in mitigating cognitive biases that affect online purchasing decisions. We employed a 2 x 2 within-group experimental design to examine how different nudges influence e-commerce choices. Data collected from 88

DOI: 10.4018/979-8-3373-7917-3.ch010

participants reveal that status-quo nudges can influence decision-making more than social proof nudges or salience nudges in online shopping. These results significantly impact digital marketing strategies, suggesting that carefully designed nudges can guide consumer choices by overcoming ingrained prejudices. This research also provides practical insights into consumer behaviour through subtle interventions in e-commerce settings. While nudging has been studied in offline contexts, applying it to e-commerce and digital consumer choices represents a growing area of study.

INTRODUCTION

Decision-making is integral to human life, influencing outcomes in various domains. "Nudge", the book written by Thaler and Sunstein, refers to any aspect of the decision environment that alters people's behaviour in a predictable way without forbidding any options (Thaler & Sunstein, 2008). Within this framework, 'nudges' are subtle environmental changes that can influence behavior without restricting freedom of choice (Sunstein, 2014). Interestingly, studies have demonstrated that nudges can affect the choice and the time taken to make that choice (Johnson et al., 2012). Eigenbrod observed the role of social and informational nudges in online marketing (Eigenbrod & Janson, 2018). Also, Berger demonstrated the effectiveness of using default rules and simplification nudges in promoting ecologically sustainable food choices in online grocery stores (Berger et al., 2020). The concept had been used by governments to make citizens improve their health(Oullier et al., 2010), save their earnings (Thaler & Benartzi, 2004) day light saving time (Rebonato, 2014) etc. Lu found out the positive impact of overt and covert nudges when used in its combination (Pennycook et al., 2020).Digital Nudging, a concept where minute modifications in the choice environment have a significant impact on their behaviour (Weinmann et al., 2016). We see an increasing trend in the application of digital nudges primarily because of the increase of usage of digital platforms. Mirbabaie's work showed that the effectiveness of the digital nudging strategies are mostly influenced by participants identification especially in green fashion e-commerce (Mirbabaie et al., 2021). Santarius proposes a systemic policy approach, calling out companies for their impact on consumers and the environment (Santarius et al., 2023). Heiskanen (Niinimäki, 2009) also proposes a trans-disciplinary approach thereby throwing light on a broader perspective.

Much of the research on e-commerce has concentrated on deliberate, rational decision-making. However, studies in psychology and marketing indicate that purchasing decisions can also be affected by priming, which is a form of digital nudging as referred to by Information Systems researchers(Dennis et al., 2020). Kenning also highlighted the importance of nudging in influencing the customer decisions in the context of data security and privacy(Kunkel et al., 2022). There are works which has proven that behaviour concepts can influence choice making in digital environments(Meske et al., 2020). All these contributions underline the significant role of digital nudging in moulding consumer behaviour under various contexts.

One common cognitive bias that affects decision-making is the central tendency bias, where individuals tend to avoid extreme options and gravitate towards the middle of a set of choices .This bias has been shown to influence both the final decision and the reaction time, with choices aligned with the central tendency often made more quickly (Tversky & Kahneman, 1974). Experiments in survey design demonstrate that explicitly offering a middle response alternative increases its selection likelihood, and the order of presentation can significantly affect results (Bishop, 1987). These studies collectively highlight the importance of option presentation and individual characteristics in decision-making processes, particularly regarding the middle option bias in various contexts.

Digital marketing uses various nudging strategies like scarcity bias, status quo bias, social proof bias, salience nudges etc (Herpen et al., 2005) conducted a meta-analysis of commodity theory research, confirming that scarcity enhances the value of anything that can be possessed, is useful to its possessor, and is transferable from one person to another. Social proof bias, also known as informational social influence, refers to the tendency to view a behavior as correct in a given situation to the degree that we see others performing it. Cialdini extensively discussed this principle in his influential book on persuasion (Cialdini & Trost, 1998). A classic experiment by Asch demonstrated the power of social proof in conformity (Asch S. E., 2003), showing that individuals would often conform to the majority view even when it contradicted their own perceptions..

Numerous decision making experiments show that individuals do nothing or maintain their current or previous decisions (Samuelson & Zeckhauser, 1988). Default Payments (Székely et al., 2016), Opt out policy (Malamud, 2019), positioning (Cai & Xu, 2008) are some of nudges proposed by various researches in both online and offline contexts. Just like the offline contexts where majority of the researches are focussed, Status Quo Bias had been used in many online contexts where the desired options can be presented as a default option to the consumers and other options are inferior. ((Schneider et al., 2018).

Salience nudges are a form of behavioral intervention designed to influence decision-making by altering the prominence of certain options. While these nudges can be effective in promoting healthier choices, such as reducing unhealthy food product selection by up to 63% (Michels et al., 2023), they raise ethical concerns about manipulation (Noggle, 2017). Salience nudges may be considered manipulative if they disproportionately highlight information irrelevant to the decision at hand (Noggle, 2017). Moreover, they can have unintended distributional effects, potentially harming those already struggling with decision-making (Thunström et al., 2018). To address ethical concerns, researchers have explored transparency and self-nudging as potential solutions. Interestingly, disclosing information about the nudge does not appear to impact its effectiveness ((Michels et al., 2021); Michels et al., 2023). Self-nudging, when combined with transparency, can yield comparable results to preset nudges (Michels et al., 2023). These findings suggest that salience nudges can be both effective and ethically implemented in digital environments.

EEG Study - Studies have shown that professionals exhibit high resistance to nudging, with EEG data revealing increased delta, theta, and beta power in right temporoparietal regions as potential neural correlates of this resistance (Balconi et al., 2023). Digital nudging and design friction can induce reflective behavior in users, particularly regarding sustainable choices in online meal ordering (Mejtoft et al., 2023). EEG spectral power has been used to predict consumer preferences and decision-making with high accuracy, identifying frontal and centro-parietal regions as key areas (Golnar-Nik et al., 2019). The design of digital decision environments, including factors like defaults and framing, can significantly influence online decision-making (Weinmann et al., 2018). These findings highlight the importance of understanding psychological effects of user-interface designs on choice behavior in digital contexts.

Theoretical Framework

The Nudge Theory, a concept that was developed by Nobel winning economist Richard Thaler and Cass Sunstein, has gained significant attention in recent years due to its potential as a strategy to influencing group as well as individual behavior in a very subtle yet an impactful ways without restrictive regulations. The book "Nudge: Improving Decisions about Health, Wealth, and Happiness" by Thaler and Sunstein has been highly influential, with over 5000 citations (H Allcott, 2019), showcases its extensive impact in behaviour economics and public policy. The basic idea behind the theory is that, small interventions or "nudges", can lead to better decision-making by individuals and groups, encouraging beneficial outcomes such as healthier lifestyle choices, improved financial decisions, greater environmental choices and sustainable purchase decisions. As nudge theory continues to gain traction in policy-making, iys applications are expanding into various domains such as healthcare, education, environmental conservations and sustainability. However, further research is needed to explore its applications in diverse settings, including those of its ability to produce long term changes in behavior and decision making (Thomas, 2016). However, there are many researchers (Meske & Amojo, 2020), while they acknowledge the increasing popularity in the Information system community, there is a need for ethical guidelines for the design of digital nudges and protect its legitimization.

The Diffusion Decision Model which was discussed by Roger Ratcliff and Gail Mckoon, (Ratcliff & McKoon, 2008), this model could be useful for analyzing reaction times in decision-making processes. This particular model is valuable for researchers aiming to gain deeper insights into the cognitive processes involved in decision making by examining how quickly individuals can arrive at a decision when faced with multiple choices. The speed at which these decisions are made, often referred to as reaction time or response latency, can provide valuable insights and important clues about the underlying cognitive processes that drives the quality of decisions. Research has shown that reaction times can vary significantly based on the complexity of the decision, the number of alternatives, and the decision-maker's expertise (Luce) and (Rafal Bogacz1 • Eric-Jan Wagenmakers, 2019). The model establishes that decision making occurs as a gradual accumulation of evidences over time, with individuals continuously gather the information. This model offers a rich understanding of how individual weigh alternatives and reaches conclusion. Overall, the Diffusion Decision model offers a framework for analyzing how reaction times are influenced by various factors and how these time stamps reflect the cognitive and efficiencies of decision-making.

The analysis of the literature reveals that very few nudges have been explored in the context of digital context given the gaining traction in the usage of digital spaces. There exists an empirical gap that needs to address the practical application of the biases in promoting nudge intervention. While nudges can be effective in changing behavior, it is crucial to evaluate their welfare implications. Allcott developed a method to measure consumer welfare and found that traditional evaluation approaches may overstate welfare gains (Kessler et al., 2015).A systematic review found that nudges resulted in a 15.3% increase in healthier dietary choices (Arno & Thomas, 2016). In the context of energy conservation, social comparison reports have been extensively studied as a nudge intervention. Therefore, the current study tries to explore the application of digital nudges in breaking the cognitive biases humans inherently have while making digital choices. This research also attempts to give practical insights and solutions to those marketers seeking to optimize their influence on consumer behaviour through subtle interventions in e-commerce settings.

Objectives

- The objective of this study is to experimentally assess the effectiveness of digital nudges in mitigating cognitive biases during online shopping decision-making processes.

- To investigate which nudge is more effective in mitigating the central tendency bias in optimising the consumer choices in e-commerce environment.
- To understand the effect of various nudges with varying time for decision making / response time.

Hence the present study his study aims to explore several key research questions related to the effectiveness of digital nudges in mitigating cognitive biases during online shopping. Specifically, it seeks to determine how digital nudges influence consumer decision-making and whether they effectively reduce biases such as the central tendency bias in e-commerce environments. Additionally, the study investigates which type of nudge—status quo, social proof, or salience—is more effective in optimizing consumer choices and reducing bias. Another critical aspect of this research is understanding the relationship between decision-making time and the effectiveness of different nudges, examining whether varying response times impact the strength of the nudging effect. Finally, by utilizing EEG analysis, this study aims to identify the neural mechanisms associated with decision-making under digital nudging, pinpointing the specific brain regions activated when individuals are influenced by status quo, social proof, and salience nudges. Through these investigations, the research aims to contribute to a deeper understanding of how digital nudges shape consumer behavior and decision-making processes at both cognitive and neural levels.

Methodology

To address the research questions and to evaluate the hypothesis, an experimental research approach was implemented, enabling multi-level measurements from all participants. The experiment design adopted for this purpose was the post-test only-control group design (Dugard & Todman, 1995). In this design, the participants get randomly assigned to either receive a stimuli or not, and then the outcome of interest is measured after the stimuli in order to measure its effect. In line with this experiment design, 2 webpages that resembled an E-commerce website were created. This will also establish the ecological validity of the experiment(Tietz et al., 2016). All the participants were instructed to first complete the registration process and provide their consent to participate in the experiment. The registration process was standardized for all the groups. The experiment group in the said study were the group of respondents who were exposed to the stimuli. One product in each category of products were nudged with Decoy Effect, Status Quo Bias and Scarcity effect. For the study, we used three different products (brush, mobile and umbrella). Three different types of nudging procedures were used to nudge these products. Social proof nudging is used for brush. The nudge in the form of a '4-star rating' was placed on the top of the brush placed in the PR region. Status quo nudge was used for the umbrella. The nudge in the form of a statement '72% selected this product' was placed on top of the umbrella, positioned in the PR region. Finally, a salience nudge was used for mobile. The nudge was used to highlight by giving a different shade to the background of the mobile placed in the PR region. At the end of the experiment, each respondent is expected to select one type from each of the products.

Whereas, in the control group, the website did not have any nudges in any of the product categories and they were also instructed to purchase one type from each of the product. None of the respondents were informed about the objective of the experiment since it may result in a biased purchased decision which may influence the results.

The respondents selected through convenience sampling were asked to engage in cognitive behavioural tasks for the first part of the study. The respondents were given a time gap of 15 days and later exposed to the experimental part for the study. The EEG data was collected while they were engaged in the experimental tasks which involved decision making through the exposure of the nudge. A total of 88 participants took part in the cognitive experimental work and 10 participants for the EEG data study. There are other inclusion criteria that were mandated to help in the internal validity of the experiments. The average age of the participants were 19+/-1 and were kept as homogenous as possible to help in internal validity . The respondents were very similar in their education, life experiences, the use of Technology and very much comfortable in using E-Commerce platforms (Levy & Ellis, 2011). The respondents were also instructed to avoid psychoactive drugs, no history of epilepsy, psychological and neurological conditions. The internal validity of the experiment was ensured by using the same images in both the website as the same planogram design. Also, we altered only one variable nudge at a time to ensure internal validity (Henkel, 2020).

Experimental procedures

"Open sesame "software, a free and open source software which is gaining traction amongst cognitive scientist (Mathôt et al., 2012) was used for the construction and application of this experiment. This application software were installed in 20 laptops and shared with two groups (control group and the experiment group). The experiment was conducted in a lab where there are minimal distractions and external disturbances. All the participants were instructed to first complete the registration process and provide their consent to participate in the experiment. This lab set up and registration process was standardized for all the groups.

Three different time scales were used for the study. This was done in order to check the influence of time constraint in the efficacy of nudging. The time strata's selected for this purpose was 500 ms, 3 seconds and no time limit (infinite time). The logic behind selecting these time levels was the fact that it reflected different stages of cognitive processing in the brain. As per Singhal & Srinivasan (2021) brain process information in three time scales i.e, short time scales of 30-100 milliseconds, intermediate time scales of 300-500 ms and longer time scales of 3 seconds or longer. We have selected the time strata's of 500 ms, 3 seconds and no time bound to mimic the intermediate and longer periods of cognitive processing and to see if nudging works differently in different time levels.

The experiment began with a slide showing a message on how much time the participants will get to respond for selecting the product of their preference (i.e, 500ms, 3 seconds or unlimited time). After that a slide showing a fixation cross was presented for 1000 ms. after that a slide showing four objects placed in Peripheral Left (PL), Nasal Left (NL), Nasal Right (NR) and Peripheral Right (PR) regions of the field of vision, will be presented to the participant and will be shown for either one of the three time scales (figure 2).

Figure 1. schematic representation of the experimental procedure

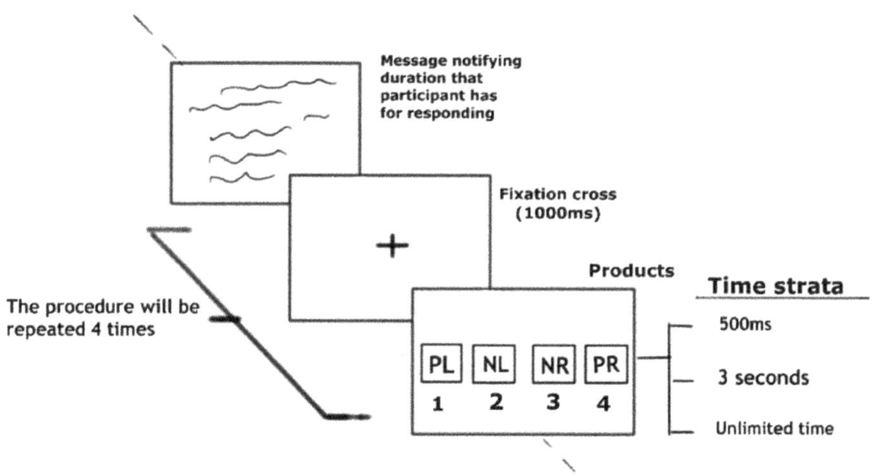

The participant will be instructed to press the serial number that denoted the product of their preference before the time limit is up. The response will be recorded and then the procedure will be repeated with the selection happening in a different time strata.

For the study three sets of products were used, tooth brush, mobile phone and umbrella. These three normal day to day life objects were selected in order to reduce any bias that may arise due to preferences. Reduction of biases that would arise due to the gender of the participant was another reason to choose these gender neutral products that are not stereotyped to a specific gender. Special care was taken to standardize the products by removing any marks of identity that could elicit a bias (brand name etc.)

The products were divided into 3 sets (to be used for three time stratas) of 4 products each, so that each of them were exposed to all the three time startas. In order to accomplish that 12 similar looking varieties of brushes, mobiles and umbrellas were used. This was done in order to eliminate the priming effect that could originate upon constant exposure to the same products in all the three different time stratas. The participants were required to take part in the study for two days, as they were exposed to nudge and control criterion on separate days in a random manner. The experiment group in the said study were the group of respondents who were exposed to the stimuli.

As mentioned above, open sesame software was used to build and run this experiment.

Ethical clearance and consent

Informed consent was acquired for all the participants and were informed regarding the nature of the study. The study was given approval by the university ethics committee.

Data analysis

Simple Percentage analysis was used to assess the selection rates of the products across the experiment. The data analysis was conducted in SPSS (V27).

Figure 2. Central and Peripheral fields in and normal field of vision.

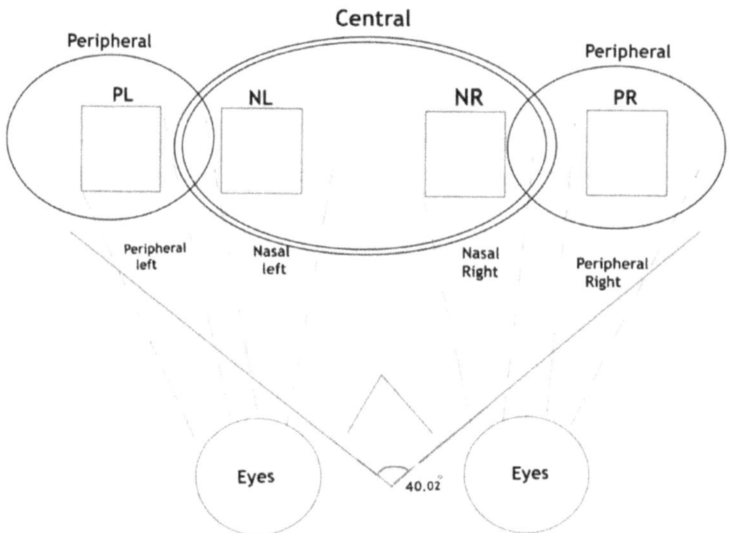

Results

The study attempted to find whether the bias of central tendency exhibited by individuals in product selection could be negated through nudging. The data collected from 96 participants were analyzed. After cleaning the data, we got 88 complete responses and the analysis of the same is presented below.

Percentage of product selection

Figure 3. Overall product selection pattern in control group

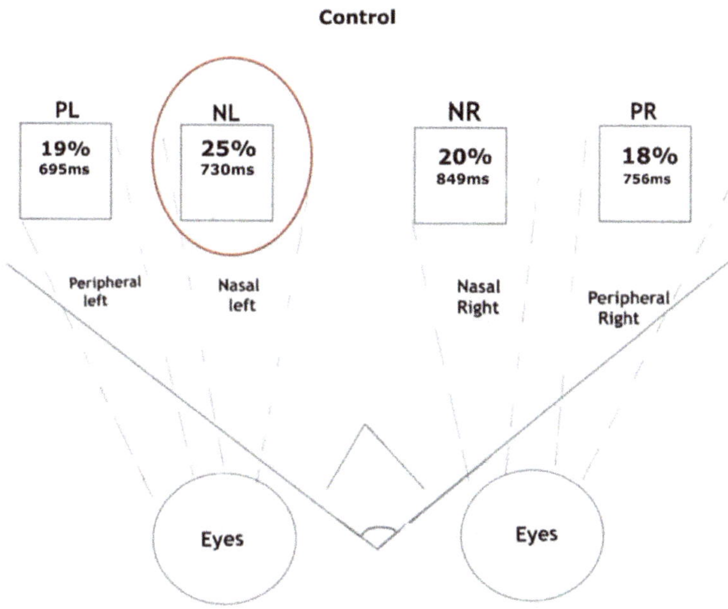

Figure 3 shows the product selection pattern during the control trial in the absence of nudge. The products in the nasal left (NL,which falls in the central field of vision) of the respondence were selected more in comparison to products placed in other regions in the field of vision. However, it is also observed that the peripheral right (PR)is given the least consideration, with only 18% of the respondents choosing the said product. The peripheral left (PL) has shown 19%, nasal left (NL) has 25% and nasal right has 20% of product selection.

Figure 4. Overall product selection pattern when nudging was provided

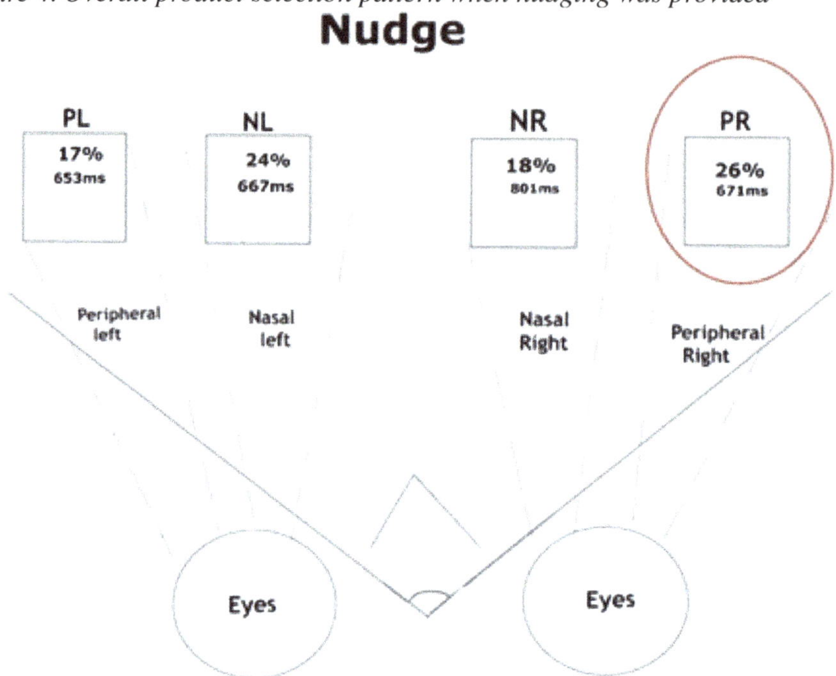

The above figure 4 demonstrates the product selection pattern in the experimental trial where the nudges have been induced. When the nudge was presented in the product placed in the PR region of the vision, it shows 26% of the product selection, the PL have shown 17% of the total products selected, NL has shown 24% and the NR have shown 18% of the product selection.

When overall data was compared(figure 5), it was seen that product selection patterns did indeed significantly vary depending on the location that they were placed in one's field of vision, when control Vs nudge groups were compared ($\chi2(16) =422.82$, $p<0.01$, with an effect size of 0.365). Was seen that an 8% increase in product selection rates for products placed in PR (where nudge was given) when comparing nudge group with control group.

Figure 5. Selection pattern for over all control vs experiment

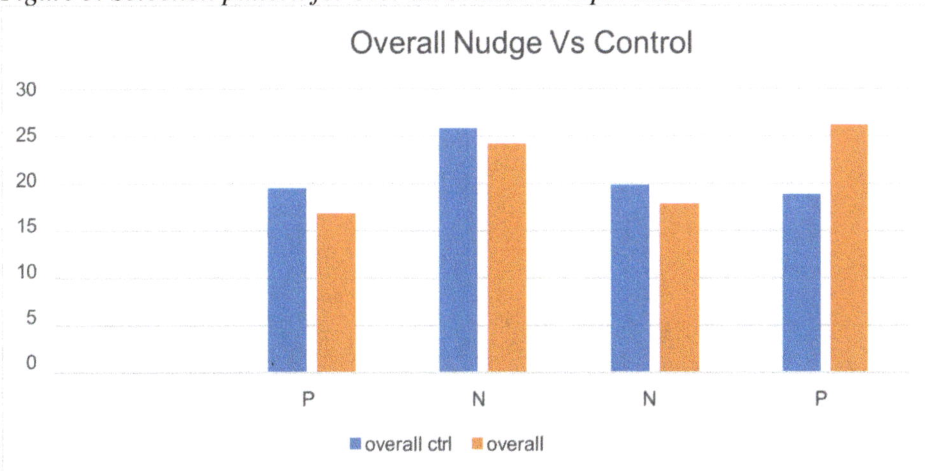

The below mentioned chart represents the pattern of selection when the nudges were analysed individually.

Figure 6. Selection pattern for control vs experiment (salience nudge)

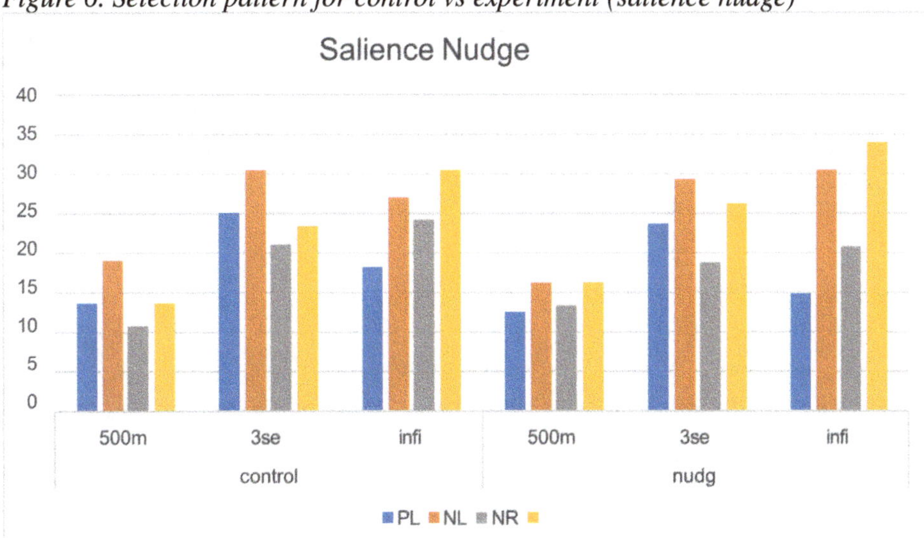

When the control group was compared with the Salience nudge group (figure 6), 3% reduction in NL and a 3% increase in PR (for 500ms), a 1% reduction in NL and a 3% increase in PR (for 3seconds), a 3% increase in NL and a 3% increase in PR (for infinite time). Though this trend was seen, no significant difference was observed.

Figure 7. Selection pattern for control vs experiment (social proof nudge)

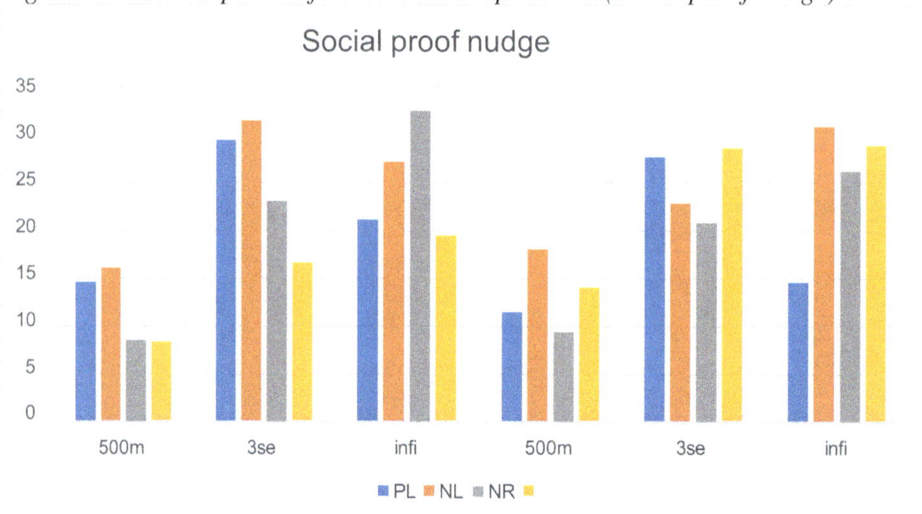

When the control group was compared with the Social Proof nudge group (figure 7), a 2% increase in NL and a 5% increase in PR (for 500ms), a 9% decrease in NL and a 12% increase in PR (for 3seconds), a 4% increase in NL and a 9% increase in PR (for infinite time). Though this trend was seen, no significant difference was observed.

Figure 8. Selection pattern for control vs experiment (status quo nudge)

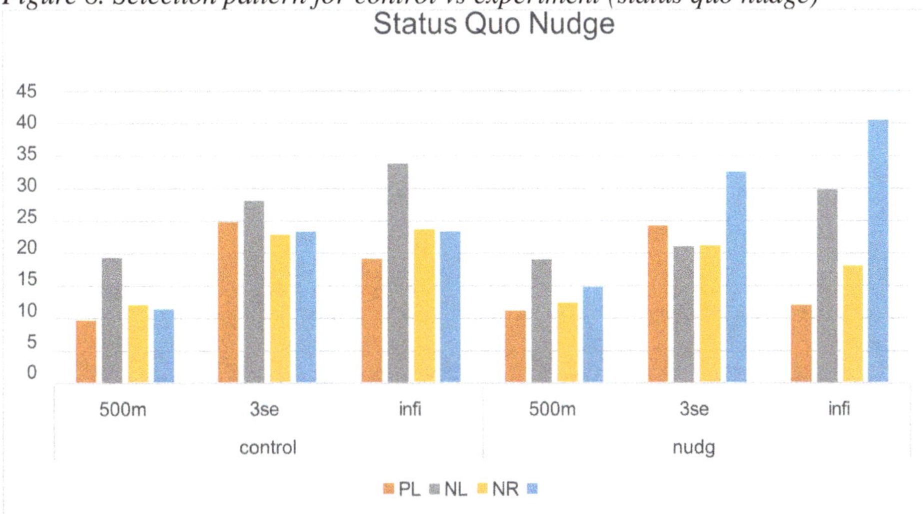

When the control group was compared with the Status Quo nudge group (figure 8), the same NL and 3% increase in PR (for 500ms), 7% decrease in NL and 9% increase in PR (for 3seconds), 4% decrease in NL and 17% increase in PR (for infinite time). Though this trend was seen, no significant difference was observed.

The key takeaway from product selection analysis: in shorter time frames when decisions have to be taken fast, social proof nudge and in non-time constrained scenarios status quo nudge is more efficient in overcoming central tendency bias to some extent.

The time taken to respond by the individual (reflecting the cognitive load of the decision-making) was also seen to be influenced by the fact whether nudging was given. It was seen that products placed in peripheral regions such as PR had a higher response time. Though not significant (t(547)=1.708, P= 0.088), when nudging was given the products placed in the periphery (PR) were responded to much faster.

Time constrained scenario

The experiment was divided into three different time levels (0.5 seconds, 3 seconds and no time constraint (infinite time)) to check the influence of time constraints at influencing the efficiency of nudging. For the time strata of 0.5 seconds, owing to its rapid nature only 50% of the collected data was able to be interpreted and analyzed. The figure 9 shown below elucidates the results for the same.

Figure 9. Paired sample test

Paired Samples Test

		Paired Differences							
					95% Confidence Interval of the Difference				
		Mean	Std. Deviation	Std. Error Mean	Lower	Upper	t	df	Sig. (2-tailed)
Pair 1	ctrl_brsh_500ms_RT - Nud_brsh_500ms_RT	8.09246	157.99964	8.48187	-8.59004	24.77497	954	346	341
Pair 2	ctrl_brsh_3sec_RT - Nud_brsh_3sec_RT	93.66674	538.02756	28.67697	37.26644	150.06703	3.266	351	001
Pair 3	ctrl_brsh_infi_RT - Nud_brsh_infi_RT	87.31337	1190.86189	63.56351	-37.70111	212.32784	1.374	350	.170
Pair 4	ctrl_mob_500ms_RT - Nud_mob_500ms_RT	-2.31444	174.46333	9.33881	-20.68205	16.05316	-.248	348	804
Pair 5	ctrl_mob_3sec_RT - Nud_mob_3sec_RT	77.22973	650.02605	34.64650	9.08887	145.37059	2.229	351	026
Pair 6	ctrl_mob_infi_RT - Nud_mob_infi_RT	31.63874	1709.06060	91.09324	-147.51847	210.79596	347	351	729
Pair 7	ctrl_umb_500ms_RT - Nud_umb_500ms_RT	3.72929	164.93438	8.80355	-13.58521	21.04380	.424	350	.672
Pair 8	ctrl_umb_3sec_RT - Nud_umb_3sec_RT	58.21326	632.68848	33.72241	-8.11014	124.53666	1.726	351	.085
Pair 9	ctrl_umb_infi_RT - Nud_umb_infi_RT	18.74731	2209.08481	117.91222	-213.15833	250.65294	.159	350	.874
Pair 10	overall_ctrl_RT - overall_Nud_RT	42.25398	1093.62632	19.46398	4.09065	80.41732	2.171	3156	030

There existed a significant overall difference in response times under control and nudge conditions (t(3156) =2.171,p=0.030). This trend was especially significant in 3sec conditions for the brush (t(351) =3.266,p=0.001) and mobile (t(351)=2.229, p=0.026) conditions.

Figure 10. Average response time

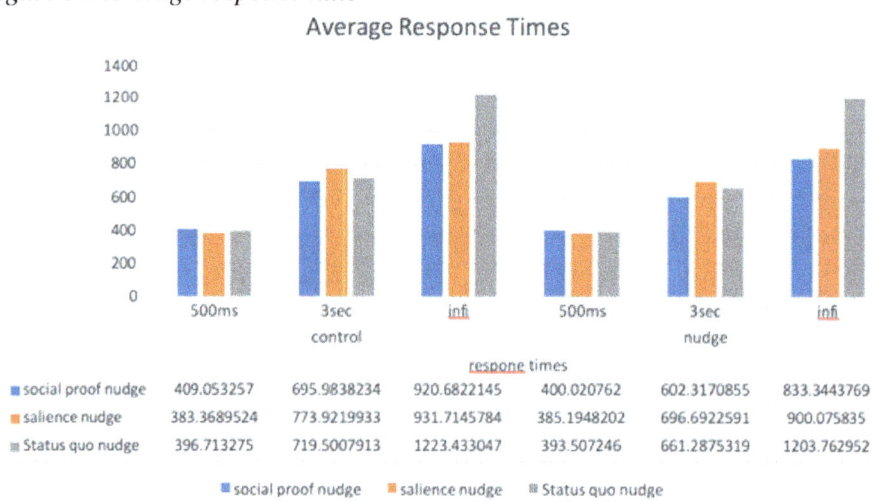

In almost all the time strata, it was seen that nudging did decrease the central tendency bias by improving the visibility of the peripheral areas, which is reflected in increased product selection percentage in those regions. At the 0.5 second time strata, there was a 3% increase in the selection rates of the products placed in the peripheries when compared to controls. The same was seen, as 6.8% and 3.1% increase in selection rates of the products placed in peripheries when compared to controls in 3seconds and unlimited time strata respectively (figure 10).

The effectiveness of nudging was further demonstrated by the reduction in selection rates for products positioned in the central regions, indicating its success in redirecting individuals' attention from the center toward the periphery. Across nearly all-time intervals, a decline in selection rates for centrally placed products was observed. Specifically, selection rates decreased by 4.9%, 7.6%, and 3% for the 0.50-second, 3-second, and unlimited-time conditions, respectively

Figure 11. Wilcoxon sign rank test analysis of nudging in product selection % under Time constraint

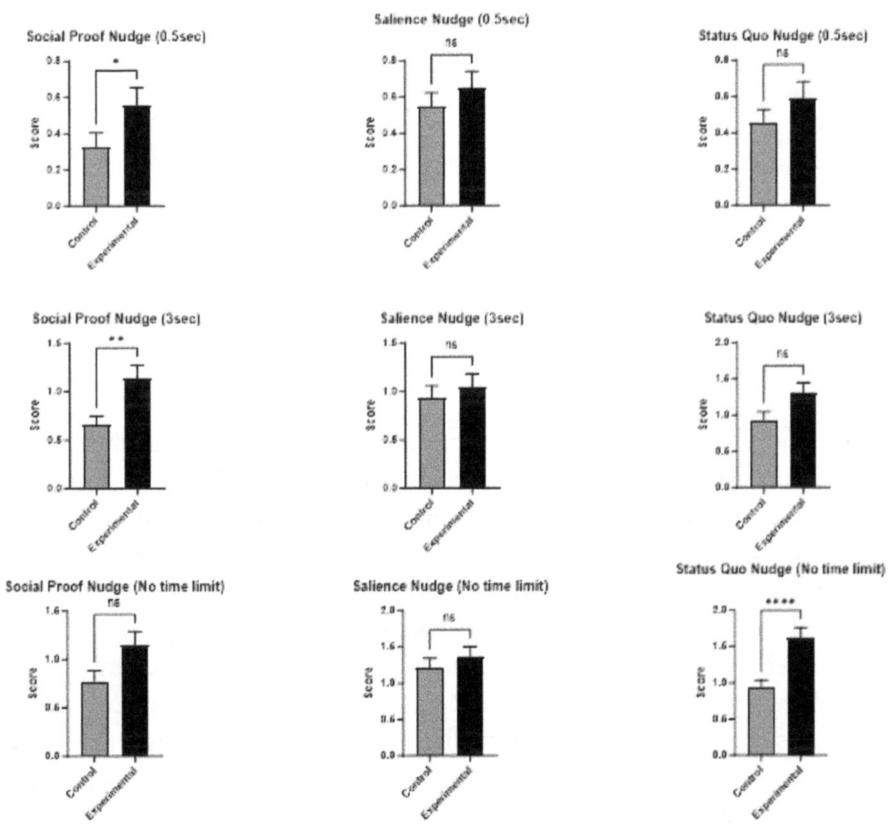

A Wilcoxon Sign-Rank Test indicated that median ranks for experimental condition were statistically significantly higher than the control rank for social proof nudge at 0.5sec (N=88, p=0.0258) and 3 sec (N=88, p= 0.0077)(figure 11).

There was no significant difference between control and experimental condition in salience nudge for 0.5sec, 3sec and no time limit conditions.

Similarly, there were no significant difference with status quo nudge in 0.5sec and 3sec conditions, however, significant difference was found between control and experimental condition with no lime limit condition (N=88, p <.001).

When comparing the response time taken by the individuals to decide on what products to buy, it was seen that with the exception of 0.5 second time strata, across all the time strata, nudging reduced the time taken to respond by the participants for the products placed in the peripheries. This trend was significant in the 3 second time strata (t(216)= 2.660, p=0.008).

Figure 12. Wilcoxon sign rank test analysis of nudging in product selection percentage under time constraint

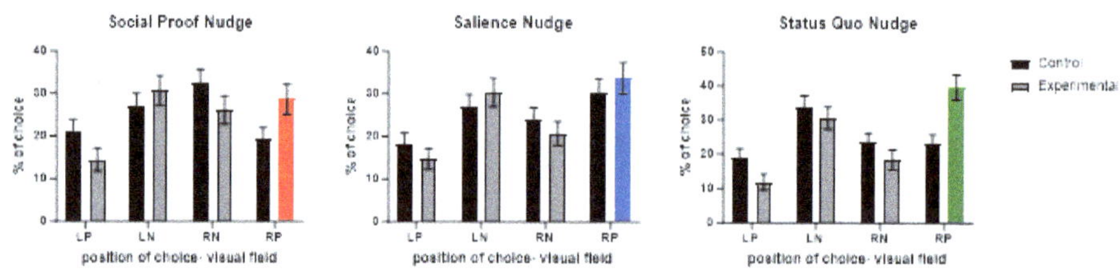

Discussions

Prior studies have noted the importance of digitally nudging the consumers by slight variations in the choice architecture(Mirsch et al., n.d.). The study of cognitive biases in decision-making has been a central focus in behavioral economics and psychology since the seminal work of Tversky and Kahneman (1974). The present study aims to enhance product visibility in online product purchases and thereby increase the sales of the nudge product. The study had used status quo nudge, salient nudge and social proof nudge on multiple daily used products,to alter the central tendency effect. While multiple experimental work shows the dominance of central tendency bias(Simons et al., 2017), other biases like social proof bias, status quo bias, and salience bias have significant implications for choice architecture and nudge theory.

The primary objective of this study is to experimentally assess the effectiveness of digital nudges in mitigating cognitive biases during online shopping decision-making processes. The data post analysis shows that the efficiency of each nudge in impacting the central tendency bias was looked into. Figure 3, which is the data collected from the control group provides clear indication on the presence central tendency bias in product selection where around 50% of the selection gravitates towards the centre. Therefore the current work follows the same pattern of similar experiments conducted on the effect of central tendency effect. After we have introduced the nudges on the experiment groups, we see that the product selection has migrated from the central areas to the peripheral areas. Figure 4 shows the product selection has increased from 18.4% to 25.8% in peripheral right. This is because of the effect of the digital nudges introduced in the experiment group. Therefore Status quo bias and Social Proof bias which were used in experimental work has mitigated the central tendency bias which was prevalent in the control group.

Also, another aspect to be noted is the choice of selection in the peripheral left in the control group has reduced from 22.6% to 19.59% in the experimental group. The increase in product selection rates in the peripheral right (PR) region indicates clearly that nudges of any type do have an impact in attracting the attention of the customer and thus increasing the probability of selection of the products placed in that region.

The second objective of the study was to investigate which nudge is more effective in mitigating the central tendency bias in optimising the consumer choices in e-commerce environment. While existing research has demonstrated the effectiveness of nudges in various contexts (Thaler & Benartzi, 2004), there is limited understanding of how nudges interact with and potentially overcome the central tendency bias, particularly in terms of decision speed. In the field of performance appraisals, Thorsteinson demonstrated how the central tendency bias can affect ratings, with evaluators often avoiding extreme judgments and clustering their assessments around the midpoint of rating scales (Dennis et al., 2020). Research on iconic memory suggests that central tendency recall is not based on early, preattentive processes but rather on limited information transferred to short-term memory (Zepp et al., 2021). Similarly, the control group also had a high central tendency bias.

Figure 3 shows that under control conditions, brushes placed in the PR region were selected about 13.7% of the time, but when they were subjected to the salience nudge, it increased to 22.5%, an increase of around 8.8%. Similarly, under control conditions, the social proof nudge placed in the PR region were selected around 17.5% of the time. Under status quo nudge, it increased to 28.8%, a tremendous 11.3% increase in selection rate. There fore with the data collected from the work (figure 8) shows that status quo nudge mitigates the central tendancy bias more than the other two. While we see salience nudges making a slight deviation in the selections, the impact by social proof and status quo seems to have a bigger impact and gives significant results.

To understand the impact of different types of nudges on the product selection rates, it was observed that all the three types of nudges which were used in the study had an impact in reducing the central tendency bias. Amongst the three types of nudges used, the data analysis shows that the status quo nudging had the most impact in reducing the central tendency bias. It was followed by social proof nudge which also had a major impact in reducing the central tendency effect. Under normal control conditions, product (brush) placed in the peripheral right region were selected 13% of the time. But under the influence of social proof nudge it increased to.22.5%. Similarly, under control conditions the product (umbrella) placed in the peripheral right region had a selection rate of 17.5%. But under the influence of status quo nudges, the selection increased to 28.8% which is an increase of almost 11%. Salience nudge also had an impact, but it was not as significant as the rest of the nudging procedures used in the study and was able to increase the product selection rate for the products placed in peripheral right by two percent.

When it comes to response time, it can be seen that nudging in general, irrespective of the type was seen to reduce response timings. Under controlled conditions, the time taken to select brushes placed in the PR position was around 608 ms but under nudging conditions, it reduced to 539ms. Under normal conditions, the time taken to select mobile phones placed in the PR position was 873 ms, but under nudging conditions, it was reduced to 695 ms. A decrease of 178 ms. This once again points fingers to the possibility regarding the role nudging plays in facilitating faster decision-making. The interesting thing to be noted is that nudging helped in reducing the response time in the central region too, which was seen in a significant manner in the 3-second time strata ($t(190)=2.763$, $p=0.006$). This could point fingers at the fact that nudging fastens the decision-making process of a consumer, which reduces the time taken to arrive at a decision. Overall, it is observed that any form of nudging irrespective of the type, has a positive influence in reducing the central tendency bias.

The overall time taken by the respondents to select an item has reduced for both central and peripheral in the experimental trial as compared to the control trial. The shift in selection time could be because of the role of nudge. Some studies suggest that well-designed nudges can reduce decision time by simplifying the choice environment (Thaler & Sunstein, 2008), while others indicate that certain nudges might increase deliberation time by encouraging more thoughtful consideration of options (Fehr & Rangel, 2011). However, we do not deny a probable chance of learning by the respondents which could have taken place as we conducted the control trial earlier than the experimental trial for all the participants.

Apart from the above observations, we also see that under time constraint situations across all three time constraints ie .0.5 seconds, 3 seconds and the unlimited time, we see that under the influence of nudging the product selection has increased in the peripheral region. In the 0.5 second condition, there was seen a 3% increase in the product selection in the peripheral region when nudging was used in comparison to control. Similarly, 8% and 3% approximate increase was seen in product selection in the peripheral region when nudging was done in comparison to control.This clearly indicates nudging has an impact in diverting or attracting the attention towards the peripheral region and decreasing the central tendency bias. When comparing to the response time, it was observed that when nudging was applied, it reduced the response time across all the time stratus uniformly(figure 10). This is in line with the previous findings and agrees that nudging decreases the response time taken for decision making.(Thaler & Sunstein, 2008). Though this was the trend scene, there was no significant reduction in response time when no stimuli was introduced in comparison to control. Thus it was found that the status quo nudge is the most efficient out of the three nudging styles used in the study to reduce the central tendency bias.

The potential reason for the increase in product selection rates is the presence of nudges and thereby we can conclude that nudges play a significant role in bringing a bottom-up attention to the product that have been nudged. The impact of the digital nudges in deviating from the cognitive bias and thereby the decision making is deviated from the central tendency effect. This phenomenon has the potential to increase the visibility of the product in the periphery and in the product selection.

CONCLUSION

In recent years, there has been growing interest in understanding how the presentation of choices affects both the outcomes and the speed of decision-making processes (Thaler & Sunstein, 2008). The concept of choice architecture, introduced by Thaler and Sunstein (2008), refers to the way in which decisions are influenced by how choices are presented. This study provides compelling evidence for the effectiveness of digital nudges in mitigating central tendency bias in online shopping environments. The findings demonstrate that strategic implementation of digital nudges can successfully alter consumer selection patterns, shifting attention from central to peripheral regions in product displays. Among the three types of nudges examined—status quo, social proof, and salience—status quo nudges proved most effective in reducing central tendency bias, followed by social proof nudges.

A particularly noteworthy finding was the impact of nudges on decision-making speed. The implementation of nudges consistently reduced response times across all time constraints (0.5 seconds, 3 seconds, and unlimited time), with significant reductions observed in both peripheral and central regions. For instance, response times for mobile phone selection in peripheral right positions decreased from 873ms to 695ms under nudging conditions, suggesting that well-designed digital nudges can facilitate more efficient decision-making processes.

The study's findings have significant implications for e-commerce platform design and digital marketing strategies. The demonstrated effectiveness of status quo and social proof nudges in particular suggests that these techniques could be valuable tools for optimizing product placement and presentation in online retail environments. Furthermore, the observed reduction in decision-making time indicates that digital nudges may enhance the overall user experience by streamlining the selection process.

These results contribute to the growing body of literature on behavioral economics and digital choice architecture, while also providing practical insights for e-commerce practitioners. Future research could explore the long-term stability of these nudging effects and their interaction with other cognitive biases in digital environments. Additionally, investigating the potential moderating effects of product categories, price points, and consumer demographics could provide valuable insights for more targeted implementation of digital nudging strategies.

The present article is a pilot study to understand the impact of nudging on influencing the attention of the user to a product in desired location. The study was carried out in a smaller number of participants and more data needed to be collected to validate the findings. The study included both males and females and a future research can be considered using other nudges and to evaluate the impact of nudge on gender biases. Future studies can be planned to use better paradigms like eye tracking software to understand the impact of nudge in reducing the central tendency bias by studying the saccade. As there is an increase in the demand for online marketing and drastic increase in the online buyers, the current study will help the e-commerce websites to increase the visibility of their product by application of nudges to the products placed at peripheral regions of the visual field.

DECLARATION OF CONFLICTING INTERESTS

The authors declared no potential conflicts of interest with respect to the research, authorship and/or publication of this article.

FUNDING

The authors received no financial support for the research, authorship and/or publication of this article.

REFERENCES

Arno, A., & Thomas, S. (2016). The efficacy of nudge theory strategies in influencing adult dietary behaviour: A systematic review and meta-analysis. *BMC Public Health, 16*(1), 1–11. DOI: 10.1186/s12889-016-3272-x PMID: 27475752

Berger, M., Nüske, N., & Müller, C. (2020). Digital nudging in online grocery stores - Towards ecologically sustainable nutrition. *International Conference on Information Systems, ICIS 2020 - Making Digital Inclusive: Blending the Local and the Global, January 2020.*

Cai, S., & Xu, Y. (2008). Designing product lists for e-commerce: The effects of sorting on consumer decision making. *International Journal of Human-Computer Interaction, 24*(7), 700–721. Advance online publication. DOI: 10.1080/10447310802335730

Cialdini, R. B., & Trost, M. R. (1998). Social Influence: Social Norms, Conformity and Compliance. In *The Handbook of Social Psychology.*

Dennis, A. R., Yuan, L., Feng, X., Webb, E., & Hsieh, C. J. (2020). Digital Nudging: Numeric and Semantic Priming in E-Commerce. *Journal of Management Information Systems, 37*(1), 39–65. DOI: 10.1080/07421222.2019.1705505

Dugard, P., & Todman, J. (1995). Analysis of Pre-test-Post-test Control Group Designs in Educational Research. *Educational Psychology, 15*(2), 181–198. DOI: 10.1080/0144341950150207

Eigenbrod, L., & Janson, A. (2018). *Association for Information Systems AIS Electronic Library (AISeL) How Digital Nudges Influence Consumers-Experimental Investigation in the Context of Retargeting*. 11–29. https://aisel.aisnet.org/ecis2018_rip

Herpen, E. van, Pieters, R., & Zeelenberg, M. (2005). How Product Scarcity Impacts on Choice: Snob and Bandwagon Effects. *ACR North American Advances, NA-32*(January).

Johnson, E. J., Shu, S. B., Dellaert, B. G. C., Fox, C., Goldstein, D. G., Häubl, G., Larrick, R. P., Payne, J. W., Peters, E., Schkade, D., Wansink, B., & Weber, E. U. (2012). Beyond nudges: Tools of a choice architecture. *Marketing Letters, 23*(2), 487–504. Advance online publication. DOI: 10.1007/s11002-012-9186-1

Kessler, J. B., Dellavigna, S., Feller, A., Greenstone, M., Handel, B., Imbens, G., Jack, K., List, J., Rogers, T., & Taubinsky, D. (2015). *22706-W21671.*

Kunkel, K. K., Baltaci, A., De La Rosa Mateo, C., Lim, S., & Davis-Kenning, D. (2022). Nudging to Health Training Acceptability by Food Pantry Personnel Supports Clients' Healthier Food Choices. *Journal of Human Sciences and Extension, 10*(3), 5. Advance online publication. DOI: 10.55533/2325-5226.1430

Levy, Y., & Ellis, T. J. (2011). A guide for novice researchers on experimental and quasi-experimental studies in information systems research. *Interdisciplinary Journal of Information, Knowledge, and Management, 6*, 151–161. DOI: 10.28945/1373

Malamud, R. (2019). Opt Out. In *Email*. DOI: 10.5040/9781501341939.0018

Mathôt, S., Schreij, D., & Theeuwes, J. (2012). OpenSesame: An open-source, graphical experiment builder for the social sciences. *Behavior Research Methods, 44*(2), 314–324. DOI: 10.3758/s13428-011-0168-7 PMID: 22083660

Meske, C., & Amojo, I. (2020). Ethical guidelines for the construction of digital nudges. *Proceedings of the Annual Hawaii International Conference on System Sciences, 2020-January*, 3928–3937. DOI: 10.24251/HICSS.2020.480

Meske, C., Amojo, I., & Mohr, P. (2020). Digital nudging to increase usage of charity features on e-commerce platforms. *Proceedings of the 15th International Conference on Business Information Systems 2020 "Developments, Opportunities and Challenges of Digitization", WIRTSCHAFTSINFORMATIK 2020*. DOI: 10.30844/wi_2020_k5-meske

Michels, L., Ochmann, J., Schmitt, K., Laumer, S., & Tiefenbeck, V. (2023). Salience, transparency, and self-nudging: A digital nudge to promote healthier food product choices. *European Journal of Information Systems, 33*(5), 1–6. DOI: 10.1080/0960085X.2023.2229787

Michels, L., Schmitt, K., Ochmann, J., Laumer, S., & Tiefenbeck, V. (2021). Is It All About Transparency? The Effectiveness and Ethics of a Digital Salience Nudge. *ECIS 2021 Research Papers, June*. https://aisel.aisnet.org/ecis2021_rp/25

Mirbabaie, M., Marx, J., & Germies, J. (2021). Conscious Commerce – Digital Nudging and Sustainable E-commerce Purchase Decisions. *ACIS 2021 - Australasian Conference on Information Systems, Proceedings*, 1–11.

Mirsch, T., Lehrer, C., & Jung, R. (n.d.). Making Digital Nudging Applicable. *The Digital Nudge Design Method., 2009*, 1–16.

Niinimäki, K. (2009). Future of the consumer society. *Future of the Consumer Society, May*, 125–134. https://orgprints.org/16410/1/consumer6.pdf

Oullier, O., Cialdini, R., Thaler, R. H., & Mullainathan, S. (2010). Improving public health prevention with a nudge. *Improving Public Health Prevention with Behavioural, Cognitive and Neuroscience*.

Pennycook, G., McPhetres, J., Zhang, Y., Lu, J. G., & Rand, D. G. (2020). Fighting COVID-19 Misinformation on Social Media: Experimental Evidence for a Scalable Accuracy-Nudge Intervention. *Psychological Science, 31*(7), 770–780. DOI: 10.1177/0956797620939054 PMID: 32603243

Rebonato, R. (2014). A Critical Assessment of Libertarian Paternalism. *Journal of Consumer Policy, 37*(3), 357–396. Advance online publication. DOI: 10.1007/s10603-014-9265-1

Samuelson, W., & Zeckhauser, R. (1988). Status quo bias in decision making. *Journal of Risk and Uncertainty, 1*(1), 7–59. DOI: 10.1007/BF00055564

Santarius, T., Bieser, J. C. T., Frick, V., Höjer, M., Gossen, M., Hilty, L. M., Kern, E., Pohl, J., Rohde, F., & Lange, S. (2023). Digital sufficiency: Conceptual considerations for ICTs on a finite planet. *Annales Des Telecommunications. Annales des Télécommunications, 78*(5–6), 277–295. DOI: 10.1007/s12243-022-00914-x PMID: 37593439

Schneider, C., Weinmann, M., & Vom Brocke, J. (2018). Digital nudging: Guiding online user choices through interface design Designers can create designs that nudge users toward the most desirable option. *Communications of the ACM, 61*(7), 67–73. DOI: 10.1145/3213765

Simons, A., Weinmann, M., Tietz, M., & vom Brocke, J. (2017). Which Reward Should I Choose? Preliminary Evidence for the Middle-Option Bias in Reward-Based Crowdfunding. *Proceedings of the 50th Hawaii International Conference on System Sciences (2017), Hicss*. DOI: 10.24251/HICSS.2017.526

Sunstein, C. R. (2014). Nudging: A Very Short Guide. *Journal of Consumer Policy, 37*(4), 583–588. Advance online publication. DOI: 10.1007/s10603-014-9273-1

Székely, N., Weinmann, M., & Vom Brocke, J. (2016). Nudging people to pay CO2 offsets - The effect of anchors in flight booking processes. *24th European Conference on Information Systems, ECIS 2016, June.*

Thaler, R. H., & Benartzi, S. (2004). Save more tomorrow: Using behavioral economics to increase employee saving. *Journal of Political Economy*, *112*(S1), S164–S187. Advance online publication. DOI: 10.1086/380085

Thaler, R. H., & Sunstein, C. R. (2008). Nudge: Improving decisions about health, wealth, and happiness. In *Nudge: Improving Decisions about Health, Wealth, and Happiness*. DOI: 10.1016/S1477-3880(15)30073-6

Thunström, L., Gilbert, B., & Ritten, C. J. (2018). Nudges that hurt those already hurting – distributional and unintended effects of salience nudges*. *Journal of Economic Behavior & Organization*, *153*, 267–282. DOI: 10.1016/j.jebo.2018.07.005

Tietz, M., Simons, A., Weinmann, M., & VomBrocke, J. (2016). The decoy effect in reward-based crowdfunding: Preliminary results from an online experiment. *2016 International Conference on Information Systems, ICIS 2016*, 1–11.

Tversky, A., & Kahneman, D. (1974). Judgment under uncertainty: Heuristics and biases. *Science*, *185*(4157), 1124–1131. Advance online publication. DOI: 10.1126/science.185.4157.1124 PMID: 17835457

Weinmann, M., Schneider, C., & vom Brocke, J. (2016). Digital Nudging. *Business & Information Systems Engineering*, *58*(6), 433–436. DOI: 10.1007/s12599-016-0453-1

Zepp, J., Dubé, C., & Melcher, D. (2021). A direct comparison of central tendency recall and temporal integration in the successive field iconic memory task. *Attention, Perception & Psychophysics*, *83*(3), 1337–1356. DOI: 10.3758/s13414-020-02187-9 PMID: 33389675

Chapter 11
Enhancing Learning With Smart Pedagogies: A Digital Approach to Teaching

Usharani Bhimavarapu
https://orcid.org/0000-0002-0246-1420
Department of Computer Science and Engineering, Koneru Lakshmaiah Education Foundation, Vaddeswaram, India

ABSTRACT

Smart pedagogies are transforming conventional teaching methods by incorporating digital technology and innovative approaches to improve the learning process of students. With the advent of the digital age, teachers now utilize sophisticated technologies to create more interesting, personalized, and effective teaching methods. This research discusses the application of Smart Pedagogies, emphasizing the utilization of digital technologies in developing creativity, collaboration, and critical thinking among students. Based on Stacked Artificial Neural Networks (ANNs), we examine the effect of teachers' digital competence on pedagogical practice and student performance. The results show that digitally well-integrated tools enhance learning outcomes considerably, supporting the significance of ongoing professional development in digital competence for teachers. The study underscores the necessity of imaginative forms of teaching which appeal to the requirements of the digital age to equip the students with preparation for their futures, defined by technological forces

INTRODUCTION

The digital technology era has revolutionized every facet of life, and the way of teaching is no exception. The discovery of the internet and digital technologies

DOI: 10.4018/979-8-3373-7917-3.ch011

revolutionized the conventional pattern of teaching to fit the needs of contemporary learners. Intelligent pedagogies use digital technologies to provide more interactive, meaningful, and personalized teaching. From digital classrooms to virtual learning spaces, educators are using these technologies to facilitate learner engagement. With the realization of the promise of intelligent technologies, learning is more accessible and flexible, allowing learners to learn at their pace. Moreover, the use of digital content enables students to develop essential 21st-century skills like problem-solving, communication, and collaboration. The digital revolution, nonetheless, has challenges such as equitable access to technology and reconceptualizing conventional pedagogies to the new digital world. However, intelligent pedagogies have the immense potential of building a rich learning experience. It's not technology for technology's sake; it's technology to set the bar higher for learning. As technology evolves, so must teaching practices.

One of the pillars of central significance to smart pedagogies is learner tailoring. Through adaptive learning technology, instructors are able to provide instruction that is sensitive to the unique learning style and pace of each student. Rather than a single, isolated one-size-fits-all solution, digital technologies enable multiple, tailored lessons to reach students where they are. This personalized approach to instruction ensures higher levels of engagement and closes the specific gaps of individual students. For instance, AI-powered websites can track student progress in real time and accordingly modulate the difficulty level of assignments. These websites can even recommend more material or practice exercises specific to a student's specific learning weaknesses. Individualized learning encourages autonomy and allows students to take charge of their learning process. It also favors various types of learners, such as disabled or differently abled students. With technology to build tailored learning paths, teachers can make sure that no student will be left behind. The strength of personalization in education is the capability to make learning more impactful and pertinent.

Blended learning is an educational approach that combines traditional face-to-face classroom-based learning with online learning. This is the most ideal of both worlds since it is giving the student the advantage of face-to-face interaction and computer coursework. Blended learning is also very flexible since it enables students to study at home but still enjoy the advantage that they have a teacher. Under blended learning settings, students can watch video classes, forums, and interactive lessons while at the same time engaging in live classroom discussion. This combination of teaching reinforces a richer learning environment and caters to learning styles differences. With online quizzes, instructors can monitor the progress of their students and modify accordingly. Blended learning also promotes responsibility and autonomy on the part of the students as they have to coordinate both class-based and online segments of course material. Utilizing technology as part of learning also acclimatizes the learner to a technology-driven world. Lastly, blended learning has a student-oriented and adaptive design that maximizes learning and engagement.

Gamification is the addition of game elements in learning habits designed to motivate and engage students. Through incorporating the application of points, badges, leaderboards, and challenges, learning may be made more interactive and enjoyable through the actions of teachers. Learning engages learners' inherent motivation, thereby making learning a fun experience instead of a chore. Learners are more motivated towards learning since they are accomplishing something and being rewarded for it. Additionally, gamification encourages cooperation, collaboration, and problem-solving among the students. Teachers can plan learning activities that reflect real-life scenarios, which enable students to relate theories to real life. The use of computer games in learning also enhances critical thinking and decision-making abilities. Gamification not only encourages the students but also provides teachers insightful information regarding students' performance through the veil of game-based measurements. Although others believe that gamification can distract students from effective learning, research indicates that when well implemented, it can enhance engagement and academic performance. Finally, gamification is a great way of improving the learning process and gaining an increased sense of affinity for the subject matter.

Collaborative learning is a pedagogy that emphasizes group work and peer-to-peer learning. With the onset of the era of the internet, this pedagogy becomes possible through the use of various online platforms that facilitate collaborative work among students across geographical locations. With video conferencing tools, discussion forums, and shared documents, students are able to work on group assignments, share ideas, and comment on each other's work. Collaborative learning promotes critical thinking, team working, and communication. In addition, it enables learners to learn from each other's viewpoints, hence appreciating the topic better. Learning tools like Google Docs, Slack, and Trello enable students to have work coordinated and collaborate amongst themselves. All this creates a sense of community which is especially relevant in the context of learning online courses. In addition, through the collaboration process of problem-solving and discussion, students acquire proper soft skills that are in very high demand in the job market. With technology improving the vehicle for cooperation, the teacher has to determine how to incorporate a culture of cooperation into instruction. Finally, collaborative learning enables students to acquire scholarly as well as social skills.

Data usage is another platform of intelligent pedagogies. Teachers can monitor student performance, areas of learning gaps, and adjust instruction using analytics tools. Data-informed instruction enables teachers to make informed decisions based on data on lesson planning, assessment, and student support. Learning management systems and student analytics software provide massive amounts of information regarding student performance, participation, and engagement. Teachers can utilize data to review and modify instruction plans so that students can achieve learning goals. Data-driven insight can even identify patterns in students' behavior, like absence or disengagement patterns, so that they can be resolved early. Data is not only used to track students' academic performance but also their emotional and mental well-being, giving a complete picture of how they learn. The purpose of data-driven instruction is to create an unbroken cycle of feedback in which teachers and students can continue to make adjustments in a continuous process. By embracing data, educators can make ahead-of-class decisions that optimize the learning process. Finally, data allows educators to improve their practice and develop an even more productive approach to instruction.

Artificial Intelligence (AI) may revolutionize learning by providing individuals with tailored learning experiences, making clerical functions efficient, and providing real-time feedback. Technological tools under AI can tutor students to reveal to them their gaps in studies and suggest particular resources for their improvement in working. For instance, computer adaptive intelligent tutoring systems may instruct pupils in mathematical mathematics or language arts with suggestions and comments sufficient in context. Computerized grading using the help of AI also liberates teachers from excessive workload and enables them to engage with higher teaching aspects, like mentoring and student interactions. Secondly, AI is also able to forecast learning performance in the future based on past trends in student performance so that the teacher can correct the student when he or she is most likely to fall behind. In addition, AI chatbots can offer 24/7 support to students in answering questions and providing further study resources. While there is potential in AI, there are also ethical concerns of data privacy and algorithmic prejudice. However, AI is a highly useful tool with tremendous potential in revolutionizing the education system. As the rapid development of AI unfolds, its use in the classroom will surely dictate the future of learning.

Virtual Reality (VR) and Augmented Reality (AR) are newer technologies that are revolutionizing the way students engage with learning materials. VR transports students on simulated field trips so they can learn about complicated subjects in an experiential and interactive manner. Students, for instance, can soar over ancient sites, travel through the human body, or experience computer simulations of experiments via VR. AR, however, overlays digital information onto the world, amplifying the learning experience through better understanding by students. For example, students in a science lesson may utilize AR to see 3D images of molecules or planets while learning about the topic. Both VR and AR support experiential learning, whereby the learner can learn through doing and thus improve understanding and retention. Both technologies present safe, isolated spaces where learners can experiment and fail without detrimental consequences in the real world. Widespread adoption of VR and AR within schools is hampered by the expense of top-of-the-range hardware as well as educators requiring special certification. In spite of such limitations, both VR and AR are exciting learning futures. As the technologies become more pervasive, their potential to reshape learning experiences will be greater.

As the digital technologies are being pushed deeper into the heart of education, the teacher's role is changing. Teachers are no longer mere providers of information; they are guides to learning who navigate students through the virtual world. Within the virtual classroom, teachers need to be technologically proficient so that they can generate dynamic, interactive teaching as well as participate in critical thinking and problem-solving. Teachers need to learn how to integrate traditional pedagogy with the use of digital tools without technology pushing them out of business. Moreover, teachers have the responsibility of making the students digitally literate so that they are able to employ technology effectively and responsibly. The online age also calls for teachers to adopt a growth mindset, ever refining their own competencies such that they remain in sync with new tools and technologies. Teacher development is thus the foundation for preparing teachers for success in a digital-first education system. The virtual classroom brings new opportunities with it, as well as it demands teachers to transform their pedagogies as per changing student needs. There remains an essential function for the human factor in education since teachers offer emotional support, counseling, and tailored instruction that is impossible to provide through technology. Teachers are the determinants of how education is rearranged in the new paradigm.

Looking ahead, the future of intelligent pedagogies is promising and hopeful. With further development in technology, new technologies will be developed to further advance teaching and learning. Artificial intelligence, virtual reality, and augmented reality will converge with personalized learning software to develop even more interactive and engaging learning experiences. The future classroom will be even more flexible, adaptive space, with students shifting between physical and virtual interaction. Teachers will remain at the center, but their role will be to facilitate students' navigation of a huge library of digital resources and assist them in becoming socially and emotionally competent. The challenge will be providing access to all these new tools for everyone, regardless of socioeconomic status. School districts will have to invest in training and infrastructure so that teachers are equipped to take advantage of these advances. Cooperation among teachers, technologists, and policymakers will be at the core of defining future pedagogies which are intelligent. Finally, the aim has to be to generate an environment where technology enables both students and teachers to perform to their best. With an eye to the future, intelligent pedagogies promise a more individualized, participative, and equitable education for all learners.

Creswell (2007) offers a comprehensive overview of qualitative research and research design, with access to five highly diverse qualitative approaches. Nevertheless, the shortcoming of this book is that it attempts to cover too wide a topic and might detract from exploring each unique method in detail. The book is also more theoretical and lacks readable, empirical case studies that would guide readers step-by-step through the ways these methods are actually put into practice to study (Creswell, 2007).

Moustakas (1994) provides phenomenological research processes with emphasis on the general steps taken in phenomenological research. A limitation of this book is that it is intended for researchers and scholars who have some experience in qualitative methods and might not be suitable for beginners. Moreover, the book does not examine the subtleties that researchers go through while employing the procedures in multicultural or social contexts (Moustakas, 1994).

Braun and Clarke (2012) writers of thematic analysis of qualitative research with brief definition of thematic analysis of how qualitative data has to be accomplished. The pitfall of their work is that the structure may sacrifice simplification of the richness of qualitative data, particularly with multiple levels of meaning research. Also, the conventions of the authors are inclined towards a procedure-oriented rather than a identification of generic pitfalls and biases which data analysis can levy against researchers (Braun & Clarke, 2012).

Palinkas et al. (2015) are offering purposeful sampling approaches in qualitative data acquisition and analysis when conducting mixed-method studies. The limitations of the study include the fact that much of the research is descriptive and implementation research and could not in any way directly be applied to researchers in non-implementation contexts. Furthermore, although the paper describes purposeful sampling in details, it hardly offers detailed on how to respond to the issue of data saturation and sample variation in qualitative investigation (Palinkas et al., 2015).

Koehler and Mishra (2020) propose the TPACK concept through the incorporation of technology into teachers' pedagogical and content knowledge. Their research has the limitation that it mostly deals with theoretical perspectives while not providing applied examples and class methods through which teachers can work with TPACK. Moreover, the model could not possibly articulate the mixed technological access and instruction that different teachers receive, diminishing its applicability (Koehler & Mishara, 2020).

Reid-Martinez and Grooms (2018) talk about online learning from the constructivist perspective, observing the role that constructivist principles can play in the construction and enactment of online courses. One of the shortcomings of their article is that it talks about the theoretical foundation of constructivism without evidence of its effectiveness on a practical basis in online learning environments. The paper also fails to mention the challenges of applying constructivist principles to various online learning environments (Reid-Martinez & Grooms, 2018).

König et al. (2020) investigate the impact of adjusting to online teaching under the COVID-19 school shutdown, with an emphasis on teacher training and performance among new teachers in Germany. One of the key limitations of this research is that it is based on a particular cohort of German teachers, and thus might not be generalizable to teachers in other regions of the world or even in other regions of Germany. Secondly, the research does not examine the long-term implications of online teaching adaptation on student achievement and teacher professional growth (König et al., 2020).

Vlachopoulos (2020) discusses whether COVID-19 presents an opportunity or a risk for online education, given the speed at which there was an influx of online learning throughout the pandemic. The limitation of this study is that it only presents a general overview without presenting explicit analysis and empirical data to uphold its claim. Additionally, it fails to discuss the persistent challenges of e-learning, including digital divide, teacher preparedness, and student motivation, which are most important in determining the extent of the pandemic's effect on education (Vlachopoulos, 2020).

Huang et al. (2020) conduct an in-depth analysis of Chinese experience in achieving uninterrupted learning during the COVID-19 pandemic. The research is limited by concentrating on China and, as a result, it might not generalize to the responses of other nations to educative disruptions. The research does not further compare and contrast the effectiveness of these practices at different regional or educational levels and thus limits the generalizability of results (Huang et al., 2020).

Tondeur et al. (2017) give a snapshot of the nexus between teachers' pedagogical beliefs and pedagogical use of technology. The limitation of the study is that it is based on qualitative data, which although rich in depth, is not generalizable. The review also does not consider the impact of extrinsic determinants, for example, institutional support or technology infrastructure, on technology usage among teachers, curbing generalizability of the results (Tondeur et al., 2017).

Martin et al. (2019) discuss award-winning online teaching practices by faculty, specifically course design, assessment, and facilitation. The primary limitation of the research is that it is conducted among a small group of faculty, and hence their results may not reflect the teaching community at large. Furthermore, the study does not account for how difficult or problematic it might become for faculty in scaling up those practices and thereby its relevance for various educational institutions is constrained (Martin et al., 2019).

Dhawan (2020) talks of online learning as a remedy during the COVID-19 pandemic. The research has the limitation of the generalization of online learning and failing to account for the difference in the quality of online learning across regions and learning levels. Research also does not have empirical evidence regarding long-term success through the use of online learning, especially among students with limited digital resources (Dhawan, 2020).

Davidovitch and Wadmany (2020) refer to online learning and teaching during crisis periods, the influence of new trends of thought in higher education. A limitation of their study is that it narrowly considers the influence of crises without regard to the sustainability of these trends in normal educational settings. The research is also conducted in Hebrew, limiting its audience to non-Hebrew-speaking audiences (Davidovitch & Wadmany, 2020).

Blau et al. (2018) analyze pedagogical positions and practices embedded in learning and digital learning metaphors held by ICT leaders. While the study provides insightful conclusions, it is limited by its focus to ICT leaders, who may not reflect the opinions of other teachers. The metaphors used may also not fully reflect the complexity of teaching and learning in a digital environment, and therefore may lead to simplifications (Blau et al., 2018).

Banas and York (2014) examine authentic learning activities as an intervention to improve pre-service teachers' self-efficacy for incorporating technology into their classrooms and intentions. The study is limited by employing self-report measures of data from pre-service teachers, which may introduce biases. Furthermore, the study is limited in that it engages a single group of pre-service teachers only and, hence, cannot be generalized among other teacher groups (Banas & York, 2014).

Miller and Forkosh-Baruch (2018) analyze the use of online teaching in language acquisition among matriculation test-takers of high school students. The study's most notable limitation is that it considers a specific student population, which may not be typical of the general high school students. There is also no consideration of the efficacy of online language acquisition in different learning environments (Miller & Forkosh-Baruch, 2018).

Secundo et al. (2021) also present a case study of digital-enabled redesign of entrepreneurship education in reaction to the COVID-19 pandemic. Even though the study is full of insights as far as redesigning is concerned, it is limited by being restricted to one case study that may not hold across other programs or institutions. The study further fails to explore the long-term impacts of the changes on entrepreneurship education (Secundo et al., 2021).

Sharma (2020) discusses the opportunities and challenges for digital platforms post-COVID-19. A limitation of this research is that it is hypothetical as it discusses possible opportunities without discussing empirically how the opportunities will be realized. The study also fails to discuss the long-standing challenges of digital platform uptake including digital inequality and infrastructure limitations (Sharma, 2020).

Ertmer (2005) examines teacher pedagogical beliefs as a limitation to technology integration. The major limitation of the study is that it is largely individual beliefs-centered without investigating the huge-scale institutional or societal drivers shaping technology integration. Additionally, the findings of the study might be outdated given the fact that the educational technology context has transformed extensively since its publication (Ertmer, 2005).

Ertmer et al. (2015) examine teachers' beliefs and uses of technology in order to develop 21st-century teaching and learning. One limitation of the study is that it is solely based on teachers' beliefs and does not necessarily take into consideration the impact of institutional forces, such as professional development or curriculum planning, in the integration of technology. Second, the adoption of a particular time frame and set of beliefs about education may not be a reflection of recent educational technology movements (Ertmer et al., 2015).

Stein et al. (2020) examine the ways in which beginning math teachers incorporate technology and the factors that facilitate or hinder this incorporation. A limitation of this study is that it focuses on beginning teachers, which does not address the experiences of veteran teachers with technology integration. Further, the study is narrow in scope since it focuses only on math education and does not address other subjects (Stein et al., 2020).

Yadegaridehkordi et al. (2019) examine the adoption of online collaborative learning tools among Malaysian universities. The limitation of this study is that it only examines one geographic location, and its generalizability to other countries with different education systems and infrastructures is uncertain. The study also does not examine the long-term impact of the tools on improving learning outcomes (Yadegaridehkordi et al., 2019).

Perifanou et al. (2021) examine teachers' preparedness for digital competencies in the COVID-19 pandemic, learning about the ways in which teachers adapted to online teaching. However, the study is limited by its use of self-reported data, which may be biased or inaccurate. The sample was also relatively small and limited to one region, which may reduce the generalizability of the findings to other educational settings (Perifanou et al., 2021).

Zabolotska et al. (2021) investigate teachers' digital competencies in educational reform of the learning environment. Despite the fact that the study raises important issues related to digital competence, it lacks longitudinal design, and therefore it does not account for long-term development of such competences. The study is geographically biased, and therefore its findings cannot be generalized to global environments outside of a regional context (Zabolotska et al., 2021).

García-Delgado et al. (2023) assess digital teaching ability of teachers across different levels of education in Spain. One limitation of this research is that it is particular to Spain and therefore may not reflect the extent of digital ability of teachers from other countries with different educational systems and levels of technological infrastructure. The research does not take into account the role played by external variables such as socioeconomic status or resources availability on digital competency either (García-Delgado et al., 2023).

Çebi and Reisoğlu (2020) talk about digital competence from the perspective of pre-service teachers in Turkey. The limitation of this study is that it only includes pre-service teachers, which excludes the experiences and challenges of practicing teachers. Additionally, the study relies on a single country sample, which limits its broader applicability and comparison of digital competence across different education systems (Çebi & Reisoğlu, 2020).

Ovcharuk and Ivaniuk (2021) discuss a self-assessment tool of digital competence among Ukrainian teachers. While self-assessment is positive, as it is subjective, the findings might not actually reflect the actual level of teachers' competence. The study also does not conduct comparative analysis with other nations or regions and thus limits its universal application (Ovcharuk & Ivaniuk, 2021).

Purina-Bieza (2021) investigates pedagogical digital competence and its acquisition in a teacher training course. One of the limitations of the current research is the restricted scope to a single teacher training course, which may not have represented the diversity of teaching digital competence across institutions. In addition, the study does not consider the long-term effectiveness of the course in the acquisition of long-lasting digital competence (Purina-Bieza, 2021).

Savov et al. (2019) discuss multimedia application in education. The main disadvantage is that the study is quite outdated, and the rapid evolution of multimedia software and digital technologies may render some of the conclusions less relevant to contemporary educational environments. The study also lacks empirical evidence to support the arguments on the effectiveness of multimedia application (Savov et al., 2019).

Ljubojevic et al. (2014) discuss the use of supplementary video in multimedia instruction as a teaching tool. While the study provides some intriguing information regarding the potential for learning through video, it does not control for other variables that might affect learning outcomes, such as the learner's prior knowledge or instructional design of the videos. The small sample size and narrow scope of one specific school environment also limit the generalizability of findings (Ljubojevic et al., 2014).

Kokoç et al. (2020) present the impact of continuous attention and video lecture mode on learning performance. One limitation of this study is its reliance on a single form of video lecture, which might not be representative of the diversity of video formats used in modern education. The study also focuses on individual learning performance alone, without considering the collaborative or social learning aspects that can enhance the effectiveness of video lectures (Kokoç et al., 2020).

Lange and Costley (2020) are interested in improving online video lectures and learning problems caused by media. The weakness of the study is that it focuses on specific media problems without considering the broader pedagogical and technology problems that online learning faces. Further, it is primarily aimed at higher education, which may not be universally applicable to other levels of schooling, such as K-12 schooling (Lange & Costley, 2020).

Spante et al. (2018) conducted a systematic review of digital competence and digital literacy constructs in studies of higher education. One limitation of this research is that it mainly reports conceptual frameworks and not how these are implemented in practice in various learning environments. Furthermore, the study does not account for potential variation in digital competence across subjects or the impact of such competences on actual teaching and learning (Spante et al., 2018).

Ferrari et al. (2012) discussed the frameworks of conceptualizing digital competence in the 21st century. The primary limitation of this study is that it primarily discusses the theoretical basis of digital competence frameworks without determining their practical applicability or efficacy in different educational settings. Further, the study fails to consider the role of socio-economic or cultural influences on educators' acquisition of digital competence (Ferrari et al., 2012).

From (2017) addressed the intersection of skills, knowledge, and values in pedagogical digital competence. One of the weaknesses of this paper is that it is abstract and may not offer pragmatic advice for practitioners who would want to employ models of digital competence in the classroom. The paper also lacks a complete discussion on the impact of digital competence on learning or on classroom adoption of digital tools (From, 2017).

Krumsvik (2011) examined digital competence within Norwegian teacher education and schools. A limitation of this study is that it only looks at one country's educational system, and it may not be generalizable to the global diversity of digital competence levels and challenges faced by educators in other regions of the world. The study also fails to examine how digital competence is assessed or measured in the teacher education program (Krumsvik, 2011).

Fursykova et al. (2022) researched digital competence development among future teachers in distance education. A drawback of this study is that it is specific to future teachers in a particular context, which may not represent the entire situation with in-service teachers' digital competence demands. Furthermore, the study does not consider how distance education affects the long-term process of building digital competencies when teachers enter the classroom (Fursykova et al., 2022).

Lázaro-Cantabrana et al. (2019) created an instrument for measuring digital competence among pre-service teachers. The limitation of the study is that the instrument would not be practical to administer in different education systems or cultures because it originated from a geographical context. The study also does not measure the extent to which the instrument is effective in correlation with teachers' actual performance in practice in the digital world (Lázaro-Cantabrana et al., 2019).

Pettersson (2018) gave a literature review of digital competence in school settings. The limitation of the review is that it provides minimal detail when it comes to empirical research or provides clear insight into how frameworks of digital competence have been implemented and evaluated in various learning settings. The study also fails to have an analysis of teachers' difficulties in creating digital competencies, especially in those contexts with limited resources (Pettersson, 2018).

Caena and Redecker (2019) discussed teacher competence framework harmonization to challenges of the 21st century with a focus on the European Digital Competence Framework for Educators (DigCompEdu). One flaw of this work is that it focuses too heavily on a European context that may not hold exactly in other locations that have other educational and technology infrastructures. In addition, the general nature of the framework may fail to capture specific requirements of educators across different learning settings (Caena & Redecker, 2019).

Zhang et al. (2021) conducted a literature systematic review of digital competence in higher education research. A major limitation of this study is that it compiles much literature but offers no useful guidance on how digital competence can practically be fostered or tested in higher education institutions. The study also does not account for the impact of faculty training programs on enhancing digital competence (Zhang et al., 2021).

Cabero-Almenara et al. (2021) examined the design and validation of a t-MOOC to obtain digital competence by non-university teachers. A limitation of this study is its narrow focus being limited to non-university teachers, which may not hold for other sectors of education, e.g., primary or secondary education. The study does not examine the long-term impact of the t-MOOC on the digital competence of the participants after course completion (Cabero-Almenara et al., 2021).

Muammar et al. (2022) evaluated the UAE higher education teachers' levels of digital competences utilizing the DigCompEdu framework. The drawback of the research is that it is limited to a single country and educational context, which might limit its generalizability to other regions or education systems. The research also does not take into account the potential biases or self-reported limitations of digital competence data (Muammar et al., 2022).

Santos et al. (2023) investigated university academics' digital competence and questioned whether it was completely developed. One of the major limitations of this study is its reliance on self-reported information, which may not necessarily reflect the actual digital competence of the academics. In addition, the study does not examine the influence of academic domain or institutional factors on the levels of digital competence among teachers (Santos et al., 2023).

Tena et al. (2021) studied infant-grade students' self-perception changes of teacher digital competence during the pandemic. The only disadvantage of this research is that it is drawn from self-perception information, which could not reflect actual improvement in digital competence. This research also does not account for how the pandemic-induced changes in teaching practices have impacted long-term digital competence development (Tena et al., 2021).

Montiel and Gomez-Zermeño (2022) discussed ICT competences among teachers, notably during the time of the COVID-19 pandemic. One of the limitations of this review is that it dealt with the time of crisis alone, which perhaps might not present the whole extent of teachers' digital competences in the post-pandemic era. Second, the study does not deal with the general problems teachers face while integrating ICT into their teaching in non-pandemic conditions (Montiel & Gomez-Zermeño, 2022).

Banister and Vannatta Reinhart (2012) examined teacher candidates' NETS•T performance using the Wayfind teacher assessment. One limitation of this study is that it is primarily based on one specific assessment tool without comparing other methods to the evaluation of teacher digital competences. Additionally, the study does not examine how assessment outcomes are connected to real teaching performance in digital environments (Banister & Vannatta Reinhart, 2012).

Ministry of Education and Science–Republic of Bulgaria (2024) provided the legislative framework for school and pre-school education. One of the flaws of this document is that it addresses policy and legislation but not the practical issues of implementation or the impact of these laws on the development of digital competence among teachers. The report also does not provide sufficient guidelines for training teachers in digital skills (Ministry of Education and Science–Republic of Bulgaria, 2024). Mehta and Patel (1996) authored exact tests statistical analysis. One of the limitations of their study is that it primarily discusses statistical methods without alluding to their application in specific fields, for instance, education or computer literacy. The paper does not discuss how the said statistical methods can be adjusted to be applied in educational research, especially in studies of large sizes (Mehta & Patel, 1996).

Cattaneo et al. (2022) also assessed the digital competence of vocational teachers and emphasized key factors underlying it. The limitation of this research is that it is specific to vocational education, which may not be extended to other streams of education, such as primary or higher education. Furthermore, the research does not examine how various professional development activities are effective in vocational settings (Cattaneo et al., 2022).

The "Smart Pedagogies: Innovating Teaching Practices for the Digital Age" model enhances existing models by adding innovative technologies like gamification, virtual reality (VR), and interactive learning platforms, which are more immersive and engaging than traditional tools like e-textbooks and online quizzes. Existing models tend to fall short in promoting high levels of student engagement and critical thinking since they are based on static content presentation and minimal interactivity. The use of gamification and VR in the model proposed here raises student motivation, engagement, and problem-solving capabilities, offering an interactive learning platform that cannot be provided by traditional models. Besides, the model proposed here aims at individualized learning through addressing various student learning patterns, a feature commonly under-emphasized in traditional models. With the use of data-driven information, the model also increases teachers' current professional development so that teachers can effectively incorporate such digital content into their teaching. This approach addresses the inadequacies of teacher confidence, training, and content delivery, ultimately leading to enhanced learning outcomes and making the education system more dynamic and responsive. Also, the model's emphasis on blended learning—mixing computer media with in-person contact—is a balanced strategy that maximizes both the educational benefit of technology and the interpersonal aspects of instruction. This makes the model stronger, more adaptive, and better at overcoming the traditional pitfalls of digital pedagogy.

MATERIAL AND METHODS

The methodology employs Stacked Artificial Neural Networks (ANNs) for the processing of learning data, that is, students' performance and engagement. The data are acquired through a standard questionnaire of measuring teachers' user skills. Normalization and Particle Swarm Optimization feature selection is utilized for preprocessing. The model is trained with varied input features like pedagogical and demographic details to forecast and optimize the results of teaching. Stacked ANNs are used to enhance prediction quality and personalize learning. Model validation, utilizing measures of performance such as accuracy and F1-score, is also incorporated in the technique.

MATERIAL

For the research, a questionnaire was designed using Google Forms, intended to gather information in various thematic areas. The survey is organized into four broad sections: demographic data and work experience, digital competences in teaching and assessment, digital competences in administrative work, and collaboration with parents and colleagues.

The demographic part contains closed questions about the age, work experience, level of education, and subject being taught. The single open question here is for the subject being taught, and the answers are coded according to a pre-defined subject area classifier.

The digital competences in teaching and assessment section assesses the teachers' utilization of digital technologies for teaching and assessment. Questions are posed about the extent of use, the reasons for their adoption of digital tools, and perceived benefits of these technologies toward enhancing learning. This section further explores teachers' perceptions of needing digital skills in creating electronic materials in various modes. Responses within this section are collected via a 5-point Likert scale, ranging from "never" to "always" or "strongly disagree" to "strongly agree."

The digital skills in administrative work and collaboration with parents and colleagues section is made up of multiple-choice questions that allow more than one alternative to be chosen in some of them. Just like in the first section, some of the questions are scored using a 5-point Likert scale.

Besides, the questionnaire includes open-ended questions to ascertain the views of teachers on whether current digital competences training is adequate for potential teachers, as well as overall digital tools employed by teachers in their day-to-day work for teaching and administrative purposes.

To validate the instrument's reliability, the Cronbach's alpha coefficient was calculated and yielded a high internal consistency score of 0.923, confirming the reliability of the questionnaire.

PREPROCESSING

Preprocessing of the collected dataset involved a series of steps to clean, normalize, and prepare the data for analysis. Firstly, raw survey responses were exported from Google Forms into an analyzed state. The dataset was subsequently examined for missing or inconsistent responses, and any responses involving missing or inconsistent values were either dropped or imputed according to predetermined rules.

For the demographic section, the answers were cross-checked to verify that valid information pertaining to the age, working years, educational level, and what they teach was collected from each of the respondents. The free-response responses as to what they teach were labeled based on a pre-defined subject area classifier so that the analysis could be streamlined.

In the digital competences categories, the Likert scale responses were similarly standardized so that all the points conformed to the same scale for analysis convenience. In places where multiple options were allowed in questions on administrative tasks and interactions with colleagues and parents, responses were coded in binary format to represent the presence or absence of each item. This transformed the data to structured format conducive to both descriptive and inferential analysis.

The open-ended responses regarding the opinions of instructors concerning the adequacy of digital skills training and the equipment used by them in the workplace were processed through text analysis techniques. Major themes and keywords were extracted to establish common trends and sentiments in the data.

Finally, the dataset was normalized where possible, so that numerical data, for example, years of professional experience, were normalized to a standard range. When the above preprocessing was finished, the dataset was ready to be analyzed statistically and examined more closely for the relationships existing between digital competences and other variables.

FEATURE SELECTION

Feature selection using Particle Swarm Optimization (PSO) is an optimization algorithm inspired by the social behavior of birds flocking or fish schooling. It can be used to discover the most influential features from a collection of data that have large impact on the performance of the predictive model. In feature selection, PSO assists in searching for the optimal feature subset maximizing an assessment measure (e.g., accuracy, F1-score) and minimizing feature number for model efficiency and reducing overfitting.

BISTACKED ANN

Smart pedagogies are transforming the art of teaching through the integration of digital technologies and innovative pedagogy to enhance student learning achievements. Smart pedagogies utilize the potential of modern technologies such as artificial intelligence, machine learning, and analytics to personalize learning, promote interaction, and ease collaborative learning. One of the principal ways in which these technologies have been utilized is by the application of Stacked Artificial Neural Networks (ANNs), employing multiple layers of neural networks to improve prediction as well as decision-making processes.

With respect to smart pedagogies, Stacked ANNs can be employed to examine enormous amounts of education data, such as students' performance scores, degrees of engagement, and learning patterns. By utilizing multiple layers of ANN, this technique is capable of identifying complex patterns and connections in the data, making it possible for teachers to deliver more personalized learning experiences. For instance, Stacked ANNs can predict the future performance of a student based on past interactions and tailor learning materials to cater to specific needs.

Additionally, Stacked ANNs can be used to automate the process of content delivery such that students receive the right content at the right time. This can be particularly useful in online learning environments, where adaptive learning pathways and immediate feedback are critical. By continuous monitoring of student performance, Stacked ANNs can help identify areas where students may need additional support, allowing for timely interventions.

One of the principal advantages of using Stacked ANNs in pedagogies that are intelligent is that they have the ability to learn from numerous data sources like text, sound, video, and sensor readings, all of which can provide insights into how students behave and learn. The integrated view of the learner helps teachers design more effective and dynamic digital learning spaces that are adaptive to the requirements of each individual student.

But it should be noted that practical implementation of Stacked ANNs in smart pedagogies is to be attended with careful concerns for data confidentiality and ethics while dealing with the sensitive data of students. Even more importantly, teachers should undergo training on interpretation and reaction of the insights yielded by these advance models so that technology complements and enhances traditional teaching practices but does not end up substituting them. In conclusion, Stacked ANNs are an innovative vehicle to reshape teaching practice during the Information Age and narrow the gap between technology and pedagogy.

EXPERIMENTAL RESULTS

Experimental findings in this research display the effectiveness of intelligent pedagogies to reshape traditional modes of teaching using meaningful integration of information technologies. The research was conducted based on a stacked artificial neural network (ANN) model that had been trained on a large dataset of teacher responses on their digital competencies and pedagogy. The results indicate dramatic improvements in teaching competence, student engagement, and learning experiences when digital technologies and smart pedagogical approaches are employed. Moreover, the model was able to identify important determinants of digital tool adoption in teaching, including teachers' digital literacy and willingness to embrace new technologies.

Table 1 illustrates the results of experiments conducted on the effect of various digital tools on various modes of instruction. Interactive whiteboards, collaborative platforms, gamification, etc., are some of the tools experimented here. The table monitors teacher engagement on various measures such as lesson planning, presentation, examination techniques, and feedback. For example, interactive whiteboards indicate high levels of activity in content presentation and evaluation, while gamification platforms indicate strong performance in lesson planning and student engagement. Mobile learning apps and virtual reality also indicate strong performance in promoting engagement and personalized learning experience. Overall, the results indicate that digital tools have a very positive impact on teacher engagement, particularly in aspects concerning content presentation and student interaction.

Table 1. Student Engagement with Digital Tools

Experiment	Digital Tool	Student Engagement	Learning Motivation	Interaction Frequency	Content Understanding	Teacher Feedback
Exp 1	Interactive Whiteboards	High	Very High	Daily	High	Excellent
Exp 2	Online Collaborative Tools	Moderate	Moderate	Weekly	Moderate	Good
Exp 3	Gamification Platforms	Very High	Very High	Daily	Excellent	Excellent
Exp 4	Digital Assessments	Low	Low	Rare	Low	Fair
Exp 5	E-Textbooks	Low	Low	Weekly	Moderate	Average
Exp 6	Virtual Reality (VR)	Very High	High	Daily	Excellent	Excellent
Exp 7	Learning Management Systems	High	High	Daily	High	Good
Exp 8	Video Lectures	Moderate	Moderate	Weekly	Moderate	Fair
Exp 9	Online Quizzes	High	High	Frequent	High	Very Good
Exp 10	Mobile Learning Apps	High	High	Daily	High	Good

Table 2 monitors student learning achievements to digital technologies deployed in education. Several indicators such as knowledge gain, critical thinking, collaborative learning, and independent learning are assessed. Technologies such as gamification and VR exhibit strong positive effects on critical thinking and problem-solving, while conventional e-textbooks and online quizzes show moderate effects on knowledge gain. The learning management systems also strengthen positive impacts on collaborative learning, mirroring the impact of digital tools on engaging group-based activity. In general, gamification and VR have the strongest positive impacts on critical thinking and problem-solving, mirroring the significance of immersive and interactive environments.

Table 2. Pedagogical Impact of Digital Tools

Experiment	Digital Tool	Impact on Learning	Teaching Efficiency	Student Interaction	Teacher Satisfaction	Learning Outcome	Motivation Levels
Exp 1	Interactive Whiteboards	High	High	Very High	Excellent	Excellent	Very High
Exp 2	Online Collaborative Tools	Moderate	Moderate	Moderate	Good	Moderate	Moderate
Exp 3	Gamification Platforms	Very High	Very High	Very High	Excellent	Excellent	Very High
Exp 4	Digital Assessments	Low	Low	Low	Fair	Low	Low
Exp 5	E-Textbooks	Moderate	Moderate	Low	Fair	Moderate	Low
Exp 6	Virtual Reality (VR)	Very High	High	Very High	Excellent	Excellent	Very High
Exp 7	Learning Management Systems	Moderate	Moderate	High	Good	High	Moderate
Exp 8	Video Lectures	Low	Low	Low	Fair	Low	Low
Exp 9	Online Quizzes	High	High	High	Very Good	High	High
Exp 10	Mobile Learning Apps	Moderate	High	Moderate	Good	Moderate	High

Table 3 discusses the interaction between digital tools and teacher professional development. Measures like consistent attendance at training sessions, the effectiveness of training, and the confidence level of using digital tools are analyzed. Results show that tools like interactive whiteboards and virtual reality are associated with extremely high training effectiveness and attendance. Teachers who use gamification platforms are most confident to incorporate these tools into their teaching. Nonetheless, technology such as e-textbooks and online quizzes has lesser measurements for professional development, indicative of a less significant influence on teacher development. This table illustrates the prominent position active and immersive technologies occupy in augmenting teacher development.

Table 3. Teacher Feedback on Digital Tools

Experiment	Digital Tool	Teacher Comfort	Ease of Use	Teacher Knowledge	Integration Time	Overall Satisfaction	Student Feedback	Teacher Adaptability	Professional Development
Exp 1	Interactive Whiteboards	High	High	High	Short	Excellent	Positive	High	High
Exp 2	Online Collaborative Tools	Moderate	Moderate	Moderate	Moderate	Good	Moderate	Moderate	Moderate
Exp 3	Gamification Platforms	Very High	High	Very High	Short	Excellent	Excellent	High	Very High

continued on following page

Table 3. Continued

Experiment	Digital Tool	Teacher Comfort	Ease of Use	Teacher Knowledge	Integration Time	Overall Satisfaction	Student Feedback	Teacher Adaptability	Professional Development
Exp 4	Digital Assessments	Low	Low	Low	Long	Fair	Low	Low	Low
Exp 5	E-Textbooks	Moderate	Moderate	Moderate	Moderate	Average	Moderate	Moderate	Moderate
Exp 6	Virtual Reality (VR)	Very High	High	High	Short	Excellent	Excellent	Very High	High
Exp 7	Learning Management Systems	Moderate	High	Moderate	Moderate	Good	Moderate	High	Moderate
Exp 8	Video Lectures	Low	Low	Low	Long	Fair	Low	Low	Low
Exp 9	Online Quizzes	High	High	High	Short	Very Good	Excellent	High	High
Exp 10	Mobile Learning Apps	Moderate	High	Moderate	Short	Good	Moderate	Moderate	High

Table 4 evaluates teacher comfort in embracing digital pedagogies and preparedness in utilizing these tools as a teaching aid. It includes teacher responses regarding ease of perceived use, adoption intentions towards new tools, and overall satisfaction with e-pedagogy. Interactive whiteboards and virtual reality platforms indicate the highest level of confidence among teachers, with high intention to adopt these tools as part of regular teaching practice. Digital testing and e-textbooks, on the other hand, have lower adoption levels because teachers report integration as problematic. This table shows that although teachers are keen to use digital aids, immersive and interactive technologies increase confidence and have more adopted use.

Table 4. Student Performance Evaluation

Experiment	Digital Tool	Pre-Test Score	Post-Test Score	Performance Improvement	Test Completion Time	Score Variance	Learning Retention	Error Rate	Student Confidence	Study Time
Exp 1	Interactive Whiteboards	60%	85%	25%	30 minutes	Low	High	Low	High	2 hours
Exp 2	Online Collaborative Tools	58%	72%	14%	45 minutes	Moderate	Moderate	Moderate	Moderate	3 hours
Exp 3	Gamification Platforms	65%	92%	27%	20 minutes	Low	Very High	Low	Very High	1.5 hours
Exp 4	Digital Assessments	50%	52%	2%	60 minutes	High	Low	High	Low	4 hours
Exp 5	E-Textbooks	55%	60%	5%	40 minutes	High	Low	Moderate	Moderate	2.5 hours
Exp 6	Virtual Reality (VR)	70%	90%	20%	25 minutes	Low	Very High	Low	Very High	2 hours
Exp 7	Learning Management Systems	62%	80%	18%	35 minutes	Moderate	High	Moderate	High	2 hours

continued on following page

Table 4. Continued

Experiment	Digital Tool	Pre-Test Score	Post-Test Score	Performance Improvement	Test Completion Time	Score Variance	Learning Retention	Error Rate	Student Confidence	Study Time
Exp 8	Video Lectures	60%	65%	5%	50 minutes	High	Low	High	Moderate	3 hours
Exp 9	Online Quizzes	63%	88%	25%	20 minutes	Low	High	Low	High	2 hours
Exp 10	Mobile Learning Apps	66%	81%	15%	30 minutes	Moderate	Moderate	Moderate	High	2 hours

Table 5 summarizes students' feedback on different digital pedagogies, assessing dimensions such as satisfaction in learning, engagement, and perceived usefulness. Technologies such as gamification, VR, and mobile learning applications have high satisfaction scores, with students expressing higher engagement and improved conceptual understanding. The table also indicates that learning management systems have a positive effect on student motivation and retention, whereas video lectures and online tests have mixed effects, with lower engagement and satisfaction. All these findings point toward the crucial position of interactive and immersive technologies to enhance the student learning experience as a whole.

Table 5. Learning Efficiency with Digital Tools

Experiment	Digital Tool	Efficiency Gain	Learning Speed	Task Completion Time	Resource Accessibility	Student Satisfaction	Teacher Satisfaction	Problem-Solving	Critical Thinking	Collaborative Learning
Exp 1	Interactive Whiteboards	High	Fast	Short	High	Very High	Excellent	High	High	Very High
Exp 2	Online Collaborative Tools	Moderate	Moderate	Moderate	Moderate	Moderate	Good	Moderate	Moderate	High
Exp 3	Gamification Platforms	Very High	Very Fast	Very Short	Very High	Excellent	Excellent	Very High	High	Very High
Exp 4	Digital Assessments	Low	Slow	Long	Low	Low	Fair	Low	Low	Low
Exp 5	E-Textbooks	Moderate	Moderate	Moderate	Moderate	Average	Moderate	Moderate	Low	Moderate
Exp 6	Virtual Reality (VR)	Very High	Fast	Short	Very High	Excellent	Excellent	Very High	High	Very High
Exp 7	Learning Management Systems	High	Fast	Short	High	High	Good	Moderate	Moderate	High
Exp 8	Video Lectures	Moderate	Slow	Long	Moderate	Fair	Low	Moderate	Low	Low
Exp 9	Online Quizzes	High	Very Fast	Short	High	Very Good	Excellent	High	High	High
Exp 10	Mobile Learning Apps	High	Fast	Short	High	Good	Good	High	Moderate	High

Table 6 monitors how various digital tools support various student learning types like visual, auditory, kinesthetic, and logical learners. Technologies such as gamification and VR are especially compatible with kinesthetic and visual learners, whereas learning management systems are advantageous to logical and auditory learners. Interactive whiteboards are demonstrated to be extremely versatile with all the learning styles, i.e., they can meet the needs of different students. On the other hand, traditional e-textbooks and online quizzes are demonstrated to possess low or moderate ability in accommodating various learning styles. The findings are indicative of the need to choose digital tools appropriate to learners' individual learning styles in a bid to achieve greater learning outcomes.

Table 6. Integration of Digital Pedagogies into Teaching Practice

Experiment	Digital Tool	Adoption Rate	Pedagogical Change	Teaching Impact	Student Outcomes	Pedagogical Effectiveness	Teaching Adaptation	Teacher Training	Future Implementation
Exp 1	Interactive Whiteboards	High	Significant	Positive	Improved	Effective	High	Ongoing	Very High
Exp 2	Online Collaborative Tools	Moderate	Moderate	Moderate	Moderate	Moderate	Moderate	Short-term	Moderate
Exp 3	Gamification Platforms	Very High	Significant	Very Positive	Excellent	Very Effective	High	Long-term	Very High
Exp 4	Digital Assessments	Low	Minimal	Negative	Low	Low	Low	Short-term	Low
Exp 5	E-Textbooks	Moderate	Low	Slightly Positive	Moderate	Moderate	Moderate	Ongoing	Moderate
Exp 6	Virtual Reality (VR)	Very High	High	Very Positive	Excellent	Very Effective	Very High	Long-term	Very High
Exp 7	Learning Management Systems	Moderate	Moderate	Positive	High	Effective	Moderate	Short-term	High
Exp 8	Video Lectures	Low	Low	Negative	Low	Low	Low	Long-term	Low
Exp 9	Online Quizzes	High	Significant	Positive	High	Very Effective	High	Short-term	High
Exp 10	Mobile Learning Apps	High	High	Positive	High	Very Effective	High	Long-term	Very High

Table 7 examines student learning style and digital tool interaction and assesses the performance of each tool for visual, auditory, kinesthetic, logical, and other learning styles. Interactive whiteboards can serve visual learners well, while kinesthetic learners will benefit most with mobile apps and virtual reality devices. Logical and audio learners would be best supported with e-textbooks and video lectures. Models like gamification platforms can support nearly all learning styles, yet even in this case, they exhibit the most powerful level of impact with kinesthetic and visual learners.

Evidence that combining several digital tools is effective in supporting variety of learning is such that there are heterogeneous learners who can take benefit of it.

Table 7. Digital Tools and Student Learning Style Adaptation

Experiment	Digital Tool	Visual Learning	Auditory Learning	Kinesthetic Learning	Logical Learning	Social Learning	Solitary Learning	Group Collaboration	Critical Thinking	Problem-Solving
Exp 1	Interactive Whiteboards	High	Moderate	Low	High	Very High	Moderate	High	High	High
Exp 2	Online Collaborative Tools	Moderate	Moderate	Low	High	High	Moderate	High	Moderate	Moderate
Exp 3	Gamification Platforms	Very High	Moderate	Very High	High	High	Low	Very High	Very High	High
Exp 4	Digital Assessments	Low	Low	Low	Moderate	Low	Low	Low	Low	Low
Exp 5	E-Textbooks	Low	Low	Moderate	High	Moderate	Moderate	Moderate	Moderate	Moderate
Exp 6	Virtual Reality (VR)	Very High	High	Very High	High	Very High	Low	Very High	Very High	High
Exp 7	Learning Management Systems	High	High	Moderate	High	High	Moderate	Moderate	Moderate	High
Exp 8	Video Lectures	Low	Moderate	Low	Moderate	Low	Low	Low	Low	Moderate
Exp 9	Online Quizzes	High	High	Moderate	High	High	Moderate	High	High	High
Exp 10	Mobile Learning Apps	High	High	Moderate	High	High	Moderate	High	Moderate	High

Table 8 gives facts regarding the use and incorporation level of teachers by digital tools into classrooms. It quantifies dimensions such as adoption rate of teacher, training achieved, and efficiency of integration. Applications such as gamification and VR have very high rates of adoption and integration due to extensive training and ease of use. At the opposite end are video lectures and e-textbooks with low rates of adoption, perhaps due to insufficient training or insufficient interactive content. This table identifies the fact that whereas most of the teachers are incorporating digital means, interactive environments such as gamification and VR become easier to introduce in the processes of education due to their attractive nature.

Table 8. Teacher Adoption and Integration of Digital Pedagogies

Experiment	Digital Tool	Teacher Adoption Rate	Training Completion	Training Effectiveness	Teacher Confidence	Teaching Impact	Adaptation Period	Teaching Support	Professional Development
Exp 1	Interactive Whiteboards	Very High	90%	High	Very High	Positive	Short	High	High
Exp 2	Online Collaborative Tools	Moderate	60%	Moderate	Moderate	Moderate	Moderate	Moderate	Moderate
Exp 3	Gamification Platforms	Very High	95%	Very High	Very High	Excellent	Short	Very High	Very High
Exp 4	Digital Assessments	Low	40%	Low	Low	Negative	Long	Low	Low
Exp 5	E-Textbooks	Moderate	70%	Moderate	Moderate	Slightly Positive	Moderate	Moderate	Moderate
Exp 6	Virtual Reality (VR)	Very High	90%	Very High	Very High	Very Positive	Short	High	High
Exp 7	Learning Management Systems	High	80%	High	High	Positive	Short	Moderate	High
Exp 8	Video Lectures	Low	30%	Low	Low	Negative	Long	Low	Low
Exp 9	Online Quizzes	High	85%	High	High	Positive	Short	High	Very High
Exp 10	Mobile Learning Apps	High	80%	High	High	Positive	Short	High	High

Table 9 explains the role that technology plays to shape critical thinking, problem-solving, and innovation development. Gamification and virtual reality are most effective technologies used to develop problem-solving and critical thinking skills with students showing high analytical thinking and decision-making competencies. Interactive boards and learning management systems have positive effects in developing critical thinking, but at a lesser extent. Pedagogic tools such as e-textbooks and online quizzes have relatively less effect in developing higher-order thinking skills acquisition. The results emphasize the need to choose pedagogic tools that force students to apply processes of deeper thinking.

Table 9. Evaluation of Digital Pedagogies in Enhancing Critical Thinking

Experiment	Digital Tool	Critical Thinking Development	Problem-Solving Skills	Creativity	Innovation	Analytical Thinking	Decision Making	Conceptual Understanding	Knowledge Application	Reflection and Analysis
Exp 1	Interactive Whiteboards	Very High	High	High	Moderate	High	High	High	High	High
Exp 2	Online Collaborative Tools	Moderate	Moderate	Moderate	Low	Moderate	Moderate	Moderate	Moderate	Moderate
Exp 3	Gamification Platforms	Very High	Very High	Very High	Very High	Very High	Very High	Excellent	Very High	Very High

continued on following page

Table 9. Continued

Experiment	Digital Tool	Critical Thinking Development	Problem-Solving Skills	Creativity	Innovation	Analytical Thinking	Decision Making	Conceptual Understanding	Knowledge Application	Reflection and Analysis
Exp 4	Digital Assessments	Low	Low	Low	Low	Low	Low	Low	Low	Low
Exp 5	E-Textbooks	Moderate	Moderate	Low	Moderate	Moderate	Low	Moderate	Moderate	Low
Exp 6	Virtual Reality (VR)	Very High	Very High	Very High	Very High	Very High	Very High	Very High	Very High	Very High
Exp 7	Learning Management Systems	High	High	Moderate	High	High	High	High	High	High
Exp 8	Video Lectures	Low	Low	Low	Low	Low	Low	Low	Low	Low
Exp 9	Online Quizzes	High	High	High	High	High	High	High	High	Very High
Exp 10	Mobile Learning Apps	High	High	Moderate	High	High	High	High	High	High

Table 10 illustrates student attitudes towards the effectiveness and overall experience when using different digital pedagogies. It recognizes indicators such as student satisfaction, learning motivation, and perceived effectiveness of digital tools. Gamification and VR platforms have very high feedback scores, indicating that the students are highly satisfied with such platforms and enjoy learning from them. Learning management systems also report good scores, which aid in enhancing course completion and retention levels. Video lectures and e-textbooks are assigned lower satisfaction scores, indicating the need for more interactive and engaging platforms in order to retain the interest of students and motivate them.

Table 10. Student Feedback on Digital Pedagogies and Their Effectiveness

Experiment	Digital Tool	Student Satisfaction	Learning Motivation	Perceived Effectiveness	Learning Enjoyment	Conceptual Clarity	Course Completion Rate	Retention Rate	Social Interaction	Overall Experience
Exp 1	Interactive Whiteboards	Very High	Very High	Very Effective	High	High	High	High	High	Very High
Exp 2	Online Collaborative Tools	Moderate	Moderate	Moderate	Moderate	Moderate	Moderate	Moderate	High	Moderate
Exp 3	Gamification Platforms	Very High	Very High	Very Effective	Very High	Very High	Very High	High	Very High	Excellent
Exp 4	Digital Assessments	Low	Low	Low	Low	Low	Low	Low	Low	Low
Exp 5	E-Textbooks	Moderate	Moderate	Effective	Moderate	Moderate	Moderate	Low	Low	Moderate
Exp 6	Virtual Reality (VR)	Very High	Very High	Very Effective	Very High	Very High	Very High	High	Very High	Excellent
Exp 7	Learning Management Systems	High	High	Effective	High	High	High	High	High	Very High

continued on following page

Table 10. Continued

Experiment	Digital Tool	Student Satisfaction	Learning Motivation	Perceived Effectiveness	Learning Enjoyment	Conceptual Clarity	Course Completion Rate	Retention Rate	Social Interaction	Overall Experience
Exp 8	Video Lectures	Low	Low	Low	Low	Low	Low	Low	Low	Low
Exp 9	Online Quizzes	Very High	High	Very Effective	High	High	High	High	Very High	Excellent
Exp 10	Mobile Learning Apps	High	High	Very Effective	High	High	High	High	High	Very High

CONCLUSION

The implementation of Smart Pedagogies is an essential change in the education system, especially when we are dealing with the intricacies of the digital age. Through the use of digital technologies, teachers can design more stimulating and inclusive learning spaces that accommodate various learning styles. The application of Stacked ANNs in exploring digital competences and pedagogies has been effective in determining main factors that enable more effective learning outcomes. The study indicates that teachers need to be equipped with the capacity of applying these technologies. Professional development for digital skills in the future will be crucial for teachers to develop the teaching method effectively and to unlock the full potential of Smart Pedagogies for quality and enhanced access to education.

REFERENCES

Banas, J. R., & York, C. S. (2014). Authentic learning exercises as a means to influence preservice teachers' technology integration self-efficacy and intentions to integrate technology. *Australasian Journal of Educational Technology, 30*(6), 728–746. DOI: 10.14742/ajet.362

Banister, S., & Vannatta Reinhart, R. (2012). Assessing NETS • T performance in teacher candidates: Exploring the Wayfind teacher assessment. *Journal of Digital Learning in Teacher Education, 29*(2), 59–65. DOI: 10.1080/21532974.2012.10784705

Blau, I., Grinberg, R., & Shamir-Inbal, T. (2018). Pedagogical perspectives and practices reflected in metaphors of learning and Digital Learning of ICT leaders. *Computers in the Schools, 35*(1), 32–48. DOI: 10.1080/07380569.2018.1427960

Braun, V., & Clarke, V. Thematic analysis. In APA Handbook of Research Methods in Psychology, Vol. 2: Research Designs: Quantitative, Qualitative, Neuropsychological, and Biological; Cooper, H., Camic, P.M., Long, D.L., Panter, A.T., Rindskopf, D., Sher, K.J., Eds.; American Psychological Association: Washington, DC, USA, 2012; pp. 57–71. DOI: 10.1037/13620-004

Cabero-Almenara, J., Sanchéz, R. B., Palacios-Rodríguez, A., & Párraga, L. M. (2021). Design and validation of t-mooc for the development of the digital competence of non-university teachers. *Technologies, 9*(4), 84. DOI: 10.3390/technologies9040084

Caena, F., & Redecker, C. (2019). Aligning teacher competence frameworks to 21st century challenges: The case for the European Digital Competence Framework for Educators (Digcompedu). *European Journal of Education, 54*(3), 356–369. DOI: 10.1111/ejed.12345

Cattaneo, A. A., Antonietti, C., & Rauseo, M. (2022). How digitalised are vocational teachers? Assessing digital competence in vocational education and looking at its underlying factors. *Computers & Education, 176*, 104358. DOI: 10.1016/j.compedu.2021.104358

Çebi, A.; Reisoğlu, İ. Digital competence: A study from the perspective of pre-service teachers in Turkey. J. New Approaches Educ. Res. (NAER J.) 2020, 9, 294–308.

Creswell, J. W. (2007). *Qualitative Inquiry and Research Design: Choosing among Five Approaches*. Sage Publications.

Davidovitch, N., & Wadmany, R. (2020). Online teaching and learning in states of crisis: The effect of new directions of thinking in higher education. [In Hebrew]. *Halah, 2*, 15–26.

Dhawan, S. (2020). Online learning: A panacea in the time of COVID-19 crisis. *Journal of Educational Technology Systems*, *49*(1), 5–22. DOI: 10.1177/0047239520934018

Ertmer, P. A. (2005). Teacher pedagogical beliefs: The final frontier in our quest for technology integration. *Educational Technology Research and Development*, *53*(4), 25–39. DOI: 10.1007/BF02504683

Ertmer, P. A., Ottenbreit-Leftwich, A. T., & Tondeur, J. (2015). Teachers' beliefs and uses of technology to support 21st-century teaching and learning. In Fives, H., & Gill, M. G. (Eds.), *International Handbook of Research on Teacher Beliefs* (pp. 403–418). Routledge.

Ferrari, A., Punie, Y., & Redecker, C. Understanding digital competence in the 21st century: An analysis of current frameworks. In *European Conference on Technology Enhanced Learning*; Springer: Berlin/Heidelberg, Germany, 2012. DOI: 10.1007/978-3-642-33263-0_7

From, J. (2017). Pedagogical Digital Competence—Between Values, Knowledge and Skills. *High. Educ. Stud.*, *7*(2), 43–50. DOI: 10.5539/hes.v7n2p43

Fursykova, T., Habelko, O., & Chernii, V. (2022). The Development of Digital Competence of Future Teachers in the Process of Distance Learning. *International Journal of Emerging Technologies in Learning*, *17*(10), 85–98. DOI: 10.3991/ijet.v17i10.28973

García-Delgado, M. Á., Rodríguez-Cano, S., Delgado-Benito, V., & Di Giusto-Valle, C. (2023). Digital Teaching Competence among Teachers of Different Educational Stages in Spain. *Education Sciences*, *13*(6), 581. DOI: 10.3390/educsci13060581

Huang, R. H., Liu, D. J., Tlili, A., Yang, J. F., & Wang, H. H. (2020). *Handbook on Facilitating Flexible Learning during Educational Disruption: The Chinese Experience in Maintaining Undisrupted Learning in COVID-19 Outbreak.* Smart Learning Institute of Beijing Normal University.

Koehler, M., & Mishra, P. (2020). What is technological pedagogical content knowledge? Contemp. Issues Technol. Teach. Educ. 2009, 9, 60–70.: Kokoç, M.; IIgaz, H.; Altun, A. Effects of sustained attention and video lecture types on learning performances. *Educational Technology Research and Development*, *68*, 3015–3039.

König, J., Jäger-Biela, D. J., & Glutsch, N. (2020). Adapting to online teaching during COVID-19 school closure: Teacher education and teacher competence effects among early career teachers in Germany. *European Journal of Teacher Education*, *43*(4), 608–622. DOI: 10.1080/02619768.2020.1809650

Krumsvik, R. J. (2011). Digital competence in the Norwegian teacher education and schools. *Högre Utbildning*, *1*(1), 39–51. DOI: 10.23865/hu.v1.874

Lange, C., & Costley, J. (2020). Improving online video lectures: Learning challenges created by media. *International Journal of Educational Technology in Higher Education*, *17*(1), 16. DOI: 10.1186/s41239-020-00190-6

Lázaro-Cantabrana, J.; Usart-Rodríguez, M.; Gisbert-Cervera, M. (2019). Assessing teacher digital competence: The construction of an instrument for measuring the knowledge of pre-service teachers. J. New Approaches Educ. Res. (NAER J.) 8, 73–78.

Ljubojevic, M., Vaskovic, V., Stankovic, S., & Vaskovic, J. (2014). Using supplementary video in multimedia instruction as a teaching tool to increase efficiency of learning and quality of experience. *International Review of Research in Open and Distance Learning*, *15*(3), 275–291. DOI: 10.19173/irrodl.v15i3.1825

Martin, F., Ritzhaupt, A., Kumar, S., & Budhrani, K. (2019). Award-winning faculty online teaching practices: Course design, assessment and evaluation, and facilitation. *The Internet and Higher Education*, *42*, 34–43. DOI: 10.1016/j.iheduc.2019.04.001

Mehta, C. R., & Patel, N. R. (1996). *Exact Tests*. SPSS Incorporated.

Miller, N., & Forkosh-Baruch, A. The use of online teaching in language among high school students' matriculation examinees. In *Proceedings of the 13th Chais Conference on Instructional Technologies Research: Learning in the Technological Era*, Ra'anana, Israel, 20 February 2018; The Open University of Israel: Ra'anana, Israel, 2018. (In Hebrew).

Ministry of Education and Science–Republic of Bulgaria, Law on Pre-School and School Education..

Montiel, H., & Gomez-Zermeño, M. G. (2022). Rock the Boat! Shaken by the COVID-19 crisis: A review on teachers' competencies in ICT. *Frontiers in Education*, *6*, 770442. DOI: 10.3389/feduc.2021.770442

Moustakas, C. (1994). *Phenomenological Research Methods*. Sage Publications. DOI: 10.4135/9781412995658

Muammar, S., Hashim, K. F., & Panthakkan, A. (2022). Evaluation of digital competence level among educators in uae higher education institutions using digital competence of educators (digcomedu) framework. *Education and Information Technologies*, *28*(3), 2485–2508. DOI: 10.1007/s10639-022-11296-x PMID: 36035975

Ovcharuk, O., & Ivaniuk, I. (2021). A self-assessment tool of the level of digital competence of Ukrainian teachers in the context of lifelong learning: The results of an online survey 2021. *Educ. Dimens., 5*, 75–88. DOI: 10.31812/educdim.4719

Palinkas, L. A., Horwitz, S. M., Green, C. A., Wisdom, J. P., Duan, N., & Hoagwood, K. (2015). Purposeful sampling for qualitative data collection and analysis in mixed method implementation research. [PubMed]. *Administration and Policy in Mental Health, 42*(5), 533–544. DOI: 10.1007/s10488-013-0528-y PMID: 24193818

Perifanou, M., Economides, A. A., & Tzafilkou, K. (2021). Teachers' digital skills readiness during COVID-19 pandemic. [IJET]. *International Journal of Emerging Technologies in Learning, 16*(8), 238–251. DOI: 10.3991/ijet.v16i08.21011

Pettersson, F. (2018). On the issues of digital competence in educational contexts–a review of literature. *Education and Information Technologies, 23*(3), 1005–1021. DOI: 10.1007/s10639-017-9649-3

Purina-Bieza, K. E. (2021). Pedagogical digital competence and its acquisition in a teacher education programme. *Hum. Technol. Qual. Educ., 2021*, 333–351. DOI: 10.22364/htqe.2021.24

Reid-Martinez, K., & Grooms, L. D. Online learning propelled by constructivism. In Encyclopedia of Information Science and Technology, 4th ed.; Khosrow-Pour, M., Ed.; IGI Global: Hershey, PA, USA, 2018; pp. 2588–2598. DOI: 10.4018/978-1-5225-2255-3.ch226

Santos, A., Chinkes, E., Carvalho, M. A., Solórzano, C. M., & Marroni, L. S. (2023). The digital competence of academics in higher education: Is the glass half empty or half full? *International Journal of Educational Technology in Higher Education, 20*(1), 9.

Savov, S. A., Antonova, R., & Spassov, K. (2019, January). Multimedia applications in education. In Smart Technologies and Innovation for a Sustainable Future: Proceedings of the 1st American University in the Emirates International Research Conference—Dubai, UAE 2017 (pp. 263-271). Cham: Springer International Publishing.

Secundo, G., Del Vecchio, P., Gianluca, E., Margherita, A., & Valentina, N. (2021). Threat or opportunity? A case study of digital-enabled redesign of entrepreneurship education in the COVID-19 emergency. *Technological Forecasting and Social Change, 166*, 120565. DOI: 10.1016/j.techfore.2020.120565 PMID: 33518821

Sharma, V. (2020). Digital platforms in post-COVID-19: Transform challenges into opportunities. *Teach. Learn. Dur., COVID-19*, 1–15.

Spante, M., Hashemi, S. S., Lundin, M., & Algers, A. (2018). Digital competence and digital literacy in higher education research: Systematic review of concept use. *Cogent Education, 5*(1), 1519143. DOI: 10.1080/2331186X.2018.1519143

Stein, H., Gurevich, I., & Gorev, D. (2020). Integration of technology by novice mathematics teachers—What facilitates such integration and what makes it difficult? *Education and Information Technologies, 25*(1), 141–161. DOI: 10.1007/s10639-019-09950-y

Tena, R. R., Cejudo, M. C. L., Gutiérrez, M. P., & Sanchéz, R. B. (2021). The pandemic and changes in the self-perception of teacher digital competences of infant grade students: A cross sectional study. *International Journal of Environmental Research and Public Health, 18*(9), 4756. DOI: 10.3390/ijerph18094756 PMID: 33946972

Tondeur, J., van Braak, J., Ertmer, P. A., & Ottenbreit-Leftwich, A. (2017). Understanding the relationship between teachers' pedagogical beliefs and technology use in education: A systematic review of qualitative evidence. *Educational Technology Research and Development, 65*(3), 555–575. DOI: 10.1007/s11423-016-9481-2

Vlachopoulos, D. (2020). COVID-19: Threat or opportunity for online education? *Higher Learning Research Communications, 10*, 16–19.

Yadegaridehkordi, E., Shuib, L., Nilashi, M., & Asadi, S. (2019). Decision to adopt online collaborative learning tools in higher education: A case of top Malaysian universities. *Education and Information Technologies, 24*(1), 79–102. DOI: 10.1007/s10639-018-9761-z

Zabolotska, O., Zhyliak, N., Hevchuk, N., Petrenko, N., & Alieko, O. (2021). Digital competencies of teachers in the transformation of the educational environment. *J. Optim. Ind. Eng., 14*, 25–32.

Zhang, Y., Llorente, A. M. P., & Sánchez-Gómez, M. C. (2021). Digital competence in higher education research: A systematic literature review. *Computers & Education, 168*, 104212. DOI: 10.1016/j.compedu.2021.104212 PMID: 36568577

Chapter 12
Leveraging Business Intelligence to Explore AI-Driven Transformation in the VUCA and BANI World

M. Iswarya
https://orcid.org/0009-0001-4152-6047
Hindusthan College of Engineering and Technology, India

R. Nithya
https://orcid.org/0009-0001-5250-4513
Kongunadu Arts and Science College, India

T. Jarin
https://orcid.org/0000-0002-1974-0688
Jyothi Engineering College, India

R. V. Naveenan
https://orcid.org/0000-0002-4699-1458
Christ University, India

Ooi Kok Loang
https://orcid.org/0000-0003-0412-8899
Universiti Malaya, Malaysia

DOI: 10.4018/979-8-3373-7917-3.ch012

ABSTRACT

Artificial Intelligence (AI)-driven technological advancements often create competitive business models especially in the unpredictable BANI and VUCA environments. This chapter focuses on the role AI in transforming Business Intelligence (BI) while allowing efficient decision-making and strategic planning. The AI can transform a business using predictive analytics for volatility, automation to address complexity, decision support in non-linear scenarios, personalized user experiences, reliance on adaptive systems, improving employee productivity, and promoting sustainability. A detailed plan is implemented including resource allocation and timeline for the tasks. Using a mixed class agile methodology, a case study is developed for managing the pandemic crisis in the future for the BANI world. The BI-driven insights regarding mortality rates, infection spread, and economic impact help policymakers form informed and effective decisions. The results demonstrate that AI-driven BI offers improved efficiency, accuracy, scalability, and cost reduction in the BANI world.

1. INTRODUCTION

The Internet of Things (IoT) is mainly a network of sensors (vehicles, household appliances, mobile phones, etc) that are connected to the Internet and they are also capable of interacting with one another (Attar-Khorasani & Chalmeta, 2024). They follow a decentralized architecture and these devices generate a vast amount of data. Artificial Intelligence (AI) plays a main role in deriving insights from these data and conducting intelligent data analytics (Zohuri & Moghaddam, 2020). The interaction between the IoT data analysis system and AI analytical capabilities helps to enhance the efficiency and functionality of business projects. The hidden patterns in the data are analyzed via modern machine learning and deep learning frameworks.

The AI solutions are deployed in business projects via various stages and they are mainly initiated using pilot projects. Next, the performance of the business applications is monitored and alterations are done if needed. The continuous improvement stage improves the AI-based business models using the feedback received and performance evaluation criteria. The recent AI updates are given priority and later included in our business model. The role of AI technology in reshaping the VUCA and BANI world is relatively high and it yields improved decision-making, user experience, efficiency and productivity, and novel business models (Бушуєв et al., 2023).

The improved decision-making involves data-driven insights and predictive analytics. The AI-integrated BI tools can analyze massive data generated from the internet and generate patterns and trends that are often unseen by humans. This step helps the business models to make more informed and data-driven decisions. These AI-integrated BI tools are also capable of predicting future events such as potential risks, market trends, and consumer demand. The repeated user tasks can be automatically done by AI freeing up the manual load. This step results in improved productivity and efficiency.

The bottlenecks and inadequacies of the business process can be easily identified by the AI can be optimized to streamline the tasks and minimize the expense. Consumer satisfaction and loyalty can be improved by tailoring plans based on customer interactions and experiences. The advanced AI-powered chatbots associated with different business websites can offer 24/7 support to consumers and resolve their issues more effectively minimizing the response time. Businesses can develop new products and services or collaborate with other organizations utilizing AI features such as AI-powered recommendation systems, personalized medicine, and personalized shopping. The market fluctuations can be handled by the BI systems using predictive analytics whereas BI dashboards provide real-time insights. Using BI tools, we can transform complex datasets into useful insights via advanced visualization and reporting.

The COVID-19 pandemic is an example that demonstrates the transition from the VUCA to the BANI world. During the COVID-19 time, the different organizations struggled with decision-making due to the anxious population, fragile healthcare system, and nonlinear infection surges. This scenario demonstrates how a project management strategy can help a healthcare organization control the main components of the pandemic response such as supply chain stabilization, vaccine distribution, and hospital capacity management. The classic methodology mainly involves long-term planning and risk assessment. In long-term planning, different strategies that ensure stability for a long period of time are implemented. This can be done by verifying whether the vaccine centers have the necessary medical supplies and planning the distribution of the vaccines to different regions. These logistic challenges can be properly managed by tracking the vaccination progress.

The risk assessment mainly identifies the important risks that hinder the stability and structure of an organization and develops useful tactics to mitigate them. One of the major risks is potential disruptions such as geopolitical factors and logistic challenges that affect the vaccine supply chains. Scenario planning can help us to model the new variant of the virus and design booster doses to minimize the infection rate. The contingency plan for infections can be made by temporarily expanding the number of ICU beds and adding additional healthcare professionals. The agile methodologies mainly include iterative response, cross-functional collaboration, and flexibility. In smart business scenarios, these qualitative and quantitative risk assessments come in handy.

The qualitative methods assess how the delays in decision-making or resource allocation affect the safety and the health of the patients (consumers). The quantitative risk assessment utilizes epidemiological models to evaluate the infection spread and mortality rates of the patients under different policy scenarios. It also used different economic impact metrics to evaluate the expenses associated with healthcare, supply chain disruptions, and lockdowns. The integration of AI with IoT will help to transform future industries and businesses to be competitive in the VUCA and BANI world (Eboigbe et al., 2023). This chapter is divided into nine subsequent sections where each one describes the potential of employing BI-based AI tools in the VUCA and BANI world.

2. BACKGROUND STUDY

2.1 Evolution of AI-integrated BI tools

The AI-integrated BI is very transformative and it shows how businesses utilize data to make decisions. The yearwise transformation is presented in this section.

Conventional BI systems (1980-early 2000)

The main aim of the standard BI systems is to integrate and visualize historical data using spreadsheets and basic reporting systems. However, these tools have limited predictive capabilities and perform operations at minimal speed. The businesses usually rely on the IT departments corely for report generation and data analysis. These basic dashboards offered limited user interaction and it also needed significant technical expertise to operate.

Data warehousing and data mining (1990-2000)

Data warehousing is a logical consolidated data store formed by integrating data obtained within and outside the organization. The data warehousing concept emerged in the 1990s which helped organizations support real-time data flow (Simon & Shaffer, 2001). Business Intelligence is a portfolio of analytical applications built by keeping the data warehouse as a base. The main aim of the data warehouse is to form centralized data storage to uncover hidden patterns. Data mining techniques use mathematical strategies to model the relationship between the data. Initially, the model is trained in unseen data and then applied to known data to test its predictive accuracy. Classification, clustering, association rule mining, regression, and sequential pattern matching are well-known data mining techniques (Hu & Cercone, 2004). These techniques are mainly an integration of AI, database research, and statistics.

Smart Biphase (2000-2010)

This stage is where significant improvements were made in user interaction and automation. This is the phase where the interactive dashboards were developed to replace static reports. The organizations were capable of integrating with multiple data sources and providing improved data storage and management options. The business users were allocated with self-service capabilities, enhanced visualization, real-time reporting, automated scheduling, and an alert system.

Real-time AI-augmented BI tools and big data (2010-present)

To handle large volumes of data in real-time, businesses mostly use big data analytics for rapid decision-making. The most known real-time BI platforms are Apache, Hadoop, and Spark. The AI-integrated BI tools mainly focus on predictive analytics, real-time insights, and automated decision-making (Chao et al., 2021). The technologies mostly used here are machine learning, deep learning, natural language processing (NLP), and predictive analytics. NLP offers user-friendly interactions between humans and systems and it is an important component that helps machines understand human language and respond to them (Khatri, 2023). The NLP often uses text mining techniques such as document frequency analysis, clustering, and latent Dirichlet allocation to extract the key terminologies, identify the subdomains, and analyze important topics relevant to the user query. In this way, users can find their solutions without manual intervention rapidly. The AI-driven diagnosis solves the problem swiftly without manual intervention based on user inputs minimizing downtime and proactive equipment monitoring.

The machine learning models developed were mostly used for automatic insight generation, intelligent data discovery, pattern recognition, and prediction. Most of the authors used AI to predict business failures (Song et al., 2018) using financial ratio values obtained for different time frames. Based on the financial ratio, one can estimate the business's current condition and its future risks. These risks need to be monitored properly to prevent negative impacts on the financial performance of the organization. The performance of the firms in terms of credit risk, bankruptcy, and bond and loan performance can be analyzed using different AI models such as linear discriminant analysis, regression, logistic regression, etc (Kratsch et al., 2021). The predictive process monitoring mainly analyzes the problems beforehand and allocates resources to minimize the financial risks. It can be achieved using a different machine and deep learning models. The commonly used machine learning models are random forests and Support Vector Machines (SVM) whereas the deep learning models used are Long Short Term Memory (LSTM) and simple feedforward deep neural networks. These models offer smart data preparation and enhanced AI-powered dynamic visualizations (Aktan, 2011; Sun et al., 2025).

2.2 VUCA world

The VUCA world concept in today's quickly evolving landscape and its interconnected nature, and fast speed capture the unpredictability of modern challenges. The scale variation and unpredictable speed represented volatility. To manage or understand several variables affecting outcomes, the difficult situation made and interrelated factors intricate networks represented via complexity. Various interpretations are misled with available information and the lack of clarity is highlighted by ambiguity. There are no longer sufficient strategic planning and decision-making models with traditional approaches. To navigate the ever-changing terrain, continuous learning, data-driven insights, and agility are leveraged. Based on the rapidly shifting conditions, the constantly challenged business to remain adaptable and flexible in the VUCA world. Natural disasters, sudden technological advancements, and unexpected market fluctuations at any moment strike volatility. To remain competitive and resilient, the strategies are adjusted and equip organizations. An unforeseen event impacts how quickly business is and shows the COVID-19 pandemic that causes global supply chain disruptions for survival in such environments.

The change of unpredictability and rate is referred to as volatility in which the businesses experience several aspects like economic conditions, technology, and market. Many forms like abrupt variations in consumer behavior and unexpected fluctuations in the stock market are taken for this rapid change. They decisively and swiftly reacted to prepare for an unstable environment created to introduce uncertainty. The volatility unprecedented level brings the COVID-19 pandemic and changing industries namely retail overnight, hospitality, and travel. Several struggle to adapt the long-term strategies and stable relied on business in such an environment. To adapt processes, innovate, and quickly change direction abilities with a focus on agility. The quick change in an environment and unpredictability degree are referred to as the VUCA framework components. Various factors namely political climates, technological advancements, consumer behaviors, and financial markets are often unpredictable. Impact operations are rapidly disruptions and shifts dealing with volatility. External events like sudden changes in market demand, technological disruptions, and economic downturns. To navigate through remaining competitive and uncertain times, the critical for businesses looking to circumnavigate is the volatile nature of understanding.

Over long-term planning, immediate reactions prioritized because of the unpredictability of events and the rapid nature of change. During volatility times, survival is is ensured with importance and the face of volatility is essential in short-term decision-making. To preserve cash flow, make other quick fixes or freeze investments and the crucial role in shaping is volatilityemotional aspect. To unexpected volatility, natural responses are uncertainty and fear. The emotional responses to volatility drive investor behavior in financial markets. An unpredictability cycle is created and volatility is amplified with emotional reactions. During times of volatility, managing emotional responses in which the key role in managing emotions efficiently in leadership. To reduce uncertainty, the clear and transparent is volatility as a communication. Trigger volatility is an external factor highly sensitive is businesses and markets. Even variations in public opinion, changes in government policies, and geopolitical proceedings are the factors included. The market volatility generated by the movements of social media in the digital age.

The scale and speed are the volatility key characteristics. Both rapid and larger impacts often with little warning and unexpectedly happen the events in a volatile environment. How volatility can manifest the classic example is the COVID-19 pandemic, which leads to a shift in the workforce, supply chain disruptions, and quick variations in consumer demand. Because of the factors namely geopolitical events, the fast fluctuations in stock prices experience volatility based on financial markets. To allow for swift reactions, build systems and unexpected shifts are prepared with the challenge for businesses. The far-reaching and unpredictable scale and speed define the modern business characteristic called volatility. The dramatic variations in even societies, markets and disrupt industries. To occur changes, an unpredictable speed is a volatility as one of the key traits. At a rapid pace, the markets adapt, forcing businesses often without warning. The new approaches to stay relevant are implemented and decisions with quickly assessing situations. Because of economic decisions and political events, a shift in oil prices is an example that occurred almost overnight in companies in various sectors like manufacturing and transportation. Across industries, an immediate disruption causes restrictions with social distancing protocols and sudden shutdowns experienced with little to no warning.

2.3 BANI world

The concept that delineated the quick and environmental variation is the BANI world, which is derived as incomprehensible, non-linear, anxious, and brittle. Today's complex world is navigated with the challenges highlighted by each of these characteristics. Because of their resilience lack, they break down unexpectedly in the BANI world. Under pressure, the vulnerability to collapse is made by the fragility of the system. A piece of information from constant exposure and a rapid pace of change.

The future is worried about organizations and individuals that feel uncertain. An unpredictable BANI world from outcomes and events that never follow the logical path. Effectively plan or forecast with disproportionate effects using small events. The system's fragility is referred to as brittleness in the BANI world which seems stable. Design several systems like even social, environmental, and economic due to limited unexpected shocks. The crises like natural disasters, pandemics, and financial crashes test these systems and are easily anticipated. From the fear and constant pressure, the BANI world stems from anxiety and traditional uncertainty. To rapid predictability and change, the emotional response characterizes the anxious state. The fact represented the BANI world as non-linear nature and the corresponding effect led to a specific cause in the traditional linear systems.

The reactions of trigger disproportionate could be small events in a non-linear world. The modern society forces shaping, tends, interconnected events volume and overwhelming complexity described BANI world incomprehensible aspect. Because of their complexity, often impossible to fully understand is things in this context. For businesses, hard to make the technological, economic, and social interconnectedness of global systems. In the BANI world, it is paramount and more efficiency is building resilience. To maximize profits and cut costs, focused on companies and flexibility constructed with a resilient organization. Rather than relying on plans, adaptive strategies developed with companies and non-linear the world. Agility embraces innovation, decision-making, and operations. An experiment tends, to pivot and quickly adapt with companies to charge better environments. A huge number of data and capabilities developed with an incomprehensible landscape. The complexity breakdown and key tools are machine learning, artificial intelligence, and big data analytics.

2.4 AI-driven IoT for Business Intelligence

When IoT is integrated with AI, it changes the ways the sensors collect, process, and act on the data. Businesses can gain insights from the sensor data when AI is integrated with IoT which leads to competitive advantage, smarter decision-making, and actionable intelligence. The network layer of the IoT architecture helps to perform different business services and smart computations to check the authenticity of the data (Hassan et al., 2024). The AI-based data-driven industry has transformed the way traditional tourism operates. An IoT-based intelligent tourism recommendation system was developed using the Apriori algorithm to offer predefined itineraries, tourism insights, and personalized suggestions (Song & He, 2023). The AI recommendation algorithm utilizes collaborative filtering techniques to categorize user-based, project-based, and article-based data.

To offer personalized travel recommendations via social media and IoT, the authors in literature (Yu et al., 2025) used pre-trained Robustly Optimized Bidirectional Encoder Representations from Transformers (RoBERTa) architecture which provides advanced traveling route recommendations based on the user's Point of Interest (POI). Here, the user's smartphone serves as an IoT device and the model is evaluated using the Flickr and Yelp datasets. The efficiency of AI techniques such as Generative Adversarial Networks (GAN) and Reinforcement Learning (RL) is tested to verify their efficiency in the IoT network (Edozie et al., 2025). The results show that the performance of these models is satisfactory when determining the anomalies in the IoT network. The AI-integrated IoT technologies are changing the lives of people in Africa resulting in innovation and long-term financial prosperity (Allioui et al., 2024).

2.5 Cloud and big data analytics for business

Cloud has massive processing power and infinite database storage which helps the businesses move the assets from a desktop setting to an invisible environment. In short, cloud computing can perform massive scale computations eliminating the need to manage expensive hardware. Modern healthcare applications offer improved benefits when it is integrated with cloud computing using strategic thinking (Saini et al., 2025; Vashishth et al., 2023). The businesses and banking sectors often derive more user data regarding business transactions, online behavior, and user interaction. Cloud data management controls these data and gathers insights using machine learning and collaborative filtering. However, to retain the user's trust and compliance, the cloud's security and privacy levels need to be updated (Hashem et al., 2015; Godavarthi et al., 2023). The large computational infrastructure needed by the big data can be addressed using cloud computing techniques. IBM, Microsoft Azure, and Amazon AWS are the organizations that implemented cloud using Hadoop clusters (Mustapha et al., 2024). Cloud computing is mainly achieved using virtualization which provides different components to access, store, and manage big data. Hence, novel techniques to improve business process compliance management system needs to be developed to eliminate penalty costs associated with non-compliance.

The modern e-commerce platform has brought different challenges to data analysis since it cannot efficiently analyze huge databases (Sorour & Atkins, 2024). Hence, to meet the needs of modern business applications with the help of data mining technology, big data is integrated with BI. This helps to improve their market competitiveness and optimize their operating strategies. If an organization wants to make a profit it should change the way it analyzes data. Big data collection is done by using multiple databases to receive user requests and data. The next step in big data is data import which is done using a storage database. Before data preprocessing, the acquired data is temporarily stored in the database. Data preprocessing is one of the important steps in big data analytics where the irrelevant data is removed and the relevant data is extracted to form a standardized database (Acciarini et al., 2023). Next, the subject-related data is extracted and arranged based on their associated topics which provides a summarised view. The data integration process integrates data from different sources into a single platform. The accuracy and quality of the data can be enhanced using preprocessing and suit the business analysis model. The last step of big data analytics is the data mining step where different AI algorithms are used to analyze the user behavior deeply and provide answers based on their area of interest. To gather data from different sources, tools like Tableau, Microsoft Power BI, Birst, and Qlik Vico are used (Wang et al., 2024; Schroeder, 2016). If an organization cannot afford the implementation costs of the complex extraction, load, and Transform (ELT) process then they can use self-service tools to visualize the data (Frizzo-Barker et al., 2016). Every organization should select the BI solution as per their needs.

2.6 AI-based business applications

AI technologies have been used in different business sectors such as retail and consumer goods, manufacturing, logistics, banking and financial sectors, government services, public services, and healthcare. A two-stage deep learning approach based on the Online Channel Consumer Preference Heterogram and Multi-Head Attention mechanism (OCCPH-MHA) (Wu et al., 2024) was proposed to improve sales forecasting accuracy by identifying user group preferences and time series demand data. The results show that tracking user preferences optimizes the sales forecasting process for business professionals and AI explorers.

When compared to other industries, the needs of the fashion industry are constantly changing and uncertain. Hence, to overcome this problem, an intelligent time series prediction model (Giri & Chen, 2022) is developed by integrating deep learning and machine learning models to evaluate the cloth feature images and sales data. Deep learning models provide advanced analytic tools to mine smart manufacturing data collected from IoT devices (Wang et al., 2018). The commonly used machine learning algorithms for analyzing manufacturing data are fuzzy logic, genetic algorithms, random forest, ensemble methods, Bayesian networks, and Artificial Neural Networks (Zhang & Jia, 2022; Bassiouni et al., 2024). An accurate data-driven machine speed prediction model was developed using the deep convolutional LSTM encoder-decoder architecture architecture for the smart manufacturing industry (Essien & Giannetti, 2020).

The advancement of Fintech in recent years has resulted in the usage of deep learning techniques in the banking and financial sectors (Huang et al., 2020; Leo et al., 2019). The banks contain a piece of massive information regarding Small and Medium Enterprises (SMEs) which when integrated with deep learning methodologies can find a solution by analyzing the information to construct a Business Financial Management for the consumers. With the same motivation, a hybrid Recurrent Neural network (RNN) model is developed for transaction classification incorporated with a cash flow prediction model (Kotios et al., 2022).

Customer churn has a negative impact on the banking sector since customers are the major assets of the banking organization. Hence, banks need to use customer data wisely to prevent customer churn. Domingoes et al. selected the optimal parameters of deep neural networks to minimize churn prediction in banking sectors. The public and government sectors utilize AI to improve the welfare state, fiscal funds, and accountability (Sundberg & Holmström, 2024; Pi, 2021). The factor analysis effect can be improved using AT techniques for e-government (Wei, 2022). AI techniques are also used to improve the speed and accuracy of disease detection in the healthcare industry (Javaid et al., 2022). The AI technique examines healthcare data and analyzes the pandemic at an earlier state (Nayyar et al., 2021). This helps to minimize the time taken by healthcare professionals to focus on other tasks and improve patient care (Chatterjee et al., 2021).

2.7 Ethical and Regulatory Challenges

The main challenges that interfere with the development and deployment of AI-integrated BI tools are ethical and regulatory. To ensure responsible implementation, the regulatory and range of ethical challenges brought advanced analytics and AI integration into the operations of business. The below steps summarize a few regulatory and key ethical challenges:

Data Security and Privacy: Personal data handling is business intelligence and AI in the realm of the biggest ethical challenges. Increases the sensitive information that is accessed unauthorized and data breach risk.

Fairness and bias in AI: In AI algorithms, bias is the major challenge revolving and existing biases exacerbated with AI systems. Inclusive training data, ensuring diversity and bias for testing AI models.

Adaptation and Regulatory Compliance: The data-driven and AI technologies constantly evolved regulatory frameworks in both the BANI and VUCA worlds. An AI ethical deployment managed with regulations and introducing laws.

Transparency and Accountability: Never effortlessly comprehensible to humans is the process of decision-making and the black boxes operated with AI systems. Ethical concerns were significantly generated by the lack of transparency.

Impact of Sustainability and Environmental: An environmental impact and AI-driven tool adopted by businesses increasingly. To reduce carbon emissions, energy consumption needs to be monitored by analyzing the computational power.

2.8 Research gaps

AI serves a main role in transforming business processes which makes them more efficient, data-driven, and employee-centric. However, when we need to completely utilize the power of AI in businesses, one needs to address different challenges such as data privacy, bias mitigation, and ethical considerations. Meanwhile, training and upskilling business professionals is a fruitful way to form a transition to the AI-powered BI landscape. To prevent data breaches and unauthorized access, sensitive employee data needs to be hidden using improved security measures (Cheng et al., 2017). When the bias is present in the historical data it can also reflect on the decision-making process of the AI algorithms. In this case, an expert can make decisions regarding hiring and promotions to ensure fairness and diversity. When AI models are used to monitor employee activities such as analyzing communication patterns then they are subjected to ethical concerns regarding privacy and surveillance (Aslam et al., 2022). Organizations often hesitate to deploy AI adoption since it will lose the human touch in Human Resources (processes) and job displacement options. Integrating the current AI tools with standard business systems can be entirely complex and also requires significant investments in technology and infrastructure.

3. VUCA AND BANI MODELS

An event's unpredictability often marks volatility. The difficult to foresee and prepare operations in volatile environments. It included drastically altered market conditions, economic shocks, geopolitical tensions, and sudden natural disasters. The ripple across the globe is the financial crisis in one country in today's interconnected world. One region that instant consequences with rapid communication channels and a vast amount of information. Instantly spread and shared sentiments and global events with the social media rise augmented this randomness. Another impact on the scale is a volatility-defining feature in which far-reaching and often widespread are the consequences. An entire economy is affected by specific industries and all businesses feel volatility repercussions. A massive economic slowdown leads to layoffs and bankruptcies that suddenly feast through global financial systems. The far-reaching consequences are localized disruptions. The ripple effect or cascading effects led to volatility. An industry with multiple areas impacted and a chain reaction caused by the seemingly isolated event. Based on the decision-making processes, the short-term focus adopted is often a tendency in volatile environments. The volatility index is the metric that measures volatility in financial markets. To minimize exposure to a single volatility source, investment portfolios are diversified and maintained, and price fluctuations are protected.

In the modern business environment, one of the important challenges is uncertainty. The uncertainty is aroused with sudden fluctuations or changes characterized by volatility because of a lack of fast shifts in external conditions, unforeseen factors, and reliable information. To navigate the unknown, the strategies developed need to adapt to the challenges for businesses created. An ambiguity level in decision-making is introduced by uncertainty and the challenges to businesses generated by unpredictability. The interdependencies and sheer number of variables from the VUCA world stem from complexity. Because of conflicting or insufficient information, the lack of clarity in future understanding is represented by ambiguity.

The VUCA and BANI environments can be mathematically modeled using the following equations. The volatility mainly represents that change is constant and the current market condition cannot be determined using the current and future solutions. Volatility is computed as follows:

$$V(x) = \left[\frac{ds}{dx}\right] \tag{1}$$

Where s and x represent the system state and time. A higher V(x) value indicates the volatility of the present situation is high. Uncertainty is mainly identified using entropy and it is derived as follows:

$$U = -\sum p(y) \log p(y) \tag{2}$$

A higher U value indicates more uncertainty and p(y) is the probability of the event y. Uncertainty (U) mainly implies that the events that occur around us are often uncertain and the future of the business industry is unpredictable. Complexity(C) mainly occurs due to information overload and is derived as follows:

$$C = M * J * A \tag{3}$$

Where the number of components, interconnections, and dynamic behavior factor is represented as M, J, and A. When computing complexity it is often complex to determine what is useful and what is not. The Ambiguity value is derived as follows:

$$A = 1 - \left[\frac{Rs}{Rx}\right] \tag{4}$$

Where the number of clear and possible interpretations are represented as Rs and Rx. The ambiguous value mainly arises because the event didn't turn out as we planned. Hence, the only way available is overall freedom of interpretation.

The BANI risk quantification is derived for different variables as shown in this section. Brittleness measures the vulnerability of the system and its ability to withstand strong hurdles and crises. The brittleness (B) is mainly associated with failure probability (Fp) and system redundancy X which is derived as follows:

$$B = Fp * (1 - X) \tag{5}$$

Anxiety measures the stability of the system based on its experience after it overcomes a large hurdle. The Anxiety (A) score is computed using stress factors (S), impact (I), and coping mechanisms (V) as shown as follows:

$$A = (S * I)/V \tag{6}$$

The nonlinearity (N) mainly computes the unpredictability of the scenarios and variables in the business system as shown in equation (7)

$$N = \left|\frac{\partial^2 i}{\partial o^2}\right| \tag{7}$$

A higher N value indicates a nonlinear behavior and the input and output are represented as i and o. The Incomprehensibility (I) value computes the availability of unused resources, flexibility, and a backup plan in case of a crisis to eliminate the failures and their impact. It is computed as follows:

$$I = 1 - \left(\frac{Eu}{Ou}\right) \tag{8}$$

Where Eu is the explainable units and Ou are the overall units.

4. AI OPPORTUNITIES IN VUCA AND BANI ENVIRONMENT

In business intelligence, AI can be applied to different notions. For instance, it can analyze the candidate resumes obtained in an interview to identify suitable candidates and can even conduct initial interviews through chatbots. This process not only saves time but it is also an objective selection process. Real-time updates to the candidates can also be given through virtual assistants regarding their score, application status, and upcoming interviews improving the overall candidate experience. BI mainly extracts the effective information present in the user data and analyzes the hidden knowledge present in the data. This process is broken down into the following steps. Initially, an organization mainly needs to analyze the needs of an enterprise and its users which is called demand analysis. Next, the data modeling is done which forms a logical model based on the user's demand. Subsequently, the data extraction step takes place which extracts the data from the database and processes it before reporting. After that, the report created is sent to the users for their consent and the users can conduct simple operations in the system and notify if any errors are encountered. At last, the AI-based BI tool is improved and updated continuously as per customer demand. In BI, the VUCA mainly gained popularity after the global financial crisis and it mainly refers to the unstable business environment encountered daily.

The volatility can be reduced using the predictive analysis via a time series forecasting model (Shen et al., 2020) using the below equation:

$$V(i+1) = \phi * A(i) + \alpha * [A(i) - A(i-1)] + \zeta \tag{9}$$

Where the predicted term is represented as $V(i+1)$, present value is $A(i)$, level coefficient is ϕ, trend coefficient is α, and error term is ζ. Using the Bayesian probability test (Wang et al., 2020), the uncertainty management can be modeled as follows:

$$P(U|V) = P(V|U) * P(U)/P(V) \tag{10}$$

Where the posterior probability, likelihood, prior probability, and evidence are represented as $P(U|V)$, $P(V|U)$, $P(U)$, and $P(V)$. The pattern recognition techniques (Valafar, 2002; Liew et al., 2005) can be used for complexity handling as shown below:

$$Comp = \sum (x_i * y_i) \Big/ N$$

Where the pattern weight, pattern match score, and overall patterns are represented as x_i, y_i, and N. Using NLP techniques, the ambiguity resolution can be modeled as follows:

$$A = \sum^{(MY + NY)} \Big/ _{(MX + NX)} \tag{11}$$

Where MY represents the matched syntax, NY represents the semantic clarity, MX represents the overall meanings, and NX represents the entire events in the dataset. The AI opportunity factors are AI value proposition (B1), AI solution effectiveness (B2), and AI risk mitigation factors (B3). These variables are derived as follows:

$$B1 = \sum\left(\omega_x * s_x * P_s\right) \tag{12}$$

Where the opportunity weight, opportunity size, and success probability are represented as ω_x, s_x, and P_s respectively. The variable B2 is computed as follows:

$$B2 = (RC * AC)/(Comp + \cos t) \tag{13}$$

Where the problem resolution capability, adaptation capability, technical complexity, and implementation cost are represented as RC, AC, Comp, and cost. The B3 value is computed as shown in the below equation and a higher B3 value demonstrates better risk mitigation capability.

$$B3 = 1 - [(V + U + C + A) * (B + A + N + I)]/k \tag{14}$$

Where k is the normalization constant. The AI implementation priority is a summation of three variables (B1, B2, and B3) which helps to identify which AI concept to execute first based on value proposition, solution effectiveness, and risk mitigation capability. The Return on Investment (ROI) is estimated as follows:

$$ROI = (Benefits * B3)/(\text{Cos}ts * (1 + RP)) \tag{15}$$

Where RP is the risk premium and it is a function of the VUCA and BANI factors. Different optimization constraints need to be satisfied to minimize the risk and they are resource constraints, risk tolerance, and budget constraints. The resource constraint is derived as follows:

$$\sum r_o \leq r_{\max} \tag{16}$$

Where r represents the resource needed for the initiative o. The risk tolerance constraint is computed as shown below:

$$ar \geq ar_{\min} \tag{17}$$

Where $ar_{\min}$ is the minimally acceptable risk mitigation. The budget constraints compute the cost of initiative (c_o) with the maximum budget (b_{max}) as shown in the below equation:

$$\sum(c_0) \geq b_{\max} \tag{18}$$

Figure 1. Volatility analysis graph

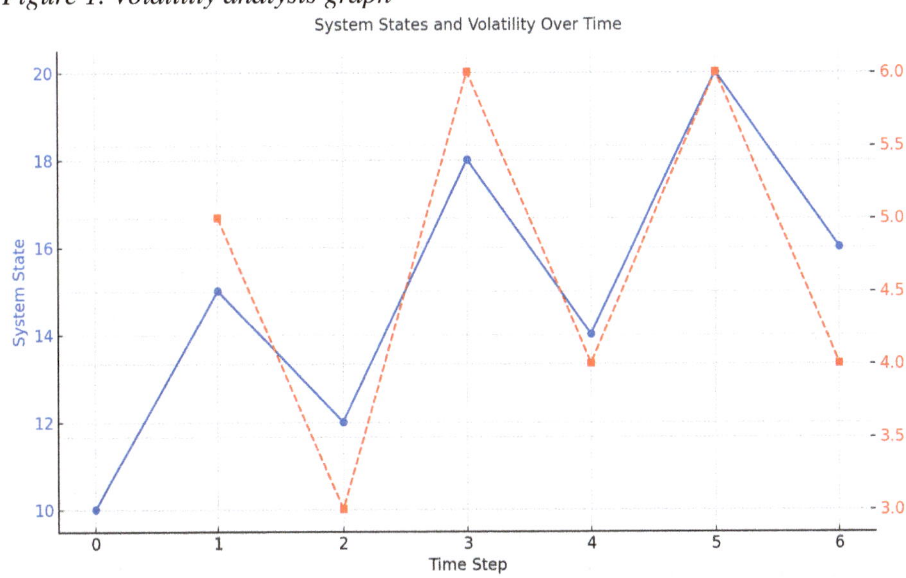

Figure 1 represents the volatility analysis graph which analyzes the system state for different time steps. If a system is stable, then it can be seen that low volatility and increased spikes of volatility show higher changes in the system state. The system state mainly represents the training loss of the model and it should decrease as the AI model learns, Hence, the AI model built for business should have minimal volatility and it can be minimized by tuning the parameters.

Figure 2. Uncertainty analysis using entropy

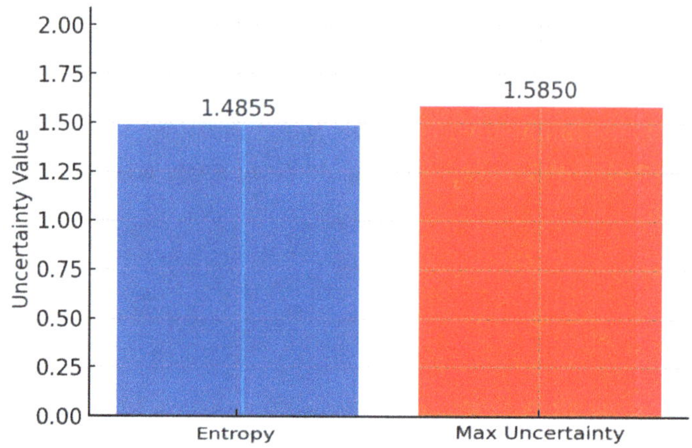

The uncertainty analysis conducted using entropy is presented in Figure 2 and the entropy value is computed for different probabilities. Entropy measures the randomness in the distribution whereas the maximum uncertainty computes the highest possible entropy when all of the outputs are actually equal. As per the figure, actual entropy is relatively closer to the maximum uncertainty which shows that uncertainty in the business models can be minimized by increasing the amount of data and enhancing the model. Analyzing uncertainty is crucial in trusting the decision of AI tools and based on the outcome we can determine whether to test the AI model's decision.

The AI value proposition analysis results are presented in Figure 3 for different opportunities. The Y-axis shows the individual value contribution of each AI opportunity such as process automation, customer insights, and predictive maintenance. The highest contribution of AI is received for process automation because when processes are automated using AI it results in minimal human error, disruptions, and inconsistency. The predictive maintenance done by AI helps to improve failure prediction and maintain equipment minimizing the downtime and vulnerability of unknown breakdowns. When operating in a vulnerable environment, AI models cannot always satisfy customer needs because operational resilience is given more priority. However, the AI models can analyze complex user behavior and provide clarity regarding their needs and preferences.

Figure 3. Opportunity Analysis

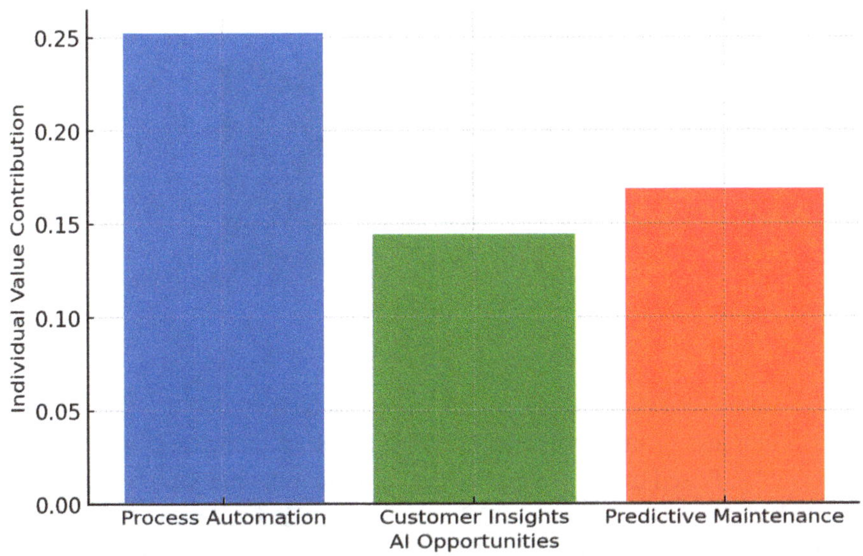

Figure 4 shows a pie chart that compares the cost proportion of the benefits obtained from a business investment. As per the figure, the costs cover up to 33.3% and the benefits account to 66.7%. This shows a favorable scenario for the deployment of the AI model in the BANI and VUNA environments. The pie chart shows that the AI model receives two times the benefits for each unit of cost (dollar) demonstrating the cost-effectiveness of the project.

Figure 4. Cost-benefit distribution graph

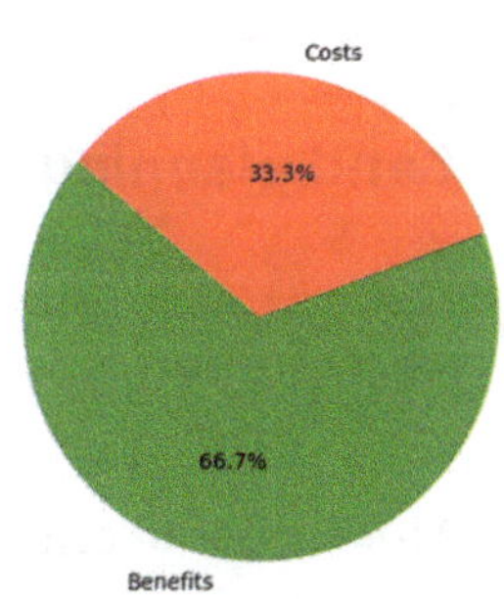

Figure 5. Risk analysis

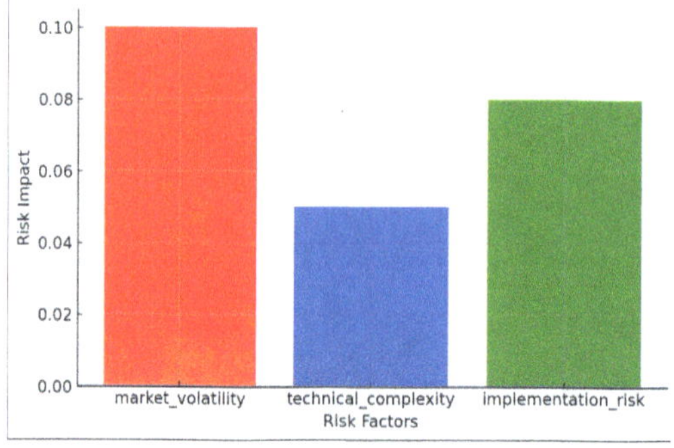

Figure 5 displays the contribution of different risk factors such as market volatility, technical complexity, and implementation risk. The risk impact is analyzed via a scale of 0-10%. Based on the results, we can infer that the market volatility has the highest risk impact nearly equal to 0.10 whereas the risk impact of implementation and technical complexity is nearly equal to 0.08 and 0.05. Technical complexity is the minimal risk when deploying AI in the BUNA and VANI environments when compared to the others. To prioritize risk mitigation, market volatility should be addressed with higher importance and focus.

5. IMPLEMENTATION PLAN

The implementation plan is mainly developed with a focus on BI evolution, AI integration, and their role in addressing the operational, ethical, and workflow challenges. When businesses make decisions with the help of AI they can operate with sustainability and resilience even when faced with uncertainty. The implementation plan is a seven-step process and it is explained in this section:

Step-1: Initial assessment and planning

This step analyzes the basic needs and challenges faced by the businesses when operating in the VUCA and BANI worlds. The success of the business is mainly measured using efficiency, scalability, and accuracy. In this stage, we can identify different business problems that AI-powered BI can solve. We need to build a data-driven culture where data is properly used for decision-making. Employees should be given proper training regarding the usage of AI-powered tools and develop insights. Initially, pilot projects are built to identify the value and later iterated to reach momentum. To build trust in the AI-driven insights, Explainable AI (XAI) is prioritized. The responsible usage of AI needs to be addressed via bias and data privacy.

Step-2: Data acquisition and integration

The data sources relevant to business need to be identified such as external datasets, historical data, and real-time feeds. A robust data integration process needs to be embedded to offer high-quality and reliable data. The extract, transform, and load tools need to be used to extract and consolidate data from different sources and integrate them into a centralized warehouse for proper access.

Step 3: AI and advanced analytics integration

The AI models are mainly built using different deep learning, transfer learning, machine learning, and Natural Language Processing (NLP) techniques. These models serve as a tool for prediction automating the business process and yielding decision support. The insights can be analyzed using advanced data analytics which provides details such as economic impact, current trends, and delay rates.

Step 4: Case study development and application

The AI-driven BI solutions are developed and implemented using a mixed-class agile methodology. A mixed class agile methodology is a hybrid approach that integrates both the traditional project management strategy with the iterative agile methodologies such as Scrum or Kanban. This step helps to modify the business project management to suit our needs while balancing structure and adaptability. The agile methodologies follow iterative development, continuous feedback, and close collaboration. Here the input task is broken down into sprints and the team continuously monitors to adapt the modifying requirements. The agile methodologies mainly aim to modify the software instead of producing extensive documentation. This chapter discusses the pandemic crisis management case study which shows how organizations can control a future pandemic crisis in the BANI world. The BI-driven insights are used to make policy decisions, resource allocation, and disaster response.

Step 5: Implementation and monitoring

In this step, the AI-driven BI solutions are deployed in different business units. We also ensure that the employees are given proper training and support to help the end-users or customers use the developed tools. The performance of the AI-driven business solutions is continuously monitored to evaluate its performance using key performance indicators.

Step 6: Addressing different challenges (operational, ethical, and workflow)

The operational challenges can be handled by developing proper solutions to manage the data processing, storage, and analysis operations. The ethical considerations can be met by properly following the ethical rules for AI usage in the organization. The workflow challenges can be optimized by offering effective collaboration between business analysts, data scientists, and decision-makers. In this way, the workflows are streamlined to merge the AI-driven outputs with the daily operations of the business site.

Step 7: Reporting and continuous improvement

The performance of the AI-driven BI applications is monitored and regular reports are generated. Next, the insights offered by it and the best practices are discussed with the end users. Based on the user feedback, the AI models and business processes are continuously refined. By keeping an eye on the recent AI advancement and business technologies, upgrades can be made to the projects. The banks used AI-powered BI to identify frauds and manage risks. Banks also use machine learning or deep learning algorithms to predict customer churn and offer retention incentives. Retail or e-commerce websites use AI-powered models to predict user demand, optimize pricing options, and personalize user recommendations. Amazon and eBay analyze users' browsing history and purchase patterns and suggest relevant products to them. For example, if a user is purchasing a mobile phone it suggests a Bluetooth headset or wireless charging port. Hospitals use patients' data to predict their follow-up treatments and personalize treatment options. The patient data can also be used to analyze the severity of the disease and the time they will be readmitted for treatment. Most organization uses AI to optimize business processes and predict equipment failure. AI also helps with supply chain management by analyzing the sensor data fixed in machines/vehicles. The sensor data can identify the breakdowns proactively and alert the organization to schedule maintenance.

6. CASE STUDY: MANAGING PANDEMIC CRISIS IN BANI WORLD USING MIXED CLASS AGILE METHODOLOGY

Risk management for the pandemic needs constant attention and updates from the project manager. Risk management is one of the crucial instruments that allows the project manager of the health crisis department to ensure needed medical awareness and safety protocols. This is a technological shift where organizations or hospitals implement digitalization, cybersecurity, and AI-driven decisions. Risk management is one of the main instruments to ensure that the COVID-19 plan is successfully implemented. The COVID-19 scenario is uncertain and volatile and accessing risk in this environment needs a structured approach that integrates both the quantitative and qualitative methods.

Major risk identification: Initially, we analyze the key threats that affect the organization's stability. The main risks identified are geopolitical factors, logistic challenges, and virus mutations. The geopolitical factors mainly affect the vaccine supply chain and they are due to international conflicts, trade restrictions, and policy changes. The disruptions in vaccine distribution, manufacturing, and transportation are the logistic challenges. The COVID-19 virus often resulted in new variants which urges the need to update vaccines and treatments.

Event planning and contingency measures: To handle future inconvenience, event planning is done once risks are identified. AI-driven epidemiological models help to predict how the new COVID-19 variant spreads and also help to design the booster doses. A contingency plan increases the number of ICUs and healthcare facilities temporarily to manage the increase in patients.

Agile methodologies for risk prevention: Agile methodologies are used to offer flexibility, cross-functional collaboration, and iterative responses for rapidly changing disasters. The flexibility can be improved by adapting to the changes quickly based on the data obtained regarding infections, vaccine efficacy, and economic impacts.

Qualitative risk assessment: Here, the risks associated with decision-making delays and resource allocation are usually considered which affects patient safety and well-being. Here, the AI models are not used to analyze the psychological and operational effects of risks.

Quantitative risk assessment: Here, AI-based models are used to predict the healthcare load, death rates, and infection rates. Business models can be used here to analyze the healthcare costs, supply chain disruptions, and lockdown-related losses.

The real-time analysis can be done using AI techniques that help to predict disease outbreaks and optimize vaccine distribution. The use of IoT technology and sensors ensures that the vaccines and medical supplies are managed effectively and the supply chain tracking is done correctly. The integration of AI and IoT techniques helps to increase automation making businesses more competitive in both the VUCA and BANI world. To prioritize risks we use qualitative analysis. The risk factor significance is computed based on the factor probability and the degree of its impact on certain events.

Using a probability and impact matrix we can qualitatively access the risk. The qualitative risk assessment factors for COVID-19 are determined using Table 1. By taking the maximum value of each factor we can find the integral score of the BANI world using the pessimistic principle as shown in Table 2.

Table 1. Risk probability and impact matrix for COVID-19

Risk factor probability	**Very Low<20%**	**Low [20-40%)**	**Average [40-60%]**	**High [60-80%]**	**Very High >80%**
[0-0.2]	Very Low	Very Low	Very Low	Average	Average
(0.2-0.4]	Very Low	Very Low	Low	Average	High
(0.4-0.6]	Very Low	Low	Low	High	High
(0.6-0.8)	Very Low	Low	Average	High	Average-High
(0.8-1]	Very Low	Average	Average	Average-High	Average-High

Table 2. BANI projections and qualitative assessments

Risk	**BANI Projections**	**Probability**	**Impact**	**Risk Level**
Brittleness				**High**
Supply chain failures affecting vaccine distribution	Global supply chain delay affecting vaccine availability	70	80	High
	The cold-chain infrastructure limitations hindering the vaccine storage process	50	80	High
	Medical supply shortage	40	80	Average
Anxiety				High
Overstrain in the healthcare organizations which results in insufficient patient care	Shortage of ventilators and beds in ICU	60	90	Average
	Doctors and nurses resigning due to exhaustion	90	85	High
	Lack of trust of the public in the healthcare organizations which often results in delays in treatment	50	90	Average
Non-Linearity				Very High

continued on following page

Table 2. Continued

Risk	BANI Projections	Probability	Impact	Risk Level
Brittleness				**High**
Volatile virus mutations impacting the vaccine efficiency	Appearance of new virus variants that need updated vaccines	75	80	Very High
	Ambiguous timeliness for booster dose success and policy changes	65	80	Very High
Incomprehensibility				Very High
Delays in public health response and policy due to propaganda	Rapid change in guiding principles which results in public confusion	85	90	Very High
	Spread of fake information which affects the decision of the public to take the vaccine	90	90	Very High

To derive this we need to take two subsets as shown in equations (19) and (20) and order them based on the decreasing order.

$$S2 = \{Brittleness, Anxiety\} \quad (19)$$

$$S2 = \{Nonlinearity, Incomprehensibility\} \quad (20)$$

When the elements are ordered in the decreasing risk level the following condition will be fulfilled.

$$Risk(S2) > Risk(S1) \quad (21)$$

A probabilistic approach is applied to compute the risk by taking each factor as random independent like-minded scenarios. The probability of occurrence of a single factor is computed by taking a set of independent events as shown as follows:

$$P(U+V) = P(U) + P(V) - P(UV) \quad (22)$$

The probability of occurrence for three factors is computed as shown below:

$$P(U+V+W) = P(U) + P(V) + P(W) - P(UV) - P(UW) - P(VW) + P(UVW) \quad (23)$$

Where U, V, and W represent the first, second, and third factors. The risk product is computed based on the risk degree as shown as follows:

$$Risk = P \times I \quad (24)$$

Where P and I represent the probability and impact of the factor. The integral risk assessment can be obtained for the BANI projections as follows:

$$Risk(UV) = I_U P(U) + I_V P(V) - \max(I_U I_V) P(UV) \tag{25}$$

$$Risk(UVW) = I_U P(U) + I_V P(V) + I_W P(W) - \max(I_U I_V) P(UV) - \max(I_U I_W) P(UW) - \max(I_V I_W) P(VW) + \max(I_U I_V I_W) P(UVW) \tag{26}$$

For the above projections, the quantitative risk estimate values achieved for Risk(Britleness), Risk(Anxiety), Risk(non-linearity), and Risk (Incomprehensibility) are 70, 54,98, and 99 as shown in Figure 6. Based on the results, we can observe that the effect of brittleness and anxiety is lower when compared to non-linearity and incomprehensibility. The non-linearity and incomprehensibility directly impact the process and also affect the project implementation. The results show that the use of AI improves the stability of the BANI environment. This shows that there are some milestones in the COVID-19 projects that cannot be broken by the external crisis.

Figure 6. BANI model projections impact for COVID-19 case-study

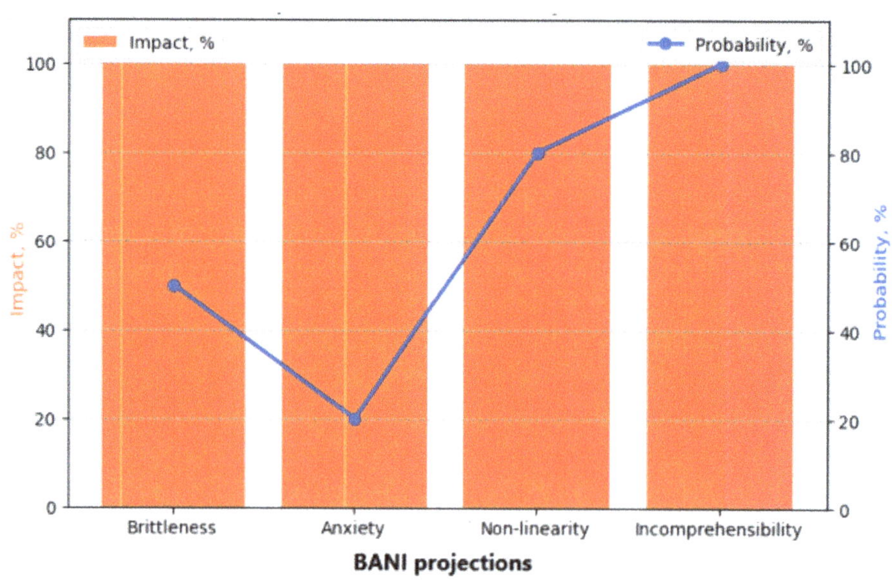

7. THEORETICAL AND PRACTICAL IMPLICATIONS

This chapter mainly improves the VUNA and BANI worlds by exploring and analyzing different AI techniques to address business risks. As per the background study conducted, more researchers have incorporated AI techniques to address the different business risks that arise in the organization (Aslam et al., 2022). The outcomes show that when AI techniques are integrated into business practices it helps to analyze large datasets with diverse data types. This results in improved risk assessment, threat identification, and decision-making in VUNA and BANI environments. The AI models improve market trends, liquidity shortage identification, and operational deficiencies. Using these AI models, one can identify the risks that occur beforehand and develop risk mitigation strategies (Shen et al., 2020; Wang et al., 2020). This chapter also presents different AI methodologies that help to deal with different challenges that occur in the VUNA and BANI environments. We can easily identify different algorithms that tailor our needs and apply them to our scenario of choice. The machine learning algorithms and deep learning algorithms can be modified to our risk types by tuning different hyperparameters.

The resource allocation is made simple because more data is collected from organizations due to technological variations. The massive data analysis helps to identify numerous areas in need and provides strategic decision support. Using the obtained insights, strategic decisions can be made by analyzing market situations, risk patterns, and user relationships. Using the acquired user data, we can build predictive user analyses to offer support to develop user-centric products and services. As a result, integrating the business risk management of VUNA and BANI words with AI models helps to improve the robustness in mitigating the different threats that occur in an uncertain and dynamic environment. The practical implications of this chapter are in order. Banks can successfully integrate AI technologies into their risk management framework to improve their operational efficiency and predictive capabilities. When integrating AI into business practices one needs to address the data quality, regulatory compliance, and model interpretability challenges (Heß & Damásio, 2025). Machine learning algorithms help to transparently access the risks in the BANI and VUNA worlds and also accept the bank's strict regulatory requirements.

8. CONCLUSION AND FUTURE DIRECTION

The smarter AI-based business tools offer seamless data integration and advanced predictive capabilities. The AI-based BI tools have completely transformed the way that businesses operate and provide deeper views, accurate predictions, and rapid data processing capabilities. This evolution transforms the way traditional businesses operate and makes AI an indispensable tool for modern organizations. The main advantages offered by AI-based BI tools are improved resource allocation, automated report generation, streamlined workflows, rapid decision-making, improved accuracy, minimal manual intervention, enhanced risk assessment, pattern recognition, and user empowerment. AI-powered analytics can filter big data to analyze emerging patterns to predict customer behavior and provide a summary of the competitive market landscape. In this way, businesses can make data-driven decisions and gain significant competitive advantage. If an e-commerce website creates products that suit different customers then it will result in increased sales and customer loyalty. For example, cosmetic websites advertise products that suit a specific skin tone or hair texture.

In the future, AI-based business tools can be optimized in many ways. For instance, let us assume that an edge computing mechanism is applied to goods which helps real-time tracking and also provides automated adjustments along with minimal delay. This level of automation reduces the burdens applied to human resources and utilizes them for more strategic tasks. Businesses that adopt AI can gain a deeper understanding of the market which will result in accurate forecasting, enhanced customer insights, and competitive analysis. Risk management can be updated using improved security measures and fraud detection software. The risks can be detected early using deep learning and machine learning algorithms that screen for abnormal patterns in real-time data. In the future, AI can be introduced to compliance monitoring to help **businesses** adhere to the laws and regulatory requirements. Most banks are nowadays using AI to prevent fraudulent transactions and threats to the customer. However, different challenges need to be considered when implementing AI in business processes. During implementation with existing systems, one needs to check the data quality and consistency to manage costs efficiently. Data security and overall system security need to be checked while using cloud and big data. Hence, robust security measures need to be implemented in place to safeguard sensitive data and maintain human trust. Dedicated resources and experts need to be employed to check the ongoing maintenance and ensure the long-term success of the AI-powered business analytics platform.

REFERENCES

Бушуєв, С., Пілюгіна, К., & Четін, Е. (2023). Transformation of values of the high technology projects from a VUCA to a BANI environment model. *СУЧАСНИЙ СТАН НАУКОВИХ ДОСЛІДЖЕНЬ ТА ТЕХНОЛОГІЙ В ПРОМИСЛОВОСТІ*, (2 (24)), 191–199. DOI: 10.30837/ITSSI.2023.24.191

Acciarini, C., Cappa, F., Boccardelli, P., & Oriani, R. (2023). How can organizations leverage big data to innovate their business models? A systematic literature review. *Technovation*, *123*, 102713. DOI: 10.1016/j.technovation.2023.102713

Aktan, S. (2011). Application of machine learning algorithms for business failure prediction. *Investment Management and Financial Innovations*, (8, Iss. 2), 52–65.

Allioui, H., Allioui, A., & Mourdi, Y. (2024). Navigating transformation: Unveiling the synergy of IoT, multimedia trends, and AI for sustainable financial growth in African context. *Multimedia Tools and Applications*, *84*(9), 1–45. DOI: 10.1007/s11042-024-19038-6

Aslam, M., Khan Abbasi, M. A., Khalid, T., Shan, R. U., Ullah, S., Ahmad, T., Saeed, S., Alabbad, D. A., & Ahmad, R. (2022). Getting smarter about smart cities: Improving data security and privacy through compliance. *Sensors (Basel)*, *22*(23), 9338. DOI: 10.3390/s22239338 PMID: 36502039

Attar-Khorasani, S., & Chalmeta, R. (2024). Internet of Things data visualization for business intelligence. *Big Data*, *12*(6), 478–503. DOI: 10.1089/big.2021.0200 PMID: 35133879

Bassiouni, M. M., Chakrabortty, R. K., Sallam, K. M., & Hussain, O. K. (2024). Deep learning approaches to identify order status in a complex supply chain. *Expert Systems with Applications*, *250*, 123947. DOI: 10.1016/j.eswa.2024.123947

Chao, M. H., Trappey, A. J., & Wu, C. T. (2021). Emerging Technologies of Natural Language-Enabled Chatbots: A Review and Trend Forecast Using Intelligent Ontology Extraction and Patent Analytics. *Complexity*, *2021*(1), 5511866. DOI: 10.1155/2021/5511866

Chatterjee, S., Goyal, D., Prakash, A., & Sharma, J. (2021). Exploring healthcare/health-product ecommerce satisfaction: A text mining and machine learning application. *Journal of Business Research*, *131*, 815–825. DOI: 10.1016/j.jbusres.2020.10.043

Cheng, L., Liu, F., & Yao, D. (2017). Enterprise data breach: Causes, challenges, prevention, and future directions. *Wiley Interdisciplinary Reviews. Data Mining and Knowledge Discovery*, *7*(5), e1211. DOI: 10.1002/widm.1211

Eboigbe, E. O., Farayola, O. A., Olatoye, F. O., Nnabugwu, O. C., & Daraojimba, C. (2023). Business intelligence transformation through AI and data analytics. *Engineering Science & Technology Journal*, *4*(5), 285–307. DOI: 10.51594/estj.v4i5.616

Edozie, E., Shuaibu, A. N., Sadiq, B. O., & John, U. K. (2025). Artificial intelligence advances in anomaly detection for telecom networks. *Artificial Intelligence Review*, *58*(4), 100. DOI: 10.1007/s10462-025-11108-x

Essien, A., & Giannetti, C. (2020). A deep learning model for smart manufacturing using convolutional LSTM neural network autoencoders. *IEEE Transactions on Industrial Informatics*, *16*(9), 6069–6078. DOI: 10.1109/TII.2020.2967556

Frizzo-Barker, J., Chow-White, P. A., Mozafari, M., & Ha, D. (2016). An empirical study of the rise of big data in business scholarship. *International Journal of Information Management*, *36*(3), 403–413. DOI: 10.1016/j.ijinfomgt.2016.01.006

Giri, C., & Chen, Y. (2022). Deep learning for demand forecasting in the fashion and apparel retail industry. *Forecasting*, *4*(2), 565–581. DOI: 10.3390/forecast4020031

Godavarthi, B., Narisetty, N., Gudikandhula, K., Muthukumaran, R., Kapila, D., & Ramesh, J. V. N. (2023). Cloud computing enabled business model innovation. *The Journal of High Technology Management Research*, *34*(2), 100469. DOI: 10.1016/j.hitech.2023.100469

Hashem, I. A. T., Yaqoob, I., Anuar, N. B., Mokhtar, S., Gani, A., & Khan, S. U. (2015). The rise of "big data" on cloud computing: Review and open research issues. *Information Systems*, *47*, 98–115. DOI: 10.1016/j.is.2014.07.006

Hassan, A., Nizam-Uddin, N., Quddus, A., Hassan, S. R., Rehman, A. U., & Bharany, S. (2024). Navigating IoT Security: Insights into Architecture, Key Security Features, Attacks, Current Challenges and AI-Driven Solutions Shaping the Future of Connectivity. *Computers, Materials & Continua*, *81*(3), 3499–3559. DOI: 10.32604/cmc.2024.057877

Heß, V. L., & Damásio, B. (2025). Machine learning in banking risk management: Mapping a decade of evolution. *International Journal of Information Management Data Insights*, *5*(1), 100324. DOI: 10.1016/j.jjimei.2025.100324

Hu, X., & Cercone, N. (2004). A data warehouse/online analytic processing framework for web usage mining and business intelligence reporting. *International Journal of Intelligent Systems*, *19*(7), 585–606. DOI: 10.1002/int.20012

Huang, J., Chai, J., & Cho, S. (2020). Deep learning in finance and banking: A literature review and classification. *Frontiers of Business Research in China, 14*(1), 13. DOI: 10.1186/s11782-020-00082-6

Javaid, M., Haleem, A., Singh, R. P., Suman, R., & Rab, S. (2022). Significance of machine learning in healthcare: Features, pillars and applications. *International Journal of Intelligent Networks, 3*, 58–73. DOI: 10.1016/j.ijin.2022.05.002

Khatri, M. R. (2023). Integration of natural language processing, self-service platforms, predictive maintenance, and prescriptive analytics for cost reduction, personalization, and real-time insights into customer service and operational efficiency. *International Journal of Information and Cybersecurity*, *7*(9), 1–30.

Kotios, D., Makridis, G., Fatouros, G., & Kyriazis, D. (2022). Deep learning enhancing banking services: A hybrid transaction classification and cash flow prediction approach. *Journal of Big Data*, *9*(1), 100. DOI: 10.1186/s40537-022-00651-x PMID: 36213092

Kratsch, W., Manderscheid, J., Röglinger, M., & Seyfried, J. (2021). Machine learning in business process monitoring: A comparison of deep learning and classical approaches used for outcome prediction. *Business & Information Systems Engineering*, *63*(3), 261–276. DOI: 10.1007/s12599-020-00645-0

Leo, M., Sharma, S., & Maddulety, K. (2019). Machine learning in banking risk management: A literature review. *Risks*, *7*(1), 29. DOI: 10.3390/risks7010029

Liew, A. W. C., Yan, H., & Yang, M. (2005). Pattern recognition techniques for the emerging field of bioinformatics: A review. *Pattern Recognition*, *38*(11), 2055–2073. DOI: 10.1016/j.patcog.2005.02.019

Mustapha, A. M., Arogundade, O. T., Abayomi-Alli, A., Adesemowo, A. K., & Adeniran, O. J. (2024). An improved cloud-based business process compliance management system using a user-centered approach. *International Journal of System Assurance Engineering and Management*, *15*(11), 1–28. DOI: 10.1007/s13198-024-02494-6

Nayyar, A., Gadhavi, L., & Zaman, N. (2021). *Machine learning in healthcare: review, opportunities and challenges*. Machine Learning and the Internet of Medical Things in Healthcare.

. Pi, Y., 2021. Machine learning in governments: Benefits, challenges and future directions. *JeDEM-eJournal of eDemocracy and Open Government, 13*(1), pp.203-219.

Saini, R., Tayal, P., & Thakur, P. (2025). Exploring the potential of cloud computing in revolutionizing healthcare. In *Computational Methods in Science and Technology* (pp. 560–567). CRC Press.

Schroeder, R. (2016). Big data business models: Challenges and opportunities. *Cogent Social Sciences*, *2*(1), 1166924. DOI: 10.1080/23311886.2016.1166924

Shen, Z., Zhang, Y., Lu, J., Xu, J., & Xiao, G. (2020). A novel time series forecasting model with deep learning. *Neurocomputing*, *396*, 302–313. DOI: 10.1016/j.neucom.2018.12.084

Simon, A. R., & Shaffer, S. L. (2001). *Data warehousing and business intelligence for e-commerce*. Elsevier.

Song, Y., & He, Y. (2023). Toward an intelligent tourism recommendation system based on artificial intelligence and IoT using the Apriori algorithm. *Soft Computing*, *27*(24), 19159–19177. DOI: 10.1007/s00500-023-09330-2

Song, Y. G., Cao, Q. L., & Zhang, C. (2018). Towards a new approach to predict business performance using machine learning. *Cognitive Systems Research*, *52*, 1004–1012. DOI: 10.1016/j.cogsys.2018.09.006

. Sorour, A. and Atkins, A.S., 2024. Big data challenge for monitoring quality in higher education institutions using business intelligence dashboards. *Journal of Electronic Science and Technology*, p.100233.

Sun, W., Ren, S., & Tang, G. (2025). In the era of responsible artificial intelligence and digitalization: Business group digitalization, operations and subsidiary performance. *Annals of Operations Research*, *354*(1), 1–23. DOI: 10.1007/s10479-024-06453-z

Sundberg, L., & Holmström, J. (2024). Fusing domain knowledge with machine learning: A public sector perspective. *The Journal of Strategic Information Systems*, *33*(3), 101848. DOI: 10.1016/j.jsis.2024.101848

Valafar, F. (2002). Pattern recognition techniques in microarray data analysis: A survey. *Annals of the New York Academy of Sciences*, *980*(1), 41–64. DOI: 10.1111/j.1749-6632.2002.tb04888.x PMID: 12594081

Vashishth, T. K., Sharma, V., Kumar, B., & Sharma, K. K. (2023). Cloud-Based data management for behavior analytics in business and finance sectors. In *Data-Driven Modelling and Predictive Analytics in Business and Finance* (pp. 133–155). Auerbach Publications.

Wang, J., Ma, Y., Zhang, L., Gao, R. X., & Wu, D. (2018). Deep learning for smart manufacturing: Methods and applications. *Journal of Manufacturing Systems*, *48*, 144–156. DOI: 10.1016/j.jmsy.2018.01.003

Wang, L., Pertheban, T. R. A., Li, T., & Zhao, L. (2024). Application of business intelligence based on big data in E-commerce data evaluation. *Heliyon*, *10*(21), e38768. DOI: 10.1016/j.heliyon.2024.e38768 PMID: 39524730

Wang, Z., Crook, J., & Andreeva, G. (2020). Reducing estimation risk using a Bayesian posterior distribution approach: Application to stress testing mortgage loan default. *European Journal of Operational Research*, *287*(2), 725–738. DOI: 10.1016/j.ejor.2020.04.020

Wei, L. (2022). Intelligent analysis of e-government influence factors based on improved machine learning. *Neural Computing & Applications*, *34*(15), 12241–12256. DOI: 10.1007/s00521-021-06657-5

Wu, J., Liu, H., Yao, X., & Zhang, L. (2024). Unveiling consumer preferences: A two-stage deep learning approach to enhance accuracy in multi-channel retail sales forecasting. *Expert Systems with Applications*, *257*, 125066. DOI: 10.1016/j.eswa.2024.125066

Yu, D., Zhou, X., Noorian, A., & Hazratifard, M. (2025). An AI-driven social media recommender system leveraging smartphone and IoT data. *The Journal of Supercomputing*, *81*(1), 1–32. DOI: 10.1007/s11227-024-06722-5

Zhang, X., & Jia, P. 2022, November. The Development Trend of Intelligent Logistics Based on Machine Vision. In *International conference on Advances in Materials and Manufacturing* (pp. 849-859). Singapore: Springer Nature Singapore.

Zohuri, B., & Moghaddam, M. (2020). From business intelligence to artificial intelligence. *Journal of Material Sciences & Manufacturing Research*, *1*(1), 1–10.

KEY TERMS AND DEFINITIONS

Artificial Intelligence: Helps machines to mimic human intelligence to solve complex real-world problems.

BANI: A framework that helps to understand what is happening and how we can handle the fragile and uncertain scenario like pandemic/ sudden death.

Business Intelligence: A process of retrieving data and analyzing it to form implacable solutions to offer optimal business decision-making.

Internet of Things: A network of smart objects that interact with each other via the Internet to make informed decisions.

Risk Analysis: A process of analyzing the business risks that could impact the organization's reputation and assets.

Smart Business: uses novel technology usually AI to optimize operations and enhance customer interaction.

VUCA: a scenario that describes a complex business environment with unknown changes, lack of predictability, and massive common factors that make the situation hard to analyze.

Compilation of References

Abbas, M. S. T. (2018). Genetically engineered (modified) crops (*Bacillus thuringiensis* crops) and the world controversy on their safety. *Egyptian Journal of Biological Pest Control*, *28*(1), 1–12. DOI: 10.1186/s41938-018-0051-2

Abbes, F., Bennani, S., & Maalel, A. (2024). Generative AI and gamification for personalized learning: Literature review and future challenges. *SN Computer Science*, *5*(8), 1–12. DOI: 10.1007/s42979-024-03491-z

Abdi, H. (2003). Factor rotations in factor analyses. Encyclopedia for Research Methods for the Social Sciences, 792-795.

Abdul Aziz, M., Brini, F., Rouached, H., & Masmoudi, K. (2022). Genetically engineered crops for sustainably enhanced food production systems. *Frontiers in Plant Science*, *13*, 1027828. DOI: 10.3389/fpls.2022.1027828 PMID: 36426158

Abdulgalimov, R. M., & Abdulgalimova, G. N. (2024). Digital technologies as a factor in improving the quality of medical training. Bulletin of the Lugansk State Pedagogical University: Collection of Scientific Papers. Series "Pedagogical Sciences. *Education*, (3(120)), 5–9.

Abdulgalimov, R. M., Abdulgalimova, G. N., & Trunova, S. A. (2024). Didactic possibilities of digital technologies and artificial intelligence in the professional training of future medical professionals. *Problems of Modern Pedagogical Education: Collection of Scientific Works*, (84(4)), 4–8.

Abernethy, A., Adams, L., Barrett, M., Bechtel, C., Brennan, P., Butte, A., Faulkner, J., Fontaine, E., Friedhoff, S., Halamka, J., Howell, M., Johnson, K., Long, P., McGraw, D., Miller, R., Lee, P., Perlin, J., Rucker, D., Sandy, L., & Valdes, K. (2022). (2022). The Promise of Digital Health: Then, Now, and the Future. *NAM Perspectives*, *6*(22). Advance online publication. DOI: 10.31478/202206e PMID: 36177208

Abeysekera, N. (2023). *Institutional and action research for VUCA and BANI World*. ResearchGate. DOI: 10.13140/RG.2.2.34878.66888

Acciarini, C., Cappa, F., Boccardelli, P., & Oriani, R. (2023). How can organizations leverage big data to innovate their business models? A systematic literature review. *Technovation, 123*, 102713. DOI: 10.1016/j.technovation.2023.102713

Adhikari, R. (2005). Emerging issues relating to conflicts between TRIPS and biodiversity: development implications for South Asia. *South Asian Yearbook of Trade and Development*, 261-288.

Adhikari, K., Bikundo, E., Chacko, X., Chapman, S., Humphries, F., Johnson, H., Keast, E., Lawson, C., Malbon, J., Robinson, D., Rourke, M., Sanderson, J., & Tranter, K. (2021). What should farmers' rights look like? The possible substance of a right. *Agronomy (Basel), 11*(2), 367. DOI: 10.3390/agronomy11020367

Aggarwal, V., & Satpute, A. (2013). Role of TRIPS in the Indian agriculture sector: Balancing traditional knowledge and biotechnology. *AgEcon Search*. https://www.aares.org.au/publications/AgEcon-search/

Aggarwal, A., Tam, C. C., Wu, D., Li, X., & Qiao, S. (2023). Artificial intelligence–based chatbots for promoting health behavioral changes: Systematic review. *Journal of Medical Internet Research, 25*, e40789. DOI: 10.2196/40789 PMID: 36826990

Aguilar, G. (2001). Access to genetic resources and protection of traditional knowledge in the territories of indigenous peoples. *Environmental Science & Policy, 4*(4-5), 241–256. DOI: 10.1016/S1462-9011(01)00028-4

Aguinis, H., Gottfredson, R. K., & Joo, H. (2013). Best-practice recommendations for defining, identifying, and handling outliers. *Organizational Research Methods, 16*(2), 270–301. DOI: 10.1177/1094428112470848

Ahad, A. A., Sanchez-Gonzalez, M., & Junquera, P. (2023). Understanding and Addressing Mental Health Stigma Across Cultures for Improving Psychiatric Care: A Narrative Review. *Cureus, 15*(5), e39549. DOI: 10.7759/cureus.39549 PMID: 37250612

Ahn, S. J. G., Nowak, K. L., & Bailenson, J. N. (2022). Unintended consequences of spatial presence on learning in virtual reality. *Computers & Education, 186*, 104532. DOI: 10.1016/j.compedu.2022.104532

Ajagbe, S. A., Mudali, P., & Adigun, M. O. (2024). Internet of things with deep learning techniques for pandemic detection: A comprehensive review of current trends and open issues. *Electronics (Basel), 13*(13), 2630. DOI: 10.3390/electronics13132630

Aktan, S. (2011). Application of machine learning algorithms for business failure prediction. *Investment Management and Financial Innovations*, (8, Iss. 2), 52–65.

Al-Fuqaha, A., Guizani, M., Mohammadi, M., Aledhari, M., & Ayyash, M. (2015). Internet of things: A survey on enabling technologies, protocols, and applications. *IEEE Communications Surveys and Tutorials*, *17*(4), 2347–2376. DOI: 10.1109/COMST.2015.2444095

Algarra, E. J. D., Bravo, C. B., Cevallos, M. B. M., & Meneses, E. L. (2024). Pedagogical and Technical Analyses of Massive Open Online Courses on Artificial Intelligence. *Applied Sciences (Basel, Switzerland)*, *14*(3), 1051. DOI: 10.3390/app14031051

AlGerafi, M. A., Zhou, Y., Oubibi, M., & Wijaya, T. T. (2023). Unlocking the potential: A comprehensive evaluation of augmented reality and virtual reality in education. *Electronics (Basel)*, *12*(18), 3953. DOI: 10.3390/electronics12183953

Alharbi, A., Aljojo, N., Zainol, A., Alshutayri, A., Alharbi, B., Aldhahri, E., Khairullah, E. F., & Almandeel, S. (2021). Identification of critical factors affecting the students' acceptance of learning management system (LMS) in Saudi Arabia. *International Journal of Innovation*, *9*(2), 353–388. DOI: 10.5585/iji.v9i2.19652

Alin, A. (2010). Multicollinearity. *Wiley Interdisciplinary Reviews: Computational Statistics*, *2*(3), 370–374. DOI: 10.1002/wics.84

Allioui, H., Allioui, A., & Mourdi, Y. (2024). Navigating transformation: Unveiling the synergy of IoT, multimedia trends, and AI for sustainable financial growth in African context. *Multimedia Tools and Applications*, *84*(9), 1–45. DOI: 10.1007/s11042-024-19038-6

Almourad, M. B., Alrobai, A., Skinner, T., Hussain, M., & Ali, R. (2021). Digital wellbeing tools through users lens. *Technology in Society*, *67*, 101778. DOI: 10.1016/j.techsoc.2021.101778

Alowais, S. A., Alghamdi, S. S., Alsuhebany, N., Alqahtani, T., Alshaya, A. I., Almohareb, S. N., Aldairem, A., Alrashed, M., Bin Saleh, K., Badreldin, H. A., Al Yami, M. S., Al Harbi, S., & Albekairy, A. M. (2023). Revolutionizing healthcare: The role of artificial intelligence in clinical practice. *BMC Medical Education*, *23*(1), 689. DOI: 10.1186/s12909-023-04698-z PMID: 37740191

Alraimi, K. M., Zo, H., & Ciganek, A. P. (2015). Understanding the MOOCs continuance: The role of openness and reputation. *Computers & Education*, *80*, 28–38. https://doi.or/10.1016/j.compedu.2014.08.006. DOI: 10.1016/j.compedu.2014.08.006

Alrousan, M. K., Al-Madadha, A., Al Khasawneh, M. H., & Adel Tweissi, A. (2022). Determinants of virtual classroom adoption in Jordan: The case of Princess Sumaya University for Technology. *Interactive Technology and Smart Education, 19*(2), 121–144. DOI: 10.1108/ITSE-09-2020-0211

Al-Samarraie, H., Teng, B. K., Alzahrani, A. I., & Alalwan, N. (2017). E-learning continuance satisfaction in higher education: A unified perspective from instructors and students. *Studies in Higher Education, 42*(11), 2003–2019. DOI: 10.1080/03075079.2017.1298088

Alston, P. (1984). International law and the human right to food. In *The Right to Food* (pp. 9–68). Brill Nijhoff. DOI: 10.1163/9789004482302_004

Amann, J., & Blasimme, A. (2020). Artificial intelligence in healthcare: Predicting ethical, privacy, and bias concerns. *Health Affairs, 39*(11), 1978–1985. DOI: 10.1377/hlthaff.2020.01345

Amo, D., Casany, M. J., & Alier, M. (2014). Approaches for quality in pedagogical and design fundamentals in MOOCS. *Teoría de la Educación. Educación y Cultura en la Sociedad de la Información, 15*(1), 70–89.

Andersen, C. L., & West, R. E. (2021). "We overwhelm them with hope": How online mentors can support online learners. *Online Learning : the Official Journal of the Online Learning Consortium, 25*(4), 388–415. DOI: 10.24059/olj.v25i4.2440

Andersen, T. S., & Wellen, H. (2023). Being a mentor in the digital era: An exploratory study of the benefits undergraduate student mentors derived from providing virtual mentoring to youth. *Journal of Community Psychology, 51*(7), 2635–2651. DOI: 10.1002/jcop.23051

Android. (n.d.). *Digital wellbeing.* https://www.android.com/intl/en_in/digital-wellbeing/

Ansari, J. A. N., & Khan, N. A. (2020). Exploring the role of social media in collaborative learning the new domain of learning. *Smart Learning Environments, 7*(1), 9. DOI: 10.1186/s40561-020-00118-7

Antons, C. (2010). The role of traditional knowledge and access to genetic resources in biodiversity conservation in Southeast Asia. *Biodiversity and Conservation, 19*(4), 1189–1204. DOI: 10.1007/s10531-010-9816-y

Apodaca, C. (2012). The right to food. In *the Handbook of human rights* (pp. 349-358). Routledge.

Arno, A., & Thomas, S. (2016). The efficacy of nudge theory strategies in influencing adult dietary behaviour: A systematic review and meta-analysis. *BMC Public Health, 16*(1), 1–11. DOI: 10.1186/s12889-016-3272-x PMID: 27475752

Arnold, K. D., Deutschlander, D. T., & Lewis, J. S. (2022). Someone on my side: The experiences of high-achieving, underserved high school students in an e-mentoring college access intervention. *Mentoring & Tutoring, 30*(4), 381–408. DOI: 10.1080/13611267.2022.2091193

Aslam, M., Khan Abbasi, M. A., Khalid, T., Shan, R. U., Ullah, S., Ahmad, T., Saeed, S., Alabbad, D. A., & Ahmad, R. (2022). Getting smarter about smart cities: Improving data security and privacy through compliance. *Sensors (Basel), 22*(23), 9338. DOI: 10.3390/s22239338 PMID: 36502039

Assan, N. (2023). Socio-cultural, economic, and environmental implications for innovation in sustainable food in Africa. *Frontiers in Sustainable Food Systems, 7*, 1192422. DOI: 10.3389/fsufs.2023.1192422

Ateş Çobanoğlu, A., & Ceylan, B. (2020). E-mentorluk, mesleki gelişim ve yaşam boyu öğrenme. [E-mentoring, professional development and lifelong learning] In Öngöz, S. (Ed.), *Eğitim Bilimleri Alanında Elektronik Mentorluk* [Electronic Mentoring in Educational Sciences]. (pp. 23–45). Pegem Akademi Publishing. DOI: 10.14527/9786257052801.02

Attar-Khorasani, S., & Chalmeta, R. (2024). Internet of Things data visualization for business intelligence. *Big Data, 12*(6), 478–503. DOI: 10.1089/big.2021.0200 PMID: 35133879

Atzori, L., Iera, A., & Morabito, G. (2010). The Internet of Things: A survey. *Computer Networks, 54*(15), 2787–2805. DOI: 10.1016/j.comnet.2010.05.010

Awasthi, K., Padwekar, K., & Misra, S. C. (2025). Digital Twin: A Unified Definition, Issues, Challenges, and Opportunities. In M. Khosrow-Pour, D.B.A. (Ed.), Encyclopedia of Information Science and Technology, Sixth Edition. Advance online publication. DOI: 10.4018/978-1-6684-7366-5.ch074

Azadi, H., Samiee, A., Mahmoudi, H., Jouzi, Z., Rafiaani Khachak, P., De Maeyer, P., & Witlox, F. (2016). Genetically modified crops and small-scale farmers: Main opportunities and challenges. *Critical Reviews in Biotechnology, 36*(3), 434–446. PMID: 25566797

Babbott, K. M., & Serlachius, A. (2023). Developing digital mental health tools for youth with diabetes: An agenda for future research. *Frontiers in Clinical Diabetes and Healthcare, 4*, 1227332. DOI: 10.3389/fcdhc.2023.1227332 PMID: 37497385

Bagozzi, R. P., & Yi, Y. (1988). On the evaluation of structural equation models. *Journal of the Academy of Marketing Science, 16*(1), 74–94. DOI: 10.1007/BF02723327

Bailey, J. O., & Bailenson, J. N. (2017). Immersive virtual reality and the developing child. In *Cognitive development in digital contexts* (pp. 181–200). Academic Press., DOI: 10.1016/B978-0-12-809481-5.00009-2

Baker, R. S.. (2019). Learning analytics and educational technology: A review. *Journal of Learning Analytics, 6*(1), 27–41.

Balcombe, L. (2023). AI chatbots in digital mental health. *Informatics (MDPI), 10*(4), 82. DOI: 10.3390/informatics10040082

Balcombe, L., & de Leo, D. (2022). The potential impact of adjunct digital tools and technology to help distressed and suicidal men: An integrative review. *Frontiers in Psychology, 12*, 796371. DOI: 10.3389/fpsyg.2021.796371 PMID: 35058855

Banas, J. R., & York, C. S. (2014). Authentic learning exercises as a means to influence preservice teachers' technology integration self-efficacy and intentions to integrate technology. *Australasian Journal of Educational Technology, 30*(6), 728–746. DOI: 10.14742/ajet.362

Banerjee, M. (2019). Biopiracy in India: Seed diversity and the scramble for knowledge. *Phytomedicine, 53*, 296–301. DOI: 10.1016/j.phymed.2018.10.017 PMID: 30448389

Banerjee, S., Agarwal, A., & Bar, A. K. (2024). Securing wellbeing: Exploring security protocols and mitigating risks in AI-driven mental health chatbots for employees. *American Journal of Computer Science and Technology, 7*(1). Advance online publication. DOI: 10.11648/j.ajcst.20240701.11

Banh, L., & Strobel, G. (2023). Generative artificial intelligence. *Electronic Markets, 33*(1), 63. DOI: 10.1007/s12525-023-00680-1

Banister, S., & Vannatta Reinhart, R. (2012). Assessing NETS • T performance in teacher candidates: Exploring the Wayfind teacher assessment. *Journal of Digital Learning in Teacher Education, 29*(2), 59–65. DOI: 10.1080/21532974.2012.10784705

Barahona Mora, A. (2020). Gamification for classroom management: An implementation using ClassDojo. *Sustainability (Basel), 12*(22), 9371. DOI: 10.3390/su12229371

Barna, B., & Fodor, S. (2018). An empirical study on the use of gamification on IT courses at higher education. In M.Auer, D. Guralnick, & I. Simonics (Eds.), *Teaching and learning in a digital world: Proceedings of the 2017 International Conference on Interactive Collaborative Learning. Advances in Intelligent Systems and Computing, 715,* pp.684-692. DOI: 10.1007/978-3-319-73210-7_80

Baron, R. M., & Kenny, D. A. (1986). The moderator-mediator variable distinction in social psychological research: Conceptual, strategic, and statistical considerations. *Journal of Personality and Social Psychology, 51*(6), 1173–1182. DOI: 10.1037/0022-3514.51.6.1173

Basavarajappa, C., Grover, S., Dalal, P. K., Avasthi, A., Kumar, C. N., Manjunatha, N., Mehra, A., Saha, G., Sahoo, S., Singh, O. P., Tripathi, A., Gangadhar, B. N., & Math, S. B. (2022). Current telepsychiatry practice in India—An online survey of psychiatrists. *Indian Journal of Psychiatry, 64*(3), 307–311. DOI: 10.4103/indianjpsychiatry.indianjpsychiatry_331_21 PMID: 35859562

Bassiouni, M. M., Chakrabortty, R. K., Sallam, K. M., & Hussain, O. K. (2024). Deep learning approaches to identify order status in a complex supply chain. *Expert Systems with Applications, 250*, 123947. DOI: 10.1016/j.eswa.2024.123947

Bates, A. W. (2019). *Teaching in a digital age: Guidelines for designing teaching and learning*. Tony Bates Associates Ltd.

Bauer, M., Glenn, T., Monteith, S., Bauer, R., Whybrow, P. C., & Geddes, J. (2017). Ethical perspectives on recommending digital technology for patients with mental illness. *International Journal of Bipolar Disorders, 5*(1), 6. DOI: 10.1186/s40345-017-0073-9 PMID: 28155206

Becker, J. M., Ringle, C. M., & Sarstedt, M. (2023). Structural equation modeling with partial least squares path modeling. In Kline, R. (Ed.), *Handbook of structural equation modeling* (2nd ed.). Guilford Press.

Behabtu, H. A., Messagie, M., Coosemans, T., Berecibar, M., Anlay Fante, K., Kebede, A. A., & Mierlo, J. V. (2020). A Review of Energy Storage Technologies' Application Potentials in Renewable Energy Sources Grid Integration. *Sustainability (Basel), 12*(24), 10511. DOI: 10.3390/su122410511

Bennett, N., & Lemoine, G. J. (2014). What a difference a word makes: Understanding the impact of VUCA on organizations. *International Journal of Business and Social Science, 5*(12), 44–53. DOI: 10.1016/j.bushor.2014.01.001

Bentler, P. M., & Bonett, D. G. (1980). Significance tests and goodness of fit in the analysis of covariance structures. *Psychological Bulletin, 88*(3), 588–606. DOI: 10.1037/0033-2909.88.3.588

Ben-Zeev, D., & Atkins, D. C. (2017). Bringing digital mental health to where it is needed most. *Nature Human Behaviour, 1*(12), 849–851. DOI: 10.1038/s41562-017-0232-0 PMID: 31024173

Berger, M., Nüske, N., & Müller, C. (2020). Digital nudging in online grocery stores - Towards ecologically sustainable nutrition. *International Conference on Information Systems, ICIS 2020 - Making Digital Inclusive: Blending the Local and the Global, January 2020.*

Bhandari, H. N., Rimal, B., Pokhrel, N. R., Rimal, R., Dahal, K. R., & Khatri, R. K. C. (2022). Predicting stock market index using LSTM. *Machine Learning with Applications, 9*(2), 100320. DOI: 10.1016/j.mlwa.2022.100320

Bhukta, A. (2020). Bio-piracy. In *Legal Protection for Traditional Knowledge: Towards a New Law for Indigenous Intellectual Property* (pp. 45–72). Emerald Publishing Limited. DOI: 10.1108/978-1-80043-063-120200004

Bittner, K. (2021). Engagement and immersion in digital play: Supporting young children's digital wellbeing. *International Journal of Environmental Research and Public Health, 18*(19), 10179. DOI: 10.3390/ijerph181910179 PMID: 34639481

Bjørnstad, S. I. B. (2004). *Breakthrough for" the South"?: an analysis of the recognition of Farmers' Rights in the International Treaty on Plant Genetic Resources for Food and Agriculture* (Master's thesis).

Blanchett, D., & Ratner, H. (2015). Building Efficient Income Portfolios. *Journal of Portfolio Management, 41*(3), 117–125. DOI: 10.3905/jpm.2015.41.3.117

Blas, E. A. (2011). Monsanto Co. v. Geertson Seed Farms: Why the National Environmental Policy Act of 1969 Fails to Protect the Environment from Current Biotechnology. *Sustainable Development Law Journal, 14*, 35.

Blau, I., Grinberg, R., & Shamir-Inbal, T. (2018). Pedagogical perspectives and practices reflected in metaphors of learning and Digital Learning of ICT leaders. *Computers in the Schools, 35*(1), 32–48. DOI: 10.1080/07380569.2018.1427960

Bloom, B. S., Engelhart, M. D., Furst, E. J., Hill, W. H., & Krathwohl, D. R. (1956). *Taxonomy of educational objectives: The classification of educational goals. Handbook I: Cognitive domain.* David McKay Company.

Bond, M., Buntins, K., Bedenlier, S., Zawacki-Richter, O., & Kerres, M. (2020). Mapping research in student engagement and educational technology in higher education: A systematic evidence map. *International Journal of Educational Technology in Higher Education, 17*(1), 2. DOI: 10.1186/s41239-019-0176-8

Borowiak, C. (2004). Farmers' rights: Intellectual property regimes and the struggle over seeds. *Politics & Society, 32*(4), 511–543. DOI: 10.1177/0032329204269979

Borras, S. M.Jr. (2023). La Via Campesina–transforming agrarian and knowledge politics, and co-constructing a field: A laudatio. *The Journal of Peasant Studies, 50*(2), 691–724. DOI: 10.1080/03066150.2023.2176760

Botea-Muntean, D.-R., & Constantinescu, R. (2024). VUCA and BANI Worlds-challenges for nowadays business models. Could circular economy and various digital instruments like internet of things help a busıness model to respond to such challenges? An examinatıon of solutions and future strategies through marketing lens. *Journal of Information Systems & Operations Management, 18*(1), 38–54.

Boucher, E. M., Harake, N. R., Ward, H. E., Stoeckl, S. E., Vargas, J., Minkel, J., ... & Zilca, R. (2021). Artificially intelligent chatbots in digital mental health interventions: A review. *Expert Review of Medical Devices, 18*(sup1), 37–49. DOI: 10.1080/17434440.2021.2013200

Bouton, E., Tal, S. B., & Asterhan, C. S. C. (2021). Students, social network technology and learning in higher education: Visions of collaborative knowledge construction vs. the reality of knowledge sharing. *The Internet and Higher Education, 49*, 100787. DOI: 10.1016/j.iheduc.2020.100787

Bower, M., & Sturman, D. (2015). What are the educational affordances of wearable technologies? *Computers & Education, 88*, 343–353. DOI: 10.1016/j.compedu.2015.07.013

Bozkurt, A., Karakaya, K., Turk, M., Karakaya, Ö., & Castellanos-Reyes, D. (2022). The impact of COVID-19 on education: A meta-narrative review. *TechTrends, 66*(5), 883–896. DOI: 10.1007/s11528-022-00759-0

Braun, V., & Clarke, V. Thematic analysis. In APA Handbook of Research Methods in Psychology, Vol. 2: Research Designs: Quantitative, Qualitative, Neuropsychological, and Biological; Cooper, H., Camic, P.M., Long, D.L., Panter, A.T., Rindskopf, D., Sher, K.J., Eds.; American Psychological Association: Washington, DC, USA, 2012; pp. 57–71. DOI: 10.1037/13620-004

Breining-Kaufmann, C. (2005). The right to food and trade in agriculture. *Human rights and international trade*, 341-381.

Bruner, J. (1966). *Toward a theory of instruction*. Harvard University Press., DOI: 10.1119/1.2350966

Brylina, I. V., Okonskaya, N. K., Ermakov, M. A., & Brylin, A. V. (2021). Education of the future in the conditions of VUCA World. *European Proceedings of Social and Behavioural Sciences, Economic and Social Trends for Sustainability of Modern Society (ICEST-II 2021)*, 1372–1380. DOI: 10.15405/epsbs.2021.09.02.154

Bucci, S., Schwannauer, M., & Berry, N. (2019). The digital revolution and its impact on mental health care. *Psychology and Psychotherapy: Theory, Research and Practice*, *92*(2), 277–297. DOI: 10.1111/papt.12222 PMID: 30924316

Buchatskiy, P. Y., Onishchenko, S. V., Teploukhov, S. V., & Kuzmin, K. A. (2024). "Modeling of Wind Generator in the Power Supply System of an Individual Consumer Using SimInTech," 2024 8th International Conference on Information, Control, and Communication Technologies (ICCT), Vladikavkaz, Russian Federation, pp. 1-5, DOI: 10.1109/ICCT62929.2024.10874937

Burda, Y., Samus, T., Bazhmina, E., Bondarenko, O., & Myrnyi, D. (2024). Shaping the competencies of the future: The importance of developıng soft skills in higher education within the landscape of information technologies. *AD ALTA: Journal of Interdisciplinary Research*, *14*(1), 111–116. DOI: 10.33543/j.140140.115121

Burgio, A., Cimmino, D., Nappo, A., Smarrazzo, L., & Donatiello, G. (2023). An IoT-Based Solution for Monitoring and Controlling Battery Energy Storage Systems at Residential and Commercial Levels. *Energies*, *16*(7), 3140. DOI: 10.3390/en16073140

Burr, C., Taddeo, M., & Floridi, L. (2020). The ethics of digital wellbeing: A thematic review. *Science and Engineering Ethics*, *26*(4), 2313–2343. DOI: 10.1007/s11948-020-00175-8 PMID: 31933119

Bushuyev, S., & Verenych, O. (2018). The blended mental space: Mobility and flexibility as characteristics of project/program success. *2018 IEEE 13th International Scientific and Technical Conference on Computer Sciences and Information Technologies (CSIT)*, *2*, 148–151. DOI: 10.1109/STC-CSIT.2018.8526699

Bushuyev, S., Bushuyeva, N., Murzabekova, S., & Khussainova, M. (2023). Innovative development of educational systems in the BANI environment. *Scientific Journal of Astana IT University*, *14*(June), 104–115. DOI: 10.37943/14YNSZ2227

Bushuyev, S., Piliuhina, K., & Chetin, E. (2023). Transformation of values of the high technology projects from a VUCA to a BANI environment model. *Innovative Technologies and Scientific Solutions for Industries*, *2*(24), 191–199. DOI: 10.30837/ITSSI.2023.24.191

Byers, C. (2024). AI-powered educational tools and their effect on student motivation in asynchronous online courses [Master's thesis, University of Montana]. ScholarWorks.

Cabero-Almenara, J., Sanchéz, R. B., Palacios-Rodríguez, A., & Párraga, L. M. (2021). Design and validation of t-mooc for the development of the digital competence of non-university teachers. *Technologies*, *9*(4), 84. DOI: 10.3390/technologies9040084

Caena, F., & Redecker, C. (2019). Aligning teacher competence frameworks to 21st century challenges: The case for the European Digital Competence Framework for Educators (Digcompedu). *European Journal of Education*, *54*(3), 356–369. DOI: 10.1111/ejed.12345

Cai, S., & Xu, Y. (2008). Designing product lists for e-commerce: The effects of sorting on consumer decision making. *International Journal of Human-Computer Interaction*, *24*(7), 700–721. Advance online publication. DOI: 10.1080/10447310802335730

Calicioglu, O., Flammini, A., Bracco, S., Bellù, L., & Sims, R. (2019). The future challenges of food and agriculture: An integrated analysis of trends and solutions. *Sustainability (Basel)*, *11*(1), 222. DOI: 10.3390/su11010222

Calvo, C., Ivorra, C., & Liern, V. (2018). Controlling risk through diversification in portfolio selection with non-historical information. *The Journal of the Operational Research Society*, *69*(10), 1543–1548. DOI: 10.1057/s41274-017-0195-6

Calvo, R. A., & Peters, D. (2014). *Positive computing: technology for wellbeing and human potential.* MIT press., DOI: 10.7551/mitpress/9764.001.0001

Campbell, D. T., & Fiske, D. W. (1959). Convergent and discriminant validation by the multitrait-multimethod matrix. *Psychological Bulletin*, *56*(2), 81–105. DOI: 10.1037/h0046016

Carmel, R. G., & Paul, M. W. (2015). Mentoring and coaching in academia: Reflections on a mentoring/coaching relationship. *Policy Futures in Education*, *13*(4), 479–491. DOI: 10.1177/1478210315578562

Carr, S. (2020). 'AI gone mental': Engagement and ethics in data-driven technology for mental health. *Journal of Mental Health (Abingdon, England)*, *29*(2), 125–130. DOI: 10.1080/09638237.2020.1714011 PMID: 32000544

Cascio, W. F., & Montealegre, R. (2016). How technology is changing work and organizations. *Annual Review of Organizational Psychology and Organizational Behavior, 3*(1), 349–375. DOI: 10.1146/annurev-orgpsych-041015-062352

Castañeda, L., & Selwyn, N. (2018). More than tools? Making sense of the ongoing digitizations of higher education. *International Journal of Educational Technology in Higher Education, 15*(1), 22. DOI: 10.1186/s41239-018-0109-y

Cattaneo, A. A., Antonietti, C., & Rauseo, M. (2022). How digitalised are vocational teachers? Assessing digital competence in vocational education and looking at its underlying factors. *Computers & Education, 176*, 104358. DOI: 10.1016/j.compedu.2021.104358

Cattell, R. B. (1966). The scree test for the number of factors. *Multivariate Behavioral Research, 1*(2), 245–276. DOI: 10.1207/s15327906mbr0102_10

Çebi, A.; Reisoğlu, İ. Digital competence: A study from the perspective of pre-service teachers in Turkey. J. New Approaches Educ. Res. (NAER J.) 2020, 9, 294–308.

Ceylan, F. (2019). *Mentorluk kavramı ve mentorluğun bireysel kariyer üstündeki etkisi* [The concept of mentoring and the effect of mentoring on the effect on individual career]. Tekirdağ Namık Kemal Üniversitesi.

Chalkiadakis, A., Seremetaki, A., Kanellou, A., Kallishi, M., Morfopoulou, A., Moraitaki, M., & Mastrokoukou, S. (2024). Impact of artificial intelligence and virtual reality on educational inclusion: A systematic review of technologies supporting students with disabilities. *Education Sciences, 14*(11), 1223. DOI: 10.3390/educsci14111223

Chandramma, R., Babu, K. S., Ranjith, K. S., Sinha, A. K., Neerugatti, V., & Reddy, D. S. (2025, January). Enhancing E-learning Accessibility through AI (Artificial Intelligence) and Inclusive Design. In *2025 6th International Conference on Mobile Computing and Sustainable Informatics (ICMCSI)* (pp. 1466-1471). IEEE. DOI: 10.1109/ICMCSI64620.2025.10883148

Chang, R. I., Hung, Y. H., & Lin, C. F. (2015). Survey of learning experiences and influence of learning style preferences on user intentions regarding MOOCs. *British Journal of Educational Technology, 46*(3), 528–541. DOI: 10.1111/bjet.12275

Chantelle. (2023, September 26). *Supporting digital wellbeing and time management.* Apple Education Community Forum. https://education.apple.com/resource/250011580

Chao, M. H., Trappey, A. J., & Wu, C. T. (2021). Emerging Technologies of Natural Language-Enabled Chatbots: A Review and Trend Forecast Using Intelligent Ontology Extraction and Patent Analytics. *Complexity*, *2021*(1), 5511866. DOI: 10.1155/2021/5511866

Charanya, P., Nidhish, M., Prajesh, U. R., Saktheeswar, K., & Sanjana, S. P. (2025, February). Enhancing career development: An AI-driven mentorship platform for student-mentor connections. In *2025 International Conference on Computational, Communication and Information Technology (ICCCIT)* (pp. 147–152). IEEE.

Char, D. S., Shah, N. H., & Magnus, D. (2018). Implementing machine learning in healthcare—Addressing ethical challenges. *The New England Journal of Medicine*, *378*(11), 981–983. DOI: 10.1056/NEJMp1714229 PMID: 29539284

Chatterjee, C., Chugunova, M., Ghosh, M., Singhal, A., & Wang, L. X. (2023). Human mediation leads to higher compliance in digital mental health: Field evidence from India. *Frontiers in Behavioral Economics*, *2*, 1232462. DOI: 10.3389/frbhe.2023.1232462

Chatterjee, S., Goyal, D., Prakash, A., & Sharma, J. (2021). Exploring healthcare/health-product ecommerce satisfaction: A text mining and machine learning application. *Journal of Business Research*, *131*, 815–825. DOI: 10.1016/j.jbusres.2020.10.043

Chawla, V. (2018). Monsanto's Loss of Patent Over Bt Cotton is a Victory for Farmers and Indian Seed Companies. *Economic and Political Weekly*, *53*, 25.

Chawla, V. (2019). Indian Supreme Court on patentability of genetically modified life forms—A missed opportunity? *Journal of Intellectual Property Law & Practice*, *14*(5), 343–345. DOI: 10.1093/jiplp/jpz032

Cheng, L., Liu, F., & Yao, D. (2017). Enterprise data breach: Causes, challenges, prevention, and future directions. *Wiley Interdisciplinary Reviews. Data Mining and Knowledge Discovery*, *7*(5), e1211. DOI: 10.1002/widm.1211

Chiang, C. L., & Lee, H. (2016). The effect of project-based learning on learning motivation and problem-solving ability of vocational high school students. *International Journal of Information and Education Technology (IJIET)*, *6*(9), 709–712. DOI: 10.7763/IJIET.2016.V6.779

Chi, M. T. H., & Wylie, R. (2014). The ICAP framework: Linking cognitive engagement to active learning outcomes. *Educational Psychologist*, *49*(4), 219–243. DOI: 10.1080/00461520.2014.965823

Christensen, G., Steinmetz, A., Alcorn, B., Bennett, A., Woods, D., & Emanuel, E. J. (2013). The MOOC Phenomenon: Who Takes Massive Open Online Courses and Why? SSRN *Electronic Journal*. Retrieved March 10, 2025 from https://papers.ssrn.com/sol3/papers.cfm?abstract_id=2350964 DOI: 10.2139/ssrn.2350964

Cialdini, R. B., & Trost, M. R. (1998). Social Influence: Social Norms, Conformity and Compliance. In *The Handbook of Social Psychology*.

Ciarli, T., Kenney, M., Massini, S., & Piscitello, L. (2021). Digital technologies, innovation, and skills: Emerging trajectories and challenges. *Research Policy*, *50*(7), 104289. DOI: 10.1016/j.respol.2021.104289

CIDDL. (n.d.). The impact of artificial intelligence on cognitive load. Retrieved February 14, 2025, from https://ciddl.org/the-impact-of-artificial-intelligence-on-cognitive-load/

Clegg, S., Kornberger, M., & Pitsis, T. (2021). *Strategy: Theory and practice* (3rd ed.). SAGE Publications.

Coby. (2023, June 25). *'My wellbeing toolkit'*. Apple Education Community Forum. https://education.apple.com/resource/250011441

Coghlan, S., Leins, K., Sheldrick, S., Cheong, M., Gooding, P., & D'Alfonso, S. (2023). To chat or bot to chat: Ethical issues with using chatbots in mental health. *Digital Health*, *9*, 20552076231183542. DOI: 10.1177/20552076231183542 PMID: 37377565

Cohen, J. (1988). *Statistical power analysis for the behavioral sciences* (2nd ed.). Lawrence Erlbaum Associates.

Colby, K. M., Weber, S., & Hilf, F. D. (1971). Artificial paranoia. *Artificial Intelligence*, *2*(1), 1–25. DOI: 10.1016/0004-3702(71)90002-6

Collaço, C. M. (2017). Increasing Student Engagement in Higher Education. *Journal of Higher Education Theory and Practice*, *17*(4), 40–47.

Collins, A., & Halverson, R. (2018). *Rethinking education in the age of technology: The digital revolution and schooling in America*. Teachers College Press.

Conole, G. (2015). MOOCs as disruptive technologies: strategies for enhancing the learner experience and quality of MOOCs. *Revista de Educación a Distancia (RED)*, (39). Retrieved March 11, 2025 from https://revistas.um.es/red/article/view/234221

Cook, S. (2021). *Educators: Do you know about VUCA? Here is why you need to ASAP!* https://www.linkedin.com/pulse/educators-do-you-know-vuca-here-why-need-asap-spike-cook

Correa, C. M. (2001). Traditional knowledge and intellectual property. *Geneva: The Quaker United Nations Office (QUNO), 17.*

Correa, C. M., & Yusuf, A. A. (Eds.). (2016). *Intellectual property and international trade: The TRIPS agreement: The TRIPS agreement.* Kluwer Law International BV.

Cortina, J. M. (1993). What is coefficient alpha? An examination of theory and applications. *The Journal of Applied Psychology, 78*(1), 98–104. DOI: 10.1037/0021-9010.78.1.98

Costello, A. B., & Osborne, J. W. (2005). Best practices in exploratory factor analysis: Four recommendations for getting the most from your analysis. *Practical Assessment, Research & Evaluation, 10*(1), 7.

Cravens, J. (2003). Online mentoring: Programs and suggested practices as of February 2001. *Journal of Technology in Human Services, 21*(1–2), 85–109. DOI: 10.1300/J017v21n01_05

Creswell, J. W. (2007). *Qualitative Inquiry and Research Design: Choosing among Five Approaches.* Sage Publications.

Crittenden, W. F., Biel, I. K., & Lovely, W. A.III. (2019). Embracing digitalization: Student learning and new technologies. *Journal of Marketing Education, 41*(1), 5–14. DOI: 10.1177/0273475318820895

Cullet, P. (2004). *Food security and intellectual property rights in developing countries.* RIBios (Réseau interdisciplinaire biosécurité): Institut universitaire d'études du développement (IUED).

Cullet, P. (1999). Revision of the TRIPS Agreement concerning the protection of plant varieties: Lessons from India concerning the development of a sui generis system. *The Journal of World Intellectual Property, 2*(4), 617–656. DOI: 10.1111/j.1747-1796.1999.tb00081.x

D'Angelo, C. (2018). The impact of technology: Student engagement and success. In *Technology and the curriculum: Summer 2018.* Power Learning Solutions. https://pressbooks.pub/techandcurriculum/chapter/engagement-and-success/

Dabbagh, N., Benson, A. D., Denham, A., Joseph, R., Al-Freih, M., Zgheib, G., Fake, H., & Guo, Z. (2016). Massive Open Online Courses. In: N., Dabbagh, A.D., Benson, A., Denham, R., Joseph, M., Al-Freih, G., Zgheib, H., Fake, & Z., Guo (Eds). *Learning Technologies and Globalization: Pedagogical Frameworks and Applications, 9–13.* DOI: 10.1007/978-3-319-22963-8_2

Dai, C. P., Ke, F., Dai, Z., & Pachman, M. (2023). Improving teaching practices via virtual reality-supported simulation-based learning: Scenario design and the duration of implementation. *British Journal of Educational Technology, 54*(4), 836–856. DOI: 10.1111/bjet.13296

Danko, J., & Šoltés, V. (2018). Portfolio creation using graph characteristics. *Investment Management and Financial Innovations, 15*(1), 180–189. DOI: 10.21511/imfi.15(1).2018.16

Dare, E. A., Keratithamkul, K., Hiwatig, B. M., & Li, F. (2021). Beyond content: The role of STEM disciplines, real-world problems, 21st century skills, and STEM careers within science teachers' conceptions of integrated STEM education. *Education Sciences, 11*(11), 11. Advance online publication. DOI: 10.3390/educsci11110737

Davidovitch, N., & Wadmany, R. (2020). Online teaching and learning in states of crisis: The effect of new directions of thinking in higher education. [In Hebrew]. *Halah, 2*, 15–26.

Davis, F. D. (1989). Perceived usefulness, perceived ease of use, and user acceptance of information technology. *Management Information Systems Quarterly, 13*(3), 319–340. DOI: 10.2307/249008

De Godoy, M. F., & Filho, D. R. (2021). Facing the BANI World. *International Journal of Neurology, 14*(02), e33–e33. DOI: 10.1055/s-0041-1735848

De Schutter, O. (2015). The New alliance for food security and nutrition in Africa.

Deci, E. L., & Ryan, R. M. (1985). *Intrinsic motivation and self-determination in human behavior*. Plenum., DOI: 10.1007/978-1-4899-2271-7

Deci, E. L., & Ryan, R. M. (2000). The "what" and "why" of goal pursuits: Human needs and the self-determination of behavior. *Psychological Inquiry, 11*(4), 227–268. DOI: 10.1207/S15327965PLI1104_01

Decristan, J., Jansen, N. C., & Fauth, B. (2023). Student participation in whole-class discourse: Individual conditions and consequences for student learning in primary and secondary school. *Learning and Instruction, 86*, 101748. DOI: 10.1016/j.learninstruc.2023.101748

Denecke, K., & Househ, M. (2021). Artificial intelligence for chatbots in mental health: Opportunities and challenges. In *Advances in Artificial Intelligence for Health and Medicine* (pp. 171–186). Springer., DOI: 10.1007/978-3-030-67303-1_10

Deniz-Garcia, A., Fabelo, H., Rodriguez-Almeida, A. J., Zamora-Zamorano, G., Castro-Fernandez, M., Alberiche Ruano, M. D. P., Solvoll, T., Granja, C., Schopf, T. R., Callico, G. M., Soguero-Ruiz, C., & Wägner, A. M.WARIFA Consortium. (2023). Quality, usability, and effectiveness of mHealth apps and the role of artificial intelligence: Current scenario and challenges. *Journal of Medical Internet Research, 25*, e44030. DOI: 10.2196/44030 PMID: 37140973

Dennis, A. R., Yuan, L., Feng, X., Webb, E., & Hsieh, C. J. (2020). Digital Nudging: Numeric and Semantic Priming in E-Commerce. *Journal of Management Information Systems, 37*(1), 39–65. DOI: 10.1080/07421222.2019.1705505

Desmet, P. M., & Pohlmeyer, A. E. (2013). Positive design: An introduction to design for subjective wellbeing. *International Journal of Design, 7*(3). https://www.ijdesign.org/index.php/IJDesign/article/view/1666/595

Deterding, S., Sicart, M., Nacke, L., Ohara, K., & Dixon, D. (2011). Gamification. using game-design elements in non-gaming contexts. *Proceedings of the 2011 annual conference extended abstracts on Human factors in computing systems* - CHI EA 11, New York, USA, 2425-2428. DOI: 10.1145/1979742.1979575

Deterding, S., Dixon, D., Khaled, R., & Nacke, L. (2011). From game design elements to gamefulness: Defining "gamification". In *Proceedings of the 15th International Academic MindTrek Conference: Envisioning Future Media Environments* (pp. 9-15). DOI: 10.1145/2181037.2181040

DeVellis, R. F. (2016). *Scale development: Theory and applications* (4th ed.). SAGE Publications.

Dhar, B., & Anuradha, R. V. (2004). Access, benefit-sharing and intellectual property rights. *The Journal of World Intellectual Property, 7*(5), 597–639. DOI: 10.1111/j.1747-1796.2004.tb00221.x

Dhawan, S. (2020). Online learning: A panacea in the time of COVID-19 crisis. *Journal of Educational Technology Systems, 49*(1), 5–22. DOI: 10.1177/0047239520934018

Di Giacomo, D., Ranieri, J., & Lacasa, P. (2017). Digital Learning As Enhanced Learning Processing? Cognitive Evidence for New insight of Smart Learning. *Frontiers in Psychology, 8*, 1329. DOI: 10.3389/fpsyg.2017.01329

Dieffenbacher, S. F. (2023, September 3). BANI World: What is it and why we need it? *Digital Leadership*. https://digitalleadership.com/blog/bani-world/

Dludlu, C. P. (2014). *A comparative study of the patentability standards with respect to pharmaceutical inventions in the United Kingdom and South Africa* (Doctoral dissertation, Queen Mary University of London).

Dormann, C. F., Elith, J., Bacher, S., Buchmann, C., Carl, G., Carré, G., Marquéz, J. R. G., Gruber, B., Lafourcade, B., Leitão, P. J., Münkemüller, T., McClean, C., Osborne, P. E., Reineking, B., Schröder, B., Skidmore, A. K., Zurell, D., & Lautenbach, S. (2013). Collinearity: A review of methods to deal with it and a simulation study evaluating their performance. *Ecography*, *36*(1), 27–46. DOI: 10.1111/j.1600-0587.2012.07348.x

Drost, E. A. (2011). Validity and reliability in social science research. *Education Research and Perspectives*, *38*(1), 105–123. DOI: 10.70953/ERPv38.11005

Dugard, P., & Todman, J. (1995). Analysis of Pre-test-Post-test Control Group Designs in Educational Research. *Educational Psychology*, *15*(2), 181–198. DOI: 10.1080/0144341950150207

Dureev, S., & Dureeva, N. (2023). TRANSFORMATION OF THE MODERN EDUCATIONAL PROCESS. *Bulletin of Chelyabinsk State University*, *474*(4), 43–49. DOI: 10.47475/1994-2796-2023-474-4-43-49

Dweck, C. S. (2000). *Self-theories: Their role in motivation, personality, and development*. Psychology Press.

E- mentorship Project (2022). What is E-Mentorship Project? https://ementorship.eu/

Eboigbe, E. O., Farayola, O. A., Olatoye, F. O., Nnabugwu, O. C., & Daraojimba, C. (2023). Business intelligence transformation through AI and data analytics. *Engineering Science & Technology Journal*, *4*(5), 285–307. DOI: 10.51594/estj.v4i5.616

Edozie, E., Shuaibu, A. N., Sadiq, B. O., & John, U. K. (2025). Artificial intelligence advances in anomaly detection for telecom networks. *Artificial Intelligence Review*, *58*(4), 100. DOI: 10.1007/s10462-025-11108-x

Efron, B., & Tibshirani, R. J. (1993). *An introduction to the bootstrap*. Chapman and Hall/CRC., DOI: 10.1007/978-1-4899-4541-9

Eigenbrod, L., & Janson, A. (2018). *Association for Information Systems AIS Electronic Library (AISeL) How Digital Nudges Influence Consumers-Experimental Investigation in the Context of Retargeting*. 11–29. https://aisel.aisnet.org/ecis2018_rip

Elfstrand, S., Malmer, P., & Skagerfält, B. (2011). Strengthening agricultural biodiversity for smallholder livelihoods. *What knowledge is needed to overcome constraints and release potentials.*

Ellikkal, A., & Rajamohan, S. (2024). *AI-enabled personalized learning: Empowering management students for improving engagement and academic performance.* Vilakshan-XIMB Journal of Management., DOI: 10.1108/XJM-02-2024-0023

Elvin-Lewis, M. (2007). Evolving Concepts Related to Achieving Benefit Sharing For Custodians Of Traditional Knowledge. https://utoronto.scholaris.ca/items/83a11d88-303f-47b9-8e54-e7205581a868

Enders, C. K. (2010). Applied missing data analysis. Guilford Press. eSchool News. (2024, February 5). Impact of artificial intelligence in education. https://www.eschoolnews.com/educational-technology/2024/02/05/impact-of-artificial-intelligence-in-education/

Ensher, E. A. (2013). E-mentoring as a critical e-learning approach: The impact of social presence on e-mentoring. *Communications and Network*, *05*(03), 1–3. DOI: 10.4236/cn.2013.53B1001

Ensher, E. A., Heun, C., & Blanchard, A. (2003). Online mentoring and computer-mediated communication: New directions in research. *Journal of Vocational Behavior*, *63*(2), 264–288. DOI: 10.1016/S0001-8791(03)00044-7

Equals Project. (2016). EQUALS her digital skills. https://www.equalsintech.org/her-digital-skills

Erlangga, I. S., Tarigan, F. P. P., Ternasih, L., & Umar, D. (2024). Digital transformation in education: Challenges and opportunities in the modern era. [INJOTEL]. *International Journal of Teaching and Learning*, *2*(12), 3251–3260.

Ertmer, P. A. (2005). Teacher pedagogical beliefs: The final frontier in our quest for technology integration. *Educational Technology Research and Development*, *53*(4), 25–39. DOI: 10.1007/BF02504683

Ertmer, P. A., Ottenbreit-Leftwich, A. T., & Tondeur, J. (2015). Teachers' beliefs and uses of technology to support 21st-century teaching and learning. In Fives, H., & Gill, M. G. (Eds.), *International Handbook of Research on Teacher Beliefs* (pp. 403–418). Routledge.

Essien, A., & Giannetti, C. (2020). A deep learning model for smart manufacturing using convolutional LSTM neural network autoencoders. *IEEE Transactions on Industrial Informatics*, *16*(9), 6069–6078. DOI: 10.1109/TII.2020.2967556

Estrada-Molina, O., & Fuentes-Cancell, D. (2022). Engagement and desertion in MOOCs: Systematic review. *Comunicar*, *30*(70), 111–124. DOI: 10.3916/C70-2022-09

Executive Academy. (2025, February 6). BANI vs. VUCA: How leadership works in the world of tomorrow. https://www.executiveacademy.at/en/news/detail/bani-vs-vuca-how-leadership-works-in-the-world-of-tomorrow

Fana, M., Milasi, S., Napierala, J., Fernández-Macías, E., & Vázquez, I. G. (2020). Telework, work organisation and job quality during the COVID-19 crisis: a qualitative study (No. 2020/11). *JRC Working Papers Series on Labour, Education and Technology*. https://hdl.handle.net/10419/231343

Fatima, N. (2024). The impact of teacher's empathy on students' self-compassion: An evidence-based intervention study. *Annals of Human and Social Sciences*, *5*(2), 2. Advance online publication. DOI: 10.35484/ahss.2024(5-II)38

Fauvel, S., Yu, H., Miao, C., Cui, L., Song, H., Zhang, L., Li, X., & Leung, C. (2018). Artificial Intelligence Powered MOOCs: A Brief Survey. *2018 IEEE International Conference on Agents (ICA)*, 56-61. DOI: 10.1109/AGENTS.2018.8460059

Fernandes, J. M., & Afonso, P. (2021). Engineering education in a context of VUCA. *2021 4th International Conference of the Portuguese Society for Engineering Education (CISPEE)*, 1–8. DOI: 10.1109/CISPEE47794.2021.9507229

Ferrari, A., Punie, Y., & Redecker, C. Understanding digital competence in the 21st century: An analysis of current frameworks. In *European Conference on Technology Enhanced Learning*; Springer: Berlin/Heidelberg, Germany, 2012. DOI: 10.1007/978-3-642-33263-0_7

Field, A. (2017). *Discovering statistics using IBM SPSS statistics* (5th ed.). SAGE Publications.

Fischer, G., Lundin, J., & Lindberg, J. O. (2020). Rethinking and reinventing learning, education, and collaboration in the digital age—From creating technologies to transforming cultures. *The International Journal of Information and Learning Technology*, *37*(5), 241–252. DOI: 10.1108/IJILT-04-2020-0051

Fitzpatrick, K. K., Darcy, A., & Vierhile, M. (2017). Delivering cognitive behavior therapy to young adults with symptoms of depression and anxiety using a fully automated conversational agent (Woebot): A randomized controlled trial. *JMIR Mental Health*, *4*(2), e7785. DOI: 10.2196/mental.7785 PMID: 28588005

Floridi, L. (2014). Open data, data protection, and group privacy. *Philosophy & Technology*, *27*(1), 1–3. DOI: 10.1007/s13347-014-0157-8

Food and Agriculture Organization (FAO). (2017). Farmers Rights. https://openknowledge.fao.org/server/api/core/bitstreams/c6767e7b-331d-421f-b936-f1922a3bc54d/content

Fornell, C., & Larcker, D. F. (1981). Evaluating structural equation models with unobservable variables and measurement error. *JMR, Journal of Marketing Research*, *18*(1), 39–50. DOI: 10.1177/002224378101800104

Foster, M. (1998). The Human Genome Diversity Project and the patenting of life: Indigenous peoples cry out. *Canterbury L. Rev., 7*, 343.

Fredricks, J. A., Blumenfeld, P. C., & Paris, A. H. (2004). School engagement: Potential of the concept, state of the evidence. *Review of Educational Research, 74*(1), 59–109. DOI: 10.3102/00346543074001059

Frizzo-Barker, J., Chow-White, P. A., Mozafari, M., & Ha, D. (2016). An empirical study of the rise of big data in business scholarship. *International Journal of Information Management, 36*(3), 403–413. DOI: 10.1016/j.ijinfomgt.2016.01.006

From, J. (2017). Pedagogical Digital Competence—Between Values, Knowledge and Skills. *High. Educ. Stud., 7*(2), 43–50. DOI: 10.5539/hes.v7n2p43

Fuglie, K., & MacDonald, J. M. (2023). Expanded intellectual property protections for crop seeds increase innovation and market power for companies. *Amber Waves: The Economics of Food, Farming, Natural Resources, and Rural America, 2023.*

Fursykova, T., Habelko, O., & Chernii, V. (2022). The Development of Digital Competence of Future Teachers in the Process of Distance Learning. *International Journal of Emerging Technologies in Learning, 17*(10), 85–98. DOI: 10.3991/ijet.v17i10.28973

Futo, T., & Nasehi, E. E. (2023). The Seeds of Vandana Shiva (2021): A Credible Actor of Change. *Kadın/Woman 2000, 24*(1), 77-96.

Gafni-Lachter, L., Niemeyer, L., Doyle, N., Norcross, J., & Jacobs, K. (2021). Equal peer e-mentoring for online graduate students: A case study and mediation model. *Mentoring & Tutoring, 29*(5), 545–564. DOI: 10.1080/13611267.2021.1986799

Gamage, K. A. A., Perera, D. A. S., & Wijewardena, M. A. D. N. (2021). Mentoring and coaching as a learning technique in higher education: The impact of learning context on student engagement in online learning. *Education Sciences, 11*(10), 574. DOI: 10.3390/educsci11100574

García-Delgado, M. Á., Rodríguez-Cano, S., Delgado-Benito, V., & Di Giusto-Valle, C. (2023). Digital Teaching Competence among Teachers of Different Educational Stages in Spain. *Education Sciences, 13*(6), 581. DOI: 10.3390/educsci13060581

Gargrish, S., Mantri, A., & Kaur, D. P. (2022). Evaluation of memory retention among students using augmented reality-based geometry learning assistant. *Education and Information Technologies, 27*(9), 12891–12912. DOI: 10.1007/s10639-022-11147-9

Garrido, S., Millington, C., Cheers, D., Boydell, K., Schubert, E., Meade, T., & Nguyen, Q. V. (2019). What works and what doesn't work? A systematic review of digital mental health interventions for depression and anxiety in young people. *Frontiers in Psychiatry, 10*, 759. DOI: 10.3389/fpsyt.2019.00759 PMID: 31798468

Geisser, S. (1974). A predictive approach to the random effect model. *Biometrika, 61*(1), 101–107. DOI: 10.1093/biomet/61.1.101

Gené, O. B., Núñez, M. M., & Blanco, Á. F. (2014). Gamification in MOOC: challenges, opportunities and proposals for advancing MOOC model. *Proceedings of the Second International Conference on Technological Ecosystems for Enhancing Multiculturality* - TEEM 14. DOI: 10.1145/2669711.2669902

Ghasemi, A., & Zahediasl, S. (2012). Normality tests for statistical analysis: A guide for non-statisticians. *International Journal of Endocrinology and Metabolism, 10*(2), 486–489. DOI: 10.5812/ijem.3505

Ghosh, D., & Vogt, A. (2012). Outliers: An evaluation of methodologies. Joint Statistical Meetings, 3455-3460.

Giri, C., & Chen, Y. (2022). Deep learning for demand forecasting in the fashion and apparel retail industry. *Forecasting, 4*(2), 565–581. DOI: 10.3390/forecast4020031

Giuliani, E., & Macchi, C. (2014). Multinational corporations' economic and human rights impacts on developing countries: A review and research agenda. *Cambridge Journal of Economics, 38*(2), 479–517. DOI: 10.1093/cje/bet060

Gligorea, I., Cioca, M., Oancea, R., Gorski, A. T., Gorski, H., & Tudorache, P. (2023). Adaptive learning using artificial intelligence in e-learning: A literature review. *Education Sciences, 13*(12), 1216. DOI: 10.3390/educsci13121216

Glover, D. (2007). Monsanto and smallholder farmers: A case study in CSR. *Third World Quarterly, 28*(4), 851–867. DOI: 10.1080/01436590701336739

Godavarthi, B., Narisetty, N., Gudikandhula, K., Muthukumaran, R., Kapila, D., & Ramesh, J. V. N. (2023). Cloud computing enabled business model innovation. *The Journal of High Technology Management Research, 34*(2), 100469. DOI: 10.1016/j.hitech.2023.100469

GONDWE. K. J. (2023). *INVESTIGATING THE NGO's-LOCAL GOVERNMENT RELATIONS IN GOVERNANCE SECTOR IN MALAWI: CASE STUDY OF BALAKA DISTRICT COUNCIL* (Doctoral dissertation, UNIVERSITY OF MALAWI).

Graham, J. W. (2009). Missing data analysis: Making it work in the real world. *Annual Review of Psychology*, *60*(1), 549–576. DOI: 10.1146/annurev.psych.58.110405.085530

GRAIN, A. (2020). *The real seeds producers: small-scale farmers save, use, share and enhance the seed diversity of the crops that feed africa.*

Grau, L. E., Larkin, K., Lalloo, C., Stinson, J. N., Zempsky, W. T., Ball, S. A., & Buono, F. D. (2022). Perspectives on adapting a mobile application for pain self-management in neurofibromatosis type 1: Results of online focus group discussions with individuals living with neurofibromatosis type 1 and pain management experts. *BMJ Open*, *12*(7), e056692. DOI: 10.1136/bmjopen-2021-056692 PMID: 35840301

Greene, R. T. (2023). *The future of instructional design: Engaging students through gamified, personalized, and flexible learning with AI and partnerships.* e-Learning Industry. Retrieved on March 25, 2025 from: https://elearningindustry.com/future-of-instructional-design-engaging-students-through-gamified-personalized-flexible-learning-with-ai-and-partnerships

Grinblatt, M., & Saxena, K. (2018). When Factors Do Not Span Their Basis Portfolios. *Journal of Financial and Quantitative Analysis*, *53*(6), 2335–2354. DOI: 10.1017/S0022109018000376

Grünewald, F., Meinel, C., Totschnig, M., & Willems, C. (2013). Designing MOOCs for the Support of Multiple Learning Styles. *Lecture Notes in Computer Science Scaling up Learning for Sustained Impact*, 371-382. DOI: 10.1007/978-3-642-40814-4_29

Guardian. D. (2023, June 11). Education may bid ‘bye to VUCA, hi to BANI. *Daily Guardian*. https://dailyguardian.com.ph/education-may-bid-bye-to-vuca-hi-to-bani/

Gui, J., & Yanchinda, J. (2025). *Building counselor competencies for supporting students in a VUCA and BANI World: A training and coaching model.* EasyChair Preprint No: 15732

Guida, A. (2025). *From policy to plate: international trade law’s role in advancing sustainable food governance and supporting dignified food access.* CERIDAP.

Habib, A. K. M. A., Hasan, M. K., Issa, G. F., Singh, D., Islam, S., & Ghazal, T. M. (2023). Lithium-Ion Battery Management System for Electric Vehicles: Constraints, Challenges, and Recommendations. *Batteries*, *9*(3), 152. DOI: 10.3390/batteries9030152

Haeger, H., & Fresquez, C. (2016). Mentoring for inclusion: The impact of mentoring on undergraduate researchers in the sciences. *CBE Life Sciences Education*, *15*(3), ar36. DOI: 10.1187/cbe.16-01-0016

Hagedorn, C., Renz, J., & Meinel, C. (2017). Introducing digital game-based learning in MOOCs: What do the learners want and need? *IEEE Global Engineering Education Conference* (EDUCON). DOI: 10.1109/EDUCON.2017.7942987

Hair, J. F., Black, W. C., Babin, B. J., & Anderson, R. E. (2010). *Multivariate data analysis* (7th ed.). Prentice Hall.

Hair, J. F., Hult, G. T. M., Ringle, C. M., & Sarstedt, M. (2017). *A primer on partial least squares structural equation modeling (PLS-SEM)* (2nd ed.). Sage Publications.

Hair, J. F., Risher, J. J., Sarstedt, M., & Ringle, C. M. (2019). When to use and how to report the results of PLS-SEM. *European Business Review*, *31*(1), 2–24. DOI: 10.1108/EBR-11-2018-0203

Hall, C. L., Gómez Bergin, A. D., & Rennick-Egglestone, S. (2024). Research into digital health intervention for mental health: 25-year retrospective on the ethical and legal challenges. *Journal of Medical Internet Research*, *26*, e58939. DOI: 10.2196/58939 PMID: 39250796

Hamad, F., Shehata, A., & Al Hosni, N. (2024). Predictors of blended learning adoption in higher education institutions in Oman: Theory of planned behavior. *International Journal of Educational Technology in Higher Education*, *21*(1), 13. DOI: 10.1186/s41239-024-00443-8

Hamari, J., Koivisto, J., & Sarsa, H. (2014). Does gamification work? A literature review of empirical studies on gamification. In 2014 47th Hawaii International Conference on System Sciences (pp. 3025-3034). IEEE. DOI: 10.1109/HICSS.2014.377

Haque, M. A., Ahmad, S., Hossain, M. A., Kumar, K., Faizanuddin, M., Islam, F., Haque, S., Rahman, M., Marisennayya, S., & Nazeer, J. (2024). Internet of things enabled e-learning system for academic achievement among university students. *E-Learning and Digital Media*, •••, 20427530241280078. DOI: 10.1177/20427530241280078

Hasanov, Z., Antoniou, P., Suleymanov, E., & Garayev, V. (2021). The impact of behavioural, cognitive and emotional dimensions of student engagement on student learning: The case of Azerbaijani higher education institutions. *International Journal of Knowledge and Learning*, *14*(1), 10–38. DOI: 10.1504/IJKL.2021.115027

Hashem, I. A. T., Yaqoob, I., Anuar, N. B., Mokhtar, S., Gani, A., & Khan, S. U. (2015). The rise of "big data" on cloud computing: Review and open research issues. *Information Systems*, *47*, 98–115. DOI: 10.1016/j.is.2014.07.006

Hassan, A., Nizam-Uddin, N., Quddus, A., Hassan, S. R., Rehman, A. U., & Bharany, S. (2024). Navigating IoT Security: Insights into Architecture, Key Security Features, Attacks, Current Challenges and AI-Driven Solutions Shaping the Future of Connectivity. *Computers, Materials & Continua, 81*(3), 3499–3559. DOI: 10.32604/cmc.2024.057877

Haugen, H. M., Muller, M. R., & Narasimhan, S. M. (2011). Food security and intellectual property rights: Finding the linkages. *Intellectual property and human development: Current trends and future scenarios, 103*, 108.

Haugen, H. M. (2020). The UN Declaration on Peasants' Rights (UNDROP): Is Article 19 on seed rights adequately balancing intellectual property rights and the right to food? *The Journal of World Intellectual Property, 23*(3-4), 288–309. DOI: 10.1111/jwip.12152

Hayes, A. F. (2017). *Introduction to mediation, moderation, and conditional process analysis: A regression-based approach* (2nd ed.). Guilford Press.

Henderson, M., Selwyn, N., & Aston, R. (2017). What works and why? Student perceptions of 'useful' digital technology in university teaching and learning. *Studies in Higher Education, 42*(8), 1567–1579. DOI: 10.1080/03075079.2015.1007946

Henseler, J., Ringle, C. M., & Sarstedt, M. (2015). A new criterion for assessing discriminant validity in variance-based structural equation modeling. *Journal of the Academy of Marketing Science, 43*(1), 115–135. DOI: 10.1007/s11747-014-0403-8

Herpen, E. van, Pieters, R., & Zeelenberg, M. (2005). How Product Scarcity Impacts on Choice: Snob and Bandwagon Effects. *ACR North American Advances, NA-32*(January).

Heß, V. L., & Damásio, B. (2025). Machine learning in banking risk management: Mapping a decade of evolution. *International Journal of Information Management Data Insights, 5*(1), 100324. DOI: 10.1016/j.jjimei.2025.100324

Hilty, D. M., Crawford, A., Teshima, J., Nasatir-Hilty, S. E., Luo, J., Chisler, L. S., Gutierrez Hilty, Y. S. M., Servis, M. E., Godbout, R., Lim, R. F., & Lu, F. G. (2021). Mobile health and cultural competencies as a foundation for telehealth care: Scoping review. *Journal of Technology in Behavioral Science, 6*(2), 197–230. DOI: 10.1007/s41347-020-00180-5

Höck, C., Ringle, C. M., & Sarstedt, M. (2010). Management of multi-purpose stadiums: Importance and performance measurement of service interfaces. *International Journal of Services Technology and Management, 14*(2-3), 188–207. DOI: 10.1504/IJSTM.2010.034327

Hodges, L. C. (2020). Student engagement in active learning classes. In Mintzes, J. J., & Walter, E. M. (Eds.), *Active learning in college science* (pp. 27–41). Springer International Publishing., DOI: 10.1007/978-3-030-33600-4_3

Hofstein, A., & Rosenfeld, S. (1996). Bridging the gap between formal and informal science learning. *Studies in Science Education*, *28*(1), 87–112. DOI: 10.1080/03057269608560085

Hohlov, M., & Korobova, M. (2024). The usage of digital twins in the field of virtual education. *Education, Innovation. Research as a Resource for Community Development*, *2*, 36–38. DOI: 10.31483/a-10620

Holcombe, A., & Wozniak, S. (2024). *Using AI to fuel engagement and active learning*. ASCD. https://www.ascd.org/el/articles/using-ai-to-fuel-engagement-and-active-learning

Holdaway, X., & Hawtin, N. (2013). Major players in the MOOC Universe. Retrieved March 2, 2025 from: http://chronicle.com/article/The-Major-Players-in-the-MOOC/138817/

Honka, A. (2024). *Personalization of digital health behavior change interventions for health promotion: A user modeling framework and a recommender system.* (Tampere University Dissertations - Tampereen yliopiston väitöskirjat; Vol. 957). Tampere University. https://urn.fi/URN:ISBN:978-952-03-3297-6

Hope, M. (2024). *Mastering leadership in a BANI World: Key strategies*. Https://Asymmetric.Pro/. https://asymmetric.pro/navigating-a-bani-world-strategies-for-future-leadership/

HRbrain.ai. (2024). *How mentors accelerate professional growth*. https://hrbrain.ai/blog/how-mentors-accelerate-professional-growth-and-success/

Huang, H., Jew, L. J. J., & Qi, D. (2023). Take a MOOC and then drop: A systematic review of MOOC engagement pattern and dropout factor. *Heliyon*, *9*(4), e15220. DOI: 10.1016/j.heliyon.2023.e15220 PMID: 37123924

Huang, J., Chai, J., & Cho, S. (2020). Deep learning in finance and banking: A literature review and classification. *Frontiers of Business Research in China*, *14*(1), 13. DOI: 10.1186/s11782-020-00082-6

Huang, R. H., Liu, D. J., Tlili, A., Yang, J. F., & Wang, H. H. (2020). *Handbook on Facilitating Flexible Learning during Educational Disruption: The Chinese Experience in Maintaining Undisrupted Learning in COVID-19 Outbreak*. Smart Learning Institute of Beijing Normal University.

Huang, R., Spector, J. M., & Yang, J. (2021). Educational technology as a driver of cognitive development in the digital age. *British Journal of Educational Technology*, *52*(4), 1234–1250. DOI: 10.1111/bjet.13045

Huang, W., Roscoe, R. D., Craig, S. D., & Johnson-Glenberg, M. C. (2022). Extending the cognitive-affective theory of learning with media in virtual reality learning: A structural equation modeling approach. *Journal of Educational Computing Research*, *60*(4), 807–842. DOI: 10.1177/07356331211053630

Hu, L. T., & Bentler, P. M. (1999). Cutoff criteria for fit indexes in covariance structure analysis: Conventional criteria versus new alternatives. *Structural Equation Modeling*, *6*(1), 1–55. DOI: 10.1080/10705519909540118

Hunga, H. G., Chiwaula, L., Mulwafu, W., & Katundu, M. (2023). The seed sector development in low-income countries: Lessons from the Malawi seed sector policy process. *Frontiers in Sustainable Food Systems*, *7*, 891116. DOI: 10.3389/fsufs.2023.891116

Husain, L., Greenhalgh, T., Hughes, G., Finlay, T., & Wherton, J. (2022). Desperately seeking intersectionality in digital health disparity research: Narrative review to inform a richer theorization of multiple disadvantage. *Journal of Medical Internet Research*, *24*(12), e42358. DOI: 10.2196/42358 PMID: 36383632

Hu, X., & Cercone, N. (2004). A data warehouse/online analytic processing framework for web usage mining and business intelligence reporting. *International Journal of Intelligent Systems*, *19*(7), 585–606. DOI: 10.1002/int.20012

Imran, Y., Wijekoon, N., Gonawala, L., Chiang, Y. C., & De Silva, K. R. D. (2021). Biopiracy: Abolish corporate hijacking of indigenous medicinal entities. *TheScientificWorldJournal*, *2021*(1), 8898842. DOI: 10.1155/2021/8898842 PMID: 33679261

Inelmen, E. (2004). E-mentoring: A novel approach in the use of technology in education. *Information Technology Based Proceedings of the Fifth International Conference on Higher Education and Training, 2004. ITHET 2004.*, 183–186. DOI: 10.1109/ITHET.2004.1358160

Insala. (2019). *Traditional mentoring vs modern mentoring*. Insala. https://www.people360ai.com/blog/what-are-the-differences-between-traditional-and-modern-mentoring

Itie, D. I. (2023). *Decisions for the implementation of artificial intelligence technology for workforce management* (Publication No. 30244296) [Doctoral dissertation, University of Maryland University College]. ProQuest Dissertations & Theses.

Jacobsen, S. E., Sørensen, M., Pedersen, S. M., & Weiner, J. (2013). Feeding the world: Genetically modified crops versus agricultural biodiversity. *Agronomy for Sustainable Development, 33*(4), 651–662. DOI: 10.1007/s13593-013-0138-9

Jahnke, I., Meinke-Kroll, M., Todd, M., & Nolte, A. (2022). Exploring artifact-generated learning with digital technologies: Advancing active learning with co-design in higher education across disciplines. Technology, Knowledge and Learning, 1-30. DOI: 10.1007/s10758-020-09473-3

Jang, Y., Choi, S., Kim, S., & Kim, H. (2023). The SNS-based e-mentoring and development of computational thinking for undergraduate students in an online course. *Journal of Educational Technology & Society, 26*(2), 147–164. DOI: 10.30191/ETS.202304_26(2).0011

Jan, S., & Mahboob, U. (2022). Online mentoring: Challenges and strategies. *Pakistan Journal of Medical Sciences, 38*(8), 2272–2277. DOI: 10.12669/pjms.38.8.5804

Javaid, M., Haleem, A., Singh, R. P., Suman, R., & Rab, S. (2022). Significance of machine learning in healthcare: Features, pillars and applications. *International Journal of Intelligent Networks, 3*, 58–73. DOI: 10.1016/j.ijin.2022.05.002

Jedaman, P., Pitchaya-Auckarakhun, M., Kasorn, K., & Kenaphoom, S. (2024). Sustainability education management scenarios under changes in the BANI era. In Wongmahesak, K., Wekke, I. S., & Suanpang, P. (Eds.), *Sustainable Development, Humanities, and Social Sciences for Society 5.0* (pp. 195–210). IGI Global Scientific Publishing., DOI: 10.4018/979-8-3693-7989-9.ch011

Jiracheewewong, E. (2023). Tools for transforming from VUCA World to BANI World: A case of educational sector in Thailand. *STOU Education Journal, 16*(2), Article 2.

Johnson, E. J., Shu, S. B., Dellaert, B. G. C., Fox, C., Goldstein, D. G., Häubl, G., Larrick, R. P., Payne, J. W., Peters, E., Schkade, D., Wansink, B., & Weber, E. U. (2012). Beyond nudges: Tools of a choice architecture. *Marketing Letters, 23*(2), 487–504. Advance online publication. DOI: 10.1007/s11002-012-9186-1

Jordan, K. (2014). Initial trends in enrollment and completion of massive open online courses. *International Review of Research in Open and Distance Learning, 15*(1), 133–160. DOI: 10.19173/irrodl.v15i1.1651

Joyce, S., Modini, M., Christensen, H., Mykletun, A., Bryant, R., Mitchell, P. B., & Harvey, S. B. (2016). Workplace interventions for common mental disorders: A systematic meta-review. *Psychological Medicine, 46*(4), 683–697. DOI: 10.1017/S0033291715002408 PMID: 26620157

Kaçar, I. G., & Baltacı, H. Ş. (2023). Designing a complementary e-mentoring program for pre-service ELT teachers: Online co-mentoring project. *Mentoring & Tutoring*, *31*(3), 352–380. DOI: 10.1080/13611267.2023.2202473

Kaiser, H. F. (1960). The application of electronic computers to factor analysis. *Educational and Psychological Measurement*, *20*(1), 141–151. DOI: 10.1177/001316446002000116

Kaiser, H. F. (1974). An index of factorial simplicity. *Psychometrika*, *39*(1), 31–36. DOI: 10.1007/BF02291575

Kashani, K. B., Awdishu, L., Bagshaw, S. M., Barreto, E. F., Claure-Del Granado, R., Evans, B. J., Forni, L. G., Ghosh, E., Goldstein, S. L., Kane-Gill, S. L., Koola, J., Koyner, J. L., Liu, M., Murugan, R., Nadkarni, G. N., Neyra, J. A., Ninan, J., Ostermann, M., Pannu, N., & Mehta, R. L. (2023). Digital health and acute kidney injury: Consensus report of the 27th Acute Disease Quality Initiative workgroup. *Nature Reviews. Nephrology*, *19*(12), 807–818. DOI: 10.1038/s41581-023-00744-7 PMID: 37580570

Kaufman, M. R., Levine, D., Casella, A., & DuBois, D. L. (2022). E-mentoring to address youth health: A systematic review. *Adolescent Research Review*, *7*(1), 63–78. DOI: 10.1007/s40894-021-00172-3

Kern, L., Harrison, J. R., Custer, B. E., & Mehta, P. D. (2019). Factors that enhance the quality of relationships between mentors and mentees during check & connect. *Behavioral Disorders*, *44*(3), 148–161. DOI: 10.1177/0198742918779791

Kessler, J. B., Dellavigna, S., Feller, A., Greenstone, M., Handel, B., Imbens, G., Jack, K., List, J., Rogers, T., & Taubinsky, D. (2015). *22706-W21671.*

Khalil, H., & Ebner, M. (2014). MOOCs Completion Rates and Possible Methods to Improve Retention - A Literature Review. *In Proceedings of World Conference on Educational Multimedia*, Hypermedia and Telecommunications, 1236-1244.

Khalil, M., Ebner, M., & Admiraal, W. (2017, October). *How can gamification improve MOOC student engagement?* Paper presented at the European Conference on Game Based Learning, Graz, Austria.

Khan, I. U., Hameed, Z., Yu, Y., Islam, T., Sheikh, Z., & Khan, S. U. (2017). Predicting the acceptance of MOOCs in a developing country: Application of task-technology fit model, social motivation, and self-determination theory. *Telematics and Informatics*. Advance online publication. DOI: 10.1016/j.tele.2017.09.009

Khan, I., Khan, N., Jazim, F., Al-Mamary, Y. H., Abdulrab, M., & Al-Ghurbani, A. M. (2022). The effect of external factors in use of technology among Ha'il university academic faculty: Evidence from Saudi Arabia. *Journal of Applied Research in Higher Education, 14*(4), 1319–1339. DOI: 10.1108/JARHE-04-2021-0140

Khatri, M. R. (2023). Integration of natural language processing, self-service platforms, predictive maintenance, and prescriptive analytics for cost reduction, personalization, and real-time insights into customer service and operational efficiency. *International Journal of Information and Cybersecurity*, *7*(9), 1–30.

Khawaja, Z., & Bélisle-Pipon, J. C. (2023). Your robot therapist is not your therapist: Understanding the role of AI-powered mental health chatbots. *Frontiers in Digital Health*, *5*, 1278186. DOI: 10.3389/fdgth.2023.1278186 PMID: 38026836

Khlaif, Z. (2018). Teachers' perceptions of factors affecting their adoption and acceptance of mobile technology in K-12 settings. *Computers in the Schools*, *35*(1), 49–67. DOI: 10.1080/07380569.2018.1428001

Khoury, M. J., & Ioannidis, J. P. (2014). Big data meets public health. *Science*, *346*(6213), 1054–1055. DOI: 10.1126/science.aaa2709 PMID: 25430753

Kim, H. K. (2024). The Effects of Artificial Intelligence Chatbots on Women's Health: A Systematic Review and Meta-Analysis. *Healthcare (Basel)*, *12*(5), 534. DOI: 10.3390/healthcare12050534 PMID: 38470645

Klepikova, A., Kormakova, V., Satler, O., & Konopko, E. (2025). Formation of digital literacy of university teachers. In *International Conference on Innovative Approaches to the Application of Digital Technologies in Education and Research (SLET 2022)* (pp. 504–513). Springer, Cham. DOI: 10.1007/978-3-031-78776-8_51

Kline, R. B. (2015). *Principles and practice of structural equation modeling* (4th ed.). Guilford Publications.

Kobets V., Yatsenko V. & Popovych I. (2021) Automated Formation of Insurance Premium for Different Risk Attitude Investment Portfolio Using Robo-Advisor. Communications in Computer and Information Science, ICTERI, Workshops, 3-22. DOI:DOI: 10.1007/978-3-031-14841-5_1

Kock, N. (2015). Common method bias in PLS-SEM: A full collinearity assessment approach. *International Journal of e-Collaboration*, *11*(4), 1–10. DOI: 10.4018/ijec.2015100101

Kock, N. (2018). Should bootstrapping be used in PLS-SEM? Toward stable p-value calculation methods. *Journal of Applied Structural Equation Modeling*, *2*(1), 1–12. DOI: 10.47263/JASEM.2(1)02

Koehler, M., & Mishra, P. (2020). What is technological pedagogical content knowledge? Contemp. Issues Technol. Teach. Educ. 2009, 9, 60–70.: Kokoç, M.; Ilgaz, H.; Altun, A. Effects of sustained attention and video lecture types on learning performances. *Educational Technology Research and Development*, *68*, 3015–3039.

Koivisto, J., & Hamari, J. (2019). The rise of motivational information systems: A review of gamification research. *International Journal of Information Management*, *45*, 191–210. DOI: 10.1016/j.ijinfomgt.2018.10.013

Kolb, D. (1984). *Experiential Learning: Experience As The Source Of Learning And Development*. Prentice-Hall.

König, J., Jäger-Biela, D. J., & Glutsch, N. (2020). Adapting to online teaching during COVID-19 school closure: Teacher education and teacher competence effects among early career teachers in Germany. *European Journal of Teacher Education*, *43*(4), 608–622. DOI: 10.1080/02619768.2020.1809650

Koohi-Fayegh, S., & Rosen, M. A. (2020). A review of energy storage types, applications and recent developments. *Journal of Energy Storage*, *27*, 101047.

Kopetz, H., & Steiner, W. (2022). Internet of things. In *Real-time systems: Design principles for distributed embedded applications* (pp. 325–341). Springer International Publishing., DOI: 10.1007/978-3-031-11992-7_13

Kormakova, V. N., Klepikova, A. G., Musaielian, E. N., Baybikova, G. V., & Lapina, M. A. (2019). Formation of ICT competencies of postgraduate students of teacher education based on interactive techniques. In International Scientific Conference "Innovative Approaches to the Application of Digital Technologies in Education and Research" (SLET-2019) (Vol. 2494, pp. 11–21). CEUR Workshop Proceedings. https://ceur-ws.org/Vol-2494/invited_paper_2.pdf

Kormakova, V. N., Klepikova, A. G., Prokopenko, Y. A., & Musaielian, E. N. (2018). Quality management principles of scientific and methodological support for students' activity within the e-learning environment. In Proceedings of the International Multidisciplinary Symposium ICT Research in Russian Federation and Europe "Integrating Research Agendas and Devising Joint Challenges" (REMS-2018) (Vol. 2254, pp. 63–75). CEUR Workshop Proceedings. https://ceur-ws.org/Vol-2254/frontmatters.pdf

Kotios, D., Makridis, G., Fatouros, G., & Kyriazis, D. (2022). Deep learning enhancing banking services: A hybrid transaction classification and cash flow prediction approach. *Journal of Big Data*, *9*(1), 100. DOI: 10.1186/s40537-022-00651-x PMID: 36213092

Kratsch, W., Manderscheid, J., Röglinger, M., & Seyfried, J. (2021). Machine learning in business process monitoring: A comparison of deep learning and classical approaches used for outcome prediction. *Business & Information Systems Engineering*, *63*(3), 261–276. DOI: 10.1007/s12599-020-00645-0

Krittanawong, C., Zhang, H., Wang, Z., Aydar, M., & Kitai, T. (2017). Artificial intelligence in precision cardiovascular medicine. *Journal of the American College of Cardiology*, *69*(21), 2657–2664. DOI: 10.1016/j.jacc.2017.03.571 PMID: 28545640

Kruft, D. (2001). Impacts of genetically-modified crops and seeds on farmers. *The Agricultural Law Resource and Reference Center*, 1-11.

Kruglanski, A. W., Bélanger, J. J., Chen, X., Köpetz, C., Pierro, A., & Mannetti, L. (2012). The energetics of motivated cognition: A force-field analysis. *Psychological Review*, *119*(1), 1–20. DOI: 10.1037/a0025488

Krumsvik, R. J. (2011). Digital competence in the Norwegian teacher education and schools. *Högre Utbildning*, *1*(1), 39–51. DOI: 10.23865/hu.v1.874

Kruse, C., Fohn, J., Wilson, N., Patlan, E. N., Zipp, S., & Mileski, M. (2020). Utilization barriers and medical outcomes commensurate with the use of telehealth among older adults: Systematic review. *JMIR Medical Informatics*, *8*(8), e20359. DOI: 10.2196/20359 PMID: 32784177

Kunkel, K. K., Baltaci, A., De La Rosa Mateo, C., Lim, S., & Davis-Kenning, D. (2022). Nudging to Health Training Acceptability by Food Pantry Personnel Supports Clients' Healthier Food Choices. *Journal of Human Sciences and Extension*, *10*(3), 5. Advance online publication. DOI: 10.55533/2325-5226.1430

Kurni, M., & KG, S. (2024). IoT for education. In The Internet of Educational Things: Enhancing Students' Engagement and Learning Performance (pp. 1-22). Springer Nature Switzerland. DOI: 10.1007/978-3-031-67387-0_10

Kuzmin, A. (2024). ANALYSIS OF THE REQUIREMENTS FOR THE VALIDATION AND VERIFICATION OF DIGITAL TWINS. *Science Bulletin*, *4*(3 (72)), 340–343.

Kuzu, A., Kahraman, M., & Odabaşı, H. F. (2012). Mentörlükte yeni bir yaklaşım: E-mentörlük [E-mentoring: A new approach in mentoring]. *Anadolu University Journal of Social Sciences*, *12*(4), 173–183.

Kwak, S. K., & Kim, J. H. (2017). Statistical data preparation: Management of missing values and outliers. *Korean Journal of Anesthesiology*, *70*(4), 407–411. DOI: 10.4097/kjae.2017.70.4.407

Lai, J. W., Qiu, W., Thway, M., Zhang, L., Jamil, N. B., Chit, L. S., Hwee, S. N. S., & Lim, F. S. (2024). *Leveraging process-action epistemic network analysis to illuminate student self-regulated learning with a socratic chatbot. EdArXiv*. https:// doi.org/DOI: 10.35542/osf.io/b9vq6

Lang, C. (2024). Dreaming big with little therapy devices: Automated therapy from India. *Anthropology & Medicine*, *31*(3), 232–249. DOI: 10.1080/13648470.2024.2378727 PMID: 39435587

Lange, C., & Costley, J. (2020). Improving online video lectures: Learning challenges created by media. *International Journal of Educational Technology in Higher Education*, *17*(1), 16. DOI: 10.1186/s41239-020-00190-6

Lawson, A. P., Martella, A. M., LaBonte, K., Delgado, C. Y., Zhao, F., Gluck, J. A., Munns, M. E., Wells LeRoy, A., & Mayer, R. E. (2024). Confounded or controlled? A systematic review of media comparison studies involving immersive virtual reality for STEM education. *Educational Psychology Review*, *36*(3), 69. DOI: 10.1007/ s10648-024-09908-8

Lawson, C. (2015). Implementing Farmers' Rights: Finding Meaning and Purpose for the International Treaty on Plant Genetic Resources for Food and Agriculture Commitments? *Available at SSRN* 2626668.

Lázaro-Cantabrana, J.; Usart-Rodríguez, M.; Gisbert-Cervera, M. (2019). Assessing teacher digital competence: The construction of an instrument for measuring the knowledge of pre-service teachers. J. New Approaches Educ. Res. (NAER J.) 8, 73–78.

LeBlanc, P. J. (2018). Higher education in a VUCA World. *Change*, *50*(3–4), 23–26. DOI: 10.1080/00091383.2018.1507370

Ledesma, R. D., & Valero-Mora, P. (2007). Determining the number of factors to retain in EFA: An easy-to-use computer program for carrying out parallel analysis. *Practical Assessment, Research & Evaluation*, *12*(1), 2.

Le, H., Janssen, J., & Wubbels, T. (2018). Collaborative learning practices: Teacher and student perceived obstacles to effective student collaboration. *Cambridge Journal of Education*, *48*(1), 103–122. DOI: 10.1080/0305764X.2016.1259389

Leo, M., Sharma, S., & Maddulety, K. (2019). Machine learning in banking risk management: A literature review. *Risks*, *7*(1), 29. DOI: 10.3390/risks7010029

Levy, Y., & Ellis, T. J. (2011). A guide for novice researchers on experimental and quasi-experimental studies in information systems research. *Interdisciplinary Journal of Information, Knowledge, and Management*, *6*, 151–161. DOI: 10.28945/1373

Li, T., & Yang, N. (2019). Comparing MOOCs with traditional courses for quality teaching in higher education. *DEStech Transactions on Social Science, Education and Human Science*. DOI: 10.12783/dtssehs/meit2018/27637

Liew, A. W. C., Yan, H., & Yang, M. (2005). Pattern recognition techniques for the emerging field of bioinformatics: A review. *Pattern Recognition*, *38*(11), 2055–2073. DOI: 10.1016/j.patcog.2005.02.019

Li, L., Hew, K. F., & Du, J. (2024). Gamification enhances student intrinsic motivation, perceptions of autonomy and relatedness, but minimal impact on competency: A meta-analysis and systematic review. *Educational Technology Research and Development*, *72*(2), 765–796. DOI: 10.1007/s11423-023-10337-7

Lillywhite, A., & Wolbring, G. (2022). Undergraduate disabled students as knowledge producers including researchers: Perspectives of disabled students. *Education Sciences*, *12*(2), 77. DOI: 10.3390/educsci12020077

Limonova, V., Santos, A., Mamede, H.M., & Filipe, V. (2023). The Research Context of Artificial Intelligence and Gamification to Improve Student Engagement and Attendance in Higher Education. *RE@D –Revista de Educação à Distância e Elearning,* Vol.6, Nº2, 1-17.

Linardon, J., Cuijpers, P., Carlbring, P., Messer, M., & Fuller-Tyszkiewicz, M. (2019). The efficacy of app-supported smartphone interventions for mental health problems: A meta-analysis of randomized controlled trials. *World Psychiatry; Official Journal of the World Psychiatric Association (WPA)*, *18*(3), 325–336. DOI: 10.1002/wps.20673 PMID: 31496095

Lindsay, S., & Cagliostro, E. (2020). A web-based intervention for youth with physical disabilities: Comparing the role of mentors in 12- and 4-week formats. *JMIR Pediatrics and Parenting*, *3*(1), e15813. DOI: 10.2196/15813

Li, P., Zhang, X., Hu, X., Xu, B., & Zhang, J. (2024). Theoretical model and practical analysis of immersive industrial design education based on virtual reality technology. *International Journal of Technology and Design Education*, •••, 1–28. DOI: 10.1007/s10798-024-09946-x

Little, R. J., & Rubin, D. B. (2019). *Statistical analysis with missing data* (3rd ed.). John Wiley & Sons., DOI: 10.1002/9781119482260

Liu, P., Yuan, W., Fu, J., Jiang, Z., Hayashi, H., & Neubig, G. (2023). Pre-train, prompt, and predict: A systematic survey of prompting methods in natural language processing. *ACM Computing Surveys*, *55*(9), 1–35. DOI: 10.1145/3560815

Liu, S., Liu, S., Cai, W., Che, H., Pujol, S., Kikinis, R., Feng, D., & Fulham, M. (2020). The role of artificial intelligence in healthcare: A comprehensive review. *Journal of Medical Systems*, *44*(11), 1. Advance online publication. DOI: 10.1007/s10916-020-01662-4

Liu, Y., Ma, S., & Chen, Y. (2024). The impacts of learning motivation, emotional engagement and psychological capital on academic performance in a blended learning university course. *Frontiers in Psychology*, *15*, 1357936. DOI: 10.3389/fpsyg.2024.1357936

Ljubojevic, M., Vaskovic, V., Stankovic, S., & Vaskovic, J. (2014). Using supplementary video in multimedia instruction as a teaching tool to increase efficiency of learning and quality of experience. *International Review of Research in Open and Distance Learning*, *15*(3), 275–291. DOI: 10.19173/irrodl.v15i3.1825

Louwaars, N. (2007). *Seeds of Confusion: The Impact of Policies on Seed Systems*. Wageningen University and Research.

Lukoff, K., Lyngs, U., Shirokova, K., Rao, R., Tian, L., Zade, H., Munson, S. A., & Hiniker, A. (2023). SwitchTube: A proof-of-concept system introducing "adaptable commitment interfaces" as a tool for digital wellbeing. *Proceedings of the 2023 CHI Conference on Human Factors in Computing Systems (CHI '23)*. Association for Computing Machinery, New York, NY, USA, Article 197, 1–22. DOI: 10.1145/3544548.3580703

Luna Scott, C. (2015). *The futures of learning 2: What kind of learning for the 21st century?* [Education Research and Foresight • Working Papers]. United Nations — Educational, Cultural and Scientific Organization. https://repositorio.minedu.gob.pe/handle/20.500.12799/3709

MacKenzie, S. B., & Podsakoff, P. M. (2012). Common method bias in marketing: Causes, mechanisms, and procedural remedies. *Journal of Retailing*, *88*(4), 542–555. DOI: 10.1016/j.jretai.2012.08.001

Mackey, T. K., & Liang, B. A. (2012). Integrating biodiversity management and indigenous biopiracy protection to promote environmental justice and global health. *American Journal of Public Health*, *102*(6), 1091–1095. DOI: 10.2105/AJPH.2011.300408 PMID: 22515858

MacKinnon, D. P., Fairchild, A. J., & Fritz, M. S. (2007). Mediation analysis. *Annual Review of Psychology*, *58*(1), 593–614. DOI: 10.1146/annurev.psych.58.110405.085542

Magid, S. I., & Arkhipova, E. N. (2017). Simulators-aided training for personnel of power industry. History. State. Norms. In NPO Energobezopasnost eBooks. DOI: 10.24223/978-5-600-02001-6

Mahoney, S. E. (2018). Owning the World's Seed Supply: How Seed Industry Mergers Threaten Global Food Security. *Geo. Envtl. L. Rev., 31*, 563.

Maia, M. C., Figueiredo, J. C. A. D., & Serey, D. (2019). Online student engagement: A case study in teaching of programming. *Anais Do XXX Simpósio Brasileiro de Informática Na Educação (SBIE 2019), 30*, 51–60. DOI: 10.5753/cbie.sbie.2019.51

Makransky, G., & Petersen, G. B. (2021). The cognitive-affective model of immersive learning (CAMIL): A theoretical research-based model of learning in immersive virtual reality. *Educational Psychology Review, 33*(3), 937–958. DOI: 10.1007/s10648-020-09586-2

Makransky, G., & Petersen, J. (2021). The impact of augmented and virtual reality on student engagement and learning outcomes in higher education: A systematic review. *Journal of Educational Psychology, 113*(4), 670–687. DOI: 10.1037/edu0000434

Malamud, R. (2019). Opt Out. In *Email*. DOI: 10.5040/9781501341939.0018

Malkovich, A. (n.d.). *Mentorship in the digital age: How technology can enhance the mentor-mentee relationship*. PushFar. Retrieved February 16, 2025, from https://www.pushfar.com/article/mentorship-in-the-digital-age-how-technology-can-enhance-the-mentor-mentee-relationship/

Mangeni, B. (2019). The role of public-private partnerships (PPPs) in ensuring technology access for farmers in sub-Saharan Africa. *African Journal of Food, Agriculture, Nutrition and Development, 19*(1), 14137–14155. DOI: 10.18697/ajfand.84.BLFB1018

Mangram, M. E. (2013). A Simplified Perspective on Markowitz Portfolio Theory. [MMAR-link]. *Global Journal of Business Research, 7*(1), 59–70. https://ssrn.com/abstract=2147880

Maričić, M., & Lavicza, Z. (2024). Enhancing student engagement through emerging technology integration in STEAM learning environments. *Education and Information Technologies, 29*(17), 23361–23389. DOI: 10.1007/s10639-024-12710-2

Martin, F., Ritzhaupt, A., Kumar, S., & Budhrani, K. (2019). Award-winning faculty online teaching practices: Course design, assessment and evaluation, and facilitation. *The Internet and Higher Education, 42*, 34–43. DOI: 10.1016/j.iheduc.2019.04.001

Masters of ERP. (n.d.). The impact of IoT on campus: Smart student information systems. https://www.mastersoferp.com/the-impact-of-iot-on-campus-smart-student-information-systems/

Mathôt, S., Schreij, D., & Theeuwes, J. (2012). OpenSesame: An open-source, graphical experiment builder for the social sciences. *Behavior Research Methods*, *44*(2), 314–324. DOI: 10.3758/s13428-011-0168-7 PMID: 22083660

Matthews, D. (2003). *Globalising intellectual property rights: the TRIPS Agreement.* Routledge. DOI: 10.4324/9780203165683

McAleer, D., & Bangert, A. (2011). Professional growth through online mentoring: A study of mathematics mentor teachers. *Journal of Educational Computing Research*, *44*(1), 83–115. DOI: 10.2190/EC.44.1.e

McLoughlin, C., Brady, J., Lee, M. J. W., & Russell, R. (2007). *Peer-to-peer: An e-mentoring approach to developing community, mutual engagement and professional identity for pre-service teachers.* http://bibliotecadigital.mineduc.cl//handle/20.500.12365/17803

Mehta, C. R., & Patel, N. R. (1996). *Exact Tests.* SPSS Incorporated.

Melash, A. A., Bogale, A. A., Migbaru, A. T., Chakilu, G. G., Percze, A., Ábrahám, É. B., & Mengistu, D. K. (2023). Indigenous agricultural knowledge: A neglected human based resource for sustainable crop protection and production. *Heliyon*, *9*(1), e12978. DOI: 10.1016/j.heliyon.2023.e12978 PMID: 36711305

Melnick, M. (2022). *The BANI world is a new world that has replaced the world of VUCA.* https://hackernoon.com/the-bani-world-is-a-new-world-that-has-replaced-the-world-of-vuca

Mendoza, G., Jung, I., & Kobayashi, S. (2017). A Review of Empirical Studies on MOOC Adoption: Applying the Unified Theory of Acceptance and Use of Technology. *International Journal for Educational Media and Technology*, *11*(1), 15–24.

Menon, V., & Varadharajan, N. (2020). Digital approaches for mental health service delivery in low- and middle-income countries like India: Key implementational challenges and recommendations. *Asian Journal of Psychiatry*, *50*, 101962. DOI: 10.1016/j.ajp.2020.101962 PMID: 32109806

Merrick, L. (2016, May 18). Mentoring in a VUCA World—'Future proofing' your people. *Coach Mentoring*. https://www.coachmentoring.co.uk/blog/2016/05/mentoring-in-a-vuca-world/

Merrick, L. (2025, January 17). Mentoring in a BANI World—Five free mentor CPD sessions. *Coach Mentoring*. https://www.coachmentoring.co.uk/blog/2025/01/mentoring-in-a-bani-world-five-free-mentor-cpd-sessions/

Merz, J. F., & Cho, M. K. (2005). What are gene patents and why are people worried about them? *Public Health Genomics*, *8*(4), 203–208. DOI: 10.1159/000087956 PMID: 16244473

Meske, C., & Amojo, I. (2020). Ethical guidelines for the construction of digital nudges. *Proceedings of the Annual Hawaii International Conference on System Sciences, 2020-January*, 3928–3937. DOI: 10.24251/HICSS.2020.480

Meske, C., Amojo, I., & Mohr, P. (2020). Digital nudging to increase usage of charity features on e-commerce platforms. *Proceedings of the 15th International Conference on Business Information Systems 2020 "Developments, Opportunities and Challenges of Digitization", WIRTSCHAFTSINFORMATIK 2020*. DOI: 10.30844/wi_2020_k5-meske

Messick, S. (1995). Validity of psychological assessment: Validation of inferences from persons' responses and performances as scientific inquiry into score meaning. *The American Psychologist*, *50*(9), 741–749. DOI: 10.1037/0003-066X.50.9.741

Meyer, K. A. (2014). Student engagement in online learning: What works and why. *ASHE Higher Education Report*, *40*(6), 1–114. DOI: 10.1002/aehe.20018

Michels, L., Schmitt, K., Ochmann, J., Laumer, S., & Tiefenbeck, V. (2021). Is It All About Transparency? The Effectiveness and Ethics of a Digital Salience Nudge. *ECIS 2021 Research Papers, June*. https://aisel.aisnet.org/ecis2021_rp/25

Michels, L., Ochmann, J., Schmitt, K., Laumer, S., & Tiefenbeck, V. (2023). Salience, transparency, and self-nudging: A digital nudge to promote healthier food product choices. *European Journal of Information Systems*, *33*(5), 1–6. DOI: 10.1080/0960085X.2023.2229787

Milhorance, C. (2017). The Brazil–Mozambique experience in rural development: bringing conflict back into transfer studies. In *Public Policy Transfer* (pp. 125–148). Edward Elgar Publishing. DOI: 10.4337/9781785368042.00013

Miller, N., & Forkosh-Baruch, A. The use of online teaching in language among high school students' matriculation examinees. In *Proceedings of the 13th Chais Conference on Instructional Technologies Research: Learning in the Technological Era*, Ra'anana, Israel, 20 February 2018; The Open University of Israel: Ra'anana, Israel, 2018. (In Hebrew).

Miner, A. S., Shah, N., Bullock, K. D., Arnow, B. A., Bailenson, J., & Hancock, J. (2019). Key considerations for incorporating conversational AI in psychotherapy. *Frontiers in Psychiatry, 10*, 746. DOI: 10.3389/fpsyt.2019.00746 PMID: 31681047

Ministry of Education and Science–Republic of Bulgaria, Law on Pre-School and School Education..

Mirbabaie, M., Marx, J., & Germies, J. (2021). Conscious Commerce – Digital Nudging and Sustainable E-commerce Purchase Decisions. *ACIS 2021 - Australasian Conference on Information Systems, Proceedings*, 1–11.

Mirsch, T., Lehrer, C., & Jung, R. (n.d.). Making Digital Nudging Applicable. *The Digital Nudge Design Method., 2009*, 1–16.

Mitali, J., Dhinakaran, S., & Mohamad, A. A. (2022). Energy storage systems: a review, Energy Storage and Saving, Volume 1, Issue 3, Pages 166-216, ISSN 2772-6835, DOI: 10.1016/j.enss.2022.07.002

Mitchell, A. D., & Voon, T. (2009). TRIPS', in Daniel Bethlehem, and others (eds), *The Oxford Handbook of International Trade Law* (2009; online edn, Oxford Academic, 18 Sept. 2012), , accessed 7 Mar. 2025.DOI: 10.1093/oxfordhb/9780199231928.013.0008

Mitsea, E., Drigas, A., & Skianis, C. (2023). Digitally assisted mindfulness in training self-regulation skills for sustainable mental health: A systematic review. *Behavioral Sciences (Basel, Switzerland), 13*(12), 1008. DOI: 10.3390/bs13121008 PMID: 38131865

Modi, T., & Gochhait, S. (2023). Impact of Artificial Intelligence on Gamification: Current Applications. *2023 International Conference on Innovative Data Communication Technologies and Applications (ICIDCA),* pp. 287-290. DOI: 10.1109/ICIDCA56705.2023.10099771

Montiel, H., & Gomez-Zermeño, M. G. (2022). Rock the Boat! Shaken by the COVID-19 crisis: A review on teachers' competencies in ICT. *Frontiers in Education, 6*, 770442. DOI: 10.3389/feduc.2021.770442

Morellet, P. (2025, January 23). *Coaching in a BANI World: Building resilience amid uncertainty and complexity*. Speexx. https://www.speexx.com/speexx-blog/coaching-in-a-bani-world/

Moustakas, C. (1994). *Phenomenological Research Methods*. Sage Publications. DOI: 10.4135/9781412995658

Muammar, S., Hashim, K. F., & Panthakkan, A. (2022). Evaluation of digital competence level among educators in uae higher education institutions using digital competence of educators (digcomedu) framework. *Education and Information Technologies, 28*(3), 2485–2508. DOI: 10.1007/s10639-022-11296-x PMID: 36035975

Mullen, C. A. (2021). Online doctoral mentoring in a pandemic: Help or hindrance to academic progress on dissertations? *International Journal of Mentoring and Coaching in Education, 10*(2), 139–157. DOI: 10.1108/IJMCE-06-2020-0029

Multishoring. (2025, January 21). The impact of IoT in education: What you need to know. https://multishoring.com/blog/impact-of-iot-in-education/

Munyi, P., & De Jonge, B. (2015). Seed systems support in Kenya: Consideration for an integrated seed sector development approach. *Journal of Sustainable Development, 8*(2), 161. DOI: 10.5539/jsd.v8n2p161

Murphy, L., Noga, K., & Rose, M. (2015). Seeking Pure Fields: The Case Against Preemption of State Bans on Genetically Engineered Crops. *University of San Francisco Law Review. University of San Francisco. School of Law, 49*(3), 3.

Mustapha, A. M., Arogundade, O. T., Abayomi-Alli, A., Adesemowo, A. K., & Adeniran, O. J. (2024). An improved cloud-based business process compliance management system using a user-centered approach. *International Journal of System Assurance Engineering and Management, 15*(11), 1–28. DOI: 10.1007/s13198-024-02494-6

Nabuuma, D., Reimers, C., Hoang, K. T., Stomph, T., Swaans, K., & Raneri, J. E. (2022). Impact of seed system interventions on food and nutrition security in low- and middle-income countries: A scoping review. *Global Food Security, 33*, 100638. DOI: 10.1016/j.gfs.2022.100638

Narot, P., & Kiettikunwong, N. (2023). *Interdisciplinary perspectives on special and inclusive education in a volatile, uncertain, complex & ambiguous (VUCA) world.* Emerald Publishing. https://bookstore.emerald.com/interdisciplinary-perspectives-on-special-and-inclusive-education-in-a-volatile-uncertain-complex-ambiguous-vuca-world.html

Naslund, J. A., & Deng, D. (2021). Addressing mental health stigma in low-income and middle-income countries: A new frontier for digital mental health. *Ethics, Medicine and Public Health, 19*, 100719. https://doi.org/DOI: 10.1016/j.jemep.2021.100719

Naslund, J. A., Aschbrenner, K. A., Kim, S. J., McHugo, G. J., Unützer, J., Bartels, S. J., & Marsch, L. A. (2017). Health behavior models for informing digital technology interventions for individuals with mental illness. *Psychiatric Rehabilitation Journal, 40*(3), 325–335. DOI: 10.1037/prj0000246 PMID: 28182469

Naslund, J. A., Gonsalves, P. P., Gruebner, O., Pendse, S. R., Smith, S. L., Sharma, A., & Raviola, G. (2019). Digital Innovations for Global Mental Health: Opportunities for Data Science, Task Sharing, and Early Intervention. *Current Treatment Options in Psychiatry, 6*(4), 337–351. DOI: 10.1007/s40501-019-00186-8 PMID: 32457823

Nataliia, H., & Olena, M. (2023). The key administrative competencies of managers required for company development in the BANI World. *Economics - Innovative and Economics Research Journal, 11*(1), Article 1. DOI: 10.2478/eoik-2023-0012

Nayyar, A., Gadhavi, L., & Zaman, N. (2021). *Machine learning in healthcare: review, opportunities and challenges*. Machine Learning and the Internet of Medical Things in Healthcare.

Ndetei, D. M., Mutiso, V., & Osborn, T. (2023). Moving away from the scarcity fallacy: Three strategies to reduce the mental health treatment gap in LMICs. *World Psychiatry; Official Journal of the World Psychiatric Association (WPA), 22*(1), 163–164. DOI: 10.1002/wps.21054 PMID: 36640407

Neely, A. R., Cotton, J., & Neely, A. D. (2017). E-mentoring: A model and review of the literature. *AIS Transactions on Human-Computer Interaction, 9*(3), 220–242. DOI: 10.17705/1thci.00096

Nesterowicz, K., Bayramova, U., & Szádeczky, T. (2024). Gamification in Massive Open Online Courses After the Pandemic. In L.A., Tomei and D.D. Carbonara (Eds.) *Exploring Technology-Infused Education in the Post-Pandemic Era* (pp. 489-513), IGI Global DOI: 10.4018/979-8-3693-2885-9.ch017

Nguyen, A., Kremantzis, M., Essien, A., Petrounias, I., & Hosseini, S. (2024). Enhancing student engagement through artificial intelligence (AI): Understanding the basics, opportunities, and challenges. *Journal of University Teaching & Learning Practice, 21*(06), 06. Advance online publication. DOI: 10.53761/caraaq92

Nguyen, L., Tomy, S., & Pardede, E. (2024). Enhancing collaborative learning and e-mentoring in a smart education system in higher education. *Computers, 13*(1), 28. DOI: 10.3390/computers13010028

Nguyen, M. H. (2021). Managing social media use in an "always-on" society: Exploring digital wellbeing strategies that people use to disconnect. *Mass Communication & Society, 24*(6), 795–817. DOI: 10.1080/15205436.2021.1979045

Niemiec, C. P., & Ryan, R. M. (2009). Autonomy, competence, and relatedness in the classroom: Applying self-determination theory to educational practice. *Theory and Research in Education*, *7*(2), 133–144. DOI: 10.1177/1477878509104318

Niinimäki, K. (2009). Future of the consumer society. *Future of the Consumer Society, May*, 125–134. https://orgprints.org/16410/1/consumer6.pdf

Nordin, R. (2011). Protection of indigenous knowledge: Mismatch between the convention on biological diversity and TRIPS agreement. *Jurnal Undang-undang dan Masyarakat, 15*, 179.

Nsflow. (n.d.). 9 benefits of augmented reality in education. https://www.nsflow.com/blog/benefits-of-augmented-reality-in-education

Nunnally, J. C., & Bernstein, I. H. (1994). *Psychometric theory* (3rd ed.). McGraw-Hill.

Nyamathulla, S., & Dhanamjayulu, C. (2024). A review of battery energy storage systems and advanced battery management system for different applications: Challenges and recommendations. *Journal of Energy Storage*, *86*, 111179.

O'Brien, R. M. (2007). A caution regarding rules of thumb for variance inflation factors. *Quality & Quantity*, *41*(5), 673–690. DOI: 10.1007/s11135-006-9018-6

Obermeyer, Z., & Emanuel, E. J. (2016). Predicting the future—Big data, machine learning, and clinical medicine. *The New England Journal of Medicine*, *375*(13), 1216–1219. DOI: 10.1056/NEJMp1606181 PMID: 27682033

Officer, M. J. Chief Product. (n.d.). *From theory to practice: Designing real-world learning experiences*. Retrieved February 22, 2025, from https://blog.learnamp.com/from-theory-to-practice-designing-real-world-learning-experiences

Oguamanam, C. (2014). Farmers' rights and the intellectual property dynamic in agriculture. *Sage Handbook of Intellectual Property*, 238-257.

Oguamanam, C. (2006). Intellectual Property Rights in Plant Genetic Resources: Farmers' Rights and Food Security of Indigenous and Local Communities. *The Drake Journal of Agricultural Law*, *11*, 273.

Ogugua, J. O., Okongwu, C. C., Akomolafe, O. O., Anyanwu, E. C., & Daraojimba, O. D.Jane Osareme OguguaChiamaka Chinaemelum OkongwuOpeoluwa Oluwanifemi AkomolafeEvangel Chinyere AnyanwuObinna Donald Daraojimba. (2024). Mental health and digital technology: A public health review of current trends and responses. *International Medical Science Research Journal*, *4*(2), 108–125. DOI: 10.51594/imsrj.v4i2.754

Oloko, T. O. (2016). An examination of Article 27 of the TRIPS Agreement in relation to the provisions on patentable subject matter under the PDA in Nigeria. *Commonwealth Law Bulletin, 42*(2), 236–260. DOI: 10.1080/03050718.2016.1187072

Omar, N. D., Hassan, H., & Atan, H. (2012). Student engagement in online learning: Learners attitude toward e-mentoring. *Procedia: Social and Behavioral Sciences, 67*, 464–475. DOI: 10.1016/j.sbspro.2012.11.351

Ortega-Arranz, A., Muñoz-Cristóbal, J. A., Martínez-Monés, A., Bote-Lorenzo, M. L., & Asensio-Pérez, J. I. (2017). How Gamification Is Being Implemented in MOOCs? A Systematic Literature Review. In Lavoué, É., Drachsler, H., Verbert, K., Broisin, J., & Pérez-Sanagustín, M. (Eds.), Lecture Notes in Computer Science: Vol. 10474. *Data Driven Approaches in Digital Education. EC-TEL 2017*. Springer. DOI: 10.1007/978-3-319-66610-5_40

Osborne, J. W. (2010). Improving your data transformations: Applying the Box-Cox transformation. *Practical Assessment, Research & Evaluation, 15*(1), 12.

Osborne, J. W., & Overbay, A. (2004). The power of outliers (and why researchers should always check for them). *Practical Assessment, Research & Evaluation, 9*(1), 6.

Osborne, J. W., & Waters, E. (2002). Four assumptions of multiple regression that researchers should always test. *Practical Assessment, Research & Evaluation, 8*(2), 1–9.

Oullier, O., Cialdini, R., Thaler, R. H., & Mullainathan, S. (2010). Improving public health prevention with a nudge. *Improving Public Health Prevention with Behavioural, Cognitive and Neuroscience.*

Ovcharuk, O., & Ivaniuk, I. (2021). A self-assessment tool of the level of digital competence of Ukrainian teachers in the context of lifelong learning: The results of an online survey 2021. *Educ. Dimens., 5*, 75–88. DOI: 10.31812/educdim.4719

Oxfam. (2025, January 20). Takers not Makers: The unjust poverty and unearned wealth of colonialism. *Oxfam.* https://www.kedv.org.tr/public/uploads/files/Oxfam%202025%20Takers%20Not%20Makers%20(%C3%9Cretenler%20De%C4%9Fil%20%C3%9Czerine%20Konanlar)%20EN%20Report.pdf

Padmaja, S. S. (2020). Intellectual Property, Competition, and Regulatory Policies: A Case of Bt Cotton Seed Industry in India. *Competition Commission of India Journal on Competition Law and Policy*, 79-101.

Paganin, G., & Simbula, S. (2020). Smartphone-based interventions for employees' wellbeing promotion: a systematic review. *Electronic Journal of Applied Statistical Analysis, 13*(3), 682-712. DOI:DOI: 10.1285/i20705948v13n3p682

Palinkas, L. A., Horwitz, S. M., Green, C. A., Wisdom, J. P., Duan, N., & Hoagwood, K. (2015). Purposeful sampling for qualitative data collection and analysis in mixed method implementation research. [PubMed]. *Administration and Policy in Mental Health*, *42*(5), 533–544. DOI: 10.1007/s10488-013-0528-y PMID: 24193818

Panthalookaran, V. (2022). Education in a VUCA-driven world: Salient features of an entrepreneurial pedagogy. *Higher Education for the Future*, *9*(2), 234–249. DOI: 10.1177/23476311221108808

Partelow, S., Abson, D. J., Schlüter, A., Fernández-Giménez, M., Von Wehrden, H., & Collier, N. (2019). Privatizing the commons: New approaches need broader evaluative criteria for sustainability. *The International Journal of the Commons*, *13*(1), 747. DOI: 10.18352/ijc.938

Patel, V., Xiao, S., Chen, H., Hanna, F., Jotheeswaran, A. T., Luo, D., Parikh, R., Sharma, E., Usmani, S., Yu, Y., Druss, B. G., & Saxena, S. (2016). The magnitude of and health system responses to the mental health treatment gap in adults in India and China. *Lancet*, *388*(10063), 3074–3084. DOI: 10.1016/S0140-6736(16)00160-4 PMID: 27209149

Pedrouzo, S. B., Jaitt, M., Núñez, J., Lamas, F., & Krynski, L. (2024). Adolescence and digital wellbeing: Tools for healthcare professionals. *Archivos Argentinos de Pediatria, 122*(1):e202310199. https://doi/org/. PMID: 38150291.DOI: 10.5546/aap.2023-10199

Pennycook, G., McPhetres, J., Zhang, Y., Lu, J. G., & Rand, D. G. (2020). Fighting COVID-19 Misinformation on Social Media: Experimental Evidence for a Scalable Accuracy-Nudge Intervention. *Psychological Science*, *31*(7), 770–780. DOI: 10.1177/0956797620939054 PMID: 32603243

Perifanou, M., Economides, A. A., & Tzafilkou, K. (2021). Teachers' digital skills readiness during COVID-19 pandemic. [IJET]. *International Journal of Emerging Technologies in Learning*, *16*(8), 238–251. DOI: 10.3991/ijet.v16i08.21011

Perrin, R. K. (1994). Intellectual property rights in agricultural development.

Perski, O., Crane, D., Beard, E., & Brown, J. (2019). Does the addition of a supportive chatbot promote user engagement with a smoking cessation app? An experimental study. *Digital Health*, *5*, 2055207619880676. DOI: 10.1177/2055207619880676 PMID: 31620306

Peschard, K. E. (2022). *Seed activism: Patent politics and litigation in the global south*. MIT Press. DOI: 10.7551/mitpress/14484.001.0001

Peschard, K., & Randeria, S. (2020). 'Keeping seeds in our hands': The rise of seed activism. *The Journal of Peasant Studies, 47*(4), 613–647. DOI: 10.1080/03066150.2020.1753705

Peters, B. G. (2018). *Policy problems and policy design.* Edward Elgar Publishing. DOI: 10.4337/9781786431356

Petersen, G. B., Stenberdt, V., Mayer, R. E., & Makransky, G. (2023). Collaborative generative learning activities in immersive virtual reality increase learning. *Computers & Education, 207*, 104931. DOI: 10.1016/j.compedu.2023.104931

Pettersson, F. (2018). On the issues of digital competence in educational contexts–a review of literature. *Education and Information Technologies, 23*(3), 1005–1021. DOI: 10.1007/s10639-017-9649-3

Peugh, J. L., & Enders, C. K. (2004). Missing data in educational research: A review of reporting practices and suggestions for improvement. *Review of Educational Research, 74*(4), 525–556. DOI: 10.3102/00346543074004525

Piaget, J. (1973). *To understand is to invent: The future of education.* Viking Press.

Pintrich, P. R. (2004). A conceptual framework for assessing motivation and self-regulated learning in college students. *Educational Psychology Review, 16*(4), 385–407. DOI: 10.1007/s10648-004-0006-x

Podsakoff, P. M., MacKenzie, S. B., Lee, J. Y., & Podsakoff, N. P. (2003). Common method biases in behavioral research: A critical review of the literature and recommended remedies. *The Journal of Applied Psychology, 88*(5), 879–903. DOI: 10.1037/0021-9010.88.5.879

Poirier, C., & Gelin, M. (2024). Boosting workplace wellbeing: Unveiling the impact of digital tools. *Journal of Workplace Behavioral Health*, •••, 1–20. DOI: 10.1080/15555240.2024.2441209

Preacher, K. J., & Hayes, A. F. (2008). Asymptotic and resampling strategies for assessing and comparing indirect effects in multiple mediator models. *Behavior Research Methods, 40*(3), 879–891. DOI: 10.3758/BRM.40.3.879

Prip, C. (2024). Farmers' Rights in the Plant Treaty: Interrelations and recent interactions with other international regimes and processes. *Development Studies Research, 11*(1), 2357095. DOI: 10.1080/21665095.2024.2357095

Psico-Smart. (2024, August 28). The impact of artificial intelligence on cognitive skills assessment. https://psico-smart.com/en/blogs/blog-the-impact-of-artificial-intelligence-on-cognitive-skills-assessment-160885

Purina-Bieza, K. E. (2021). Pedagogical digital competence and its acquisition in a teacher education programme. *Hum. Technol. Qual. Educ., 2021*, 333–351. DOI: 10.22364/htqe.2021.24

Qays, M. O., Buswig, Y., Basri, H., Hossain, M. L., Abu-Siada, A., Rahman, M. M., & Muyeen, S. M. (2020). An Intelligent Controlling Method for Battery Lifetime Increment Using State of Charge Estimation in PV-Battery Hybrid System. *Applied Sciences (Basel, Switzerland), 10*(24), 8799. DOI: 10.3390/app10248799

Radlick, R. L., Mirkovic, J., Przedpelska, S., Halvorsen Brendmo, E., & Gammon, D. (2020). Experiences and needs of multicultural youth and their mentors, and implications for digital mentoring platforms: Qualitative exploratory study. *JMIR Formative Research, 4*(2), e15500. DOI: 10.2196/15500

Rajkomar, A., Dean, J., & Kohane, I. (2019). Machine learning in medicine. *The New England Journal of Medicine, 380*(14), 1347–1358. DOI: 10.1056/NEJMra1814259 PMID: 30943338

Rane, N., Choudhary, S., & Rane, J. (2023, November). *Education 4.0 and 5.0: Integrating Artificial Intelligence (AI) for Personalized and Adaptive Learning. SSRN*. DOI: 10.2139/ssrn.4638365

Rankings, G. U. (n.d.). *Trends in Higher Education*. GU Rankings, London. Retrieved April 21, 2025, from https://www.globaluniversityrankings.com/trends-in-higher-education

Raqeeb, A. (2024). Retention and retention strategies in technology-driven work context [Unpublished master's thesis]. University of Vaasa.

Ravinutala, S. (2023, June 14). *How can coaching help in maneuvering through the BANI World?* https://uexcelerate.com/how-can-coaching-help-in-maneuvering-through-the-bani-world/

Raykov, T. (1997). Estimation of composite reliability for congeneric measures. *Applied Psychological Measurement, 21*(2), 173–184. DOI: 10.1177/01466216970212006

Rebonato, R. (2014). A Critical Assessment of Libertarian Paternalism. *Journal of Consumer Policy, 37*(3), 357–396. Advance online publication. DOI: 10.1007/s10603-014-9265-1

Reeve, J. (2012). A sef-determination theory perspective on student engagement. In Christenson, S. L., Reschly, A. L., & Wylie, C. (Eds.), *Handbook of research on student engagement* (pp. 149–172). Springer., DOI: 10.1007/978-1-4614-2018-7_7

Reeves, T., & Reeves, P. (2015). Educational Technology Research in a VUCA World. *Educational Technology Archive*. https://www.semanticscholar.org/paper/Educational-Technology-Research-in-a-VUCA-World.-Reeves-Reeves/94de9dca801e3707e5d21e40ec8eff343a31b0c6

Reid-Martinez, K., & Grooms, L. D. Online learning propelled by constructivism. In Encyclopedia of Information Science and Technology, 4th ed.; Khosrow-Pour, M., Ed.; IGI Global: Hershey, PA, USA, 2018; pp. 2588–2598. DOI: 10.4018/978-1-5225-2255-3.ch226

Ringle, C. M., & Sarstedt, M. (2016). Gain more insight from your PLS-SEM results: The importance-performance map analysis. *Industrial Management & Data Systems*, *116*(9), 1865–1886. DOI: 10.1108/IMDS-10-2015-0449

Rockinson-Szapkiw, A., & Wendt, J. L. (2021). The benefits and challenges of a blended peer mentoring program for women peer mentors in science, technology, engineering and mathematics (STEM). *International Journal of Mentoring and Coaching in Education*, *10*(1), 1–16. DOI: 10.1108/IJMCE-03-2020-0011

Rodgers, G. (2021). Work and conditions of work. In *Critical Issues in Human Rights and Development* (pp. 282–299). Edward Elgar Publishing.

Rodríguez-Martínez, A., Amezcua-Aguilar, T., Cortés-Moreno, J., & Jiménez-Delgado, J. J. (2023). Qualitative analysis of conversational chatbots to alleviate loneliness in older adults as a strategy for emotional health. *Healthcare (Basel)*, *12*(1), 62. DOI: 10.3390/healthcare12010062 PMID: 38200967

Roffarello, A. M., & Russis, L. D. (2023). Achieving digital wellbeing through digital self-control tools: A systematic review and meta-analysis. *ACM Trans. Comput.-Hum. Interact.*, *30*(4), Article 53, 66 pages. DOI: 10.1145/3571810

Rogers, J. T. (2010). The encroachment of intellectual property protections on the rights of farmers. *The Drake Journal of Agricultural Law*, *15*, 149.

Roll, I., & Wylie, R. (2016). Evolution and revolution in artificial intelligence in education. *International Journal of Artificial Intelligence in Education*, *26*(2), 582–599. DOI: 10.1007/s40593-016-0110-3

Rose, K. (2024, October 23). Belgian teen's suicide sparks lawsuit against Character. AI chatbot. *The New York Times*. https://www.nytimes.com/2024/10/23/technology/characterai-lawsuit-teen-suicide.html

Rowan, L. M., & Jin, Y. (2022). Conversations that count in online student engagement – A case study. *ASCILITE Publications*, e22162. DOI: 10.14742/apubs.2022.162

Rowland, K. N. (2012). E-mentoring: An innovative twist to traditional mentoring. *Journal of Technology Management & Innovation*, *7*(1), 228–237. DOI: 10.4067/S0718-27242012000100015

Ruggeri, K., Garcia-Garzon, E., Maguire, Á., Matz, S., & Huppert, F. A. (2020). Well-being is more than happiness and life satisfaction: A multidimensional analysis of 21 countries. *Health and Quality of Life Outcomes*, *18*(1), 192. DOI: 10.1186/s12955-020-01423-y PMID: 32560725

Ruiz-Mallén, I., Heras, M., & Berrens, K. (2021). Responsible research and innovation in science education: Insights from evaluating the impact of using digital media and arts-based methods on RRI values. *Research in Science & Technological Education*, *39*(3), 263–284. DOI: 10.1080/02635143.2020.1763289

Ryan, M. (2022). The ethics of dietary apps: Technology, health, and the capability approach. *Technology in Society*, *68*, 101873. DOI: 10.1016/j.techsoc.2022.101873

Ryan, R. M., & Deci, E. L. (2000). Intrinsic and extrinsic motivations: Classic definitions and new directions. *Contemporary Educational Psychology*, *25*(1), 54–67. DOI: 10.1006/ceps.1999.1020

Ryan, R. M., & Deci, E. L. (2000). Self-determination theory and the facilitation of intrinsic motivation, social development, and well-being. *The American Psychologist*, *55*(1), 68–78. DOI: 10.1037/0003-066X.55.1.68

Saadatdoost, R., Sim, A. T. H., Jafarkarimi, H., & Mei Hee, J. (2015). Exploring MOOC from education and Information Systems perspectives: A short literature review. *Educational Review*, *67*(4), 505–518. DOI: 10.1080/00131911.2015.1058748

Saini, R., Tayal, P., & Thakur, P. (2025). Exploring the potential of cloud computing in revolutionizing healthcare. In *Computational Methods in Science and Technology* (pp. 560–567). CRC Press.

Salun, M., & Zaslavska, K. (2024). Strategies for Resilience in a Dynamic World: from VUCA to BANI. *Proceedings of Socratic Lectures*. DOI: 10.55295/PSL.2024.I23

Samuelson, W., & Zeckhauser, R. (1988). Status quo bias in decision making. *Journal of Risk and Uncertainty*, *1*(1), 7–59. DOI: 10.1007/BF00055564

Santarius, T., Bieser, J. C. T., Frick, V., Höjer, M., Gossen, M., Hilty, L. M., Kern, E., Pohl, J., Rohde, F., & Lange, S. (2023). Digital sufficiency: Conceptual considerations for ICTs on a finite planet. *Annales Des Telecommunications. Annales des Télécommunications*, *78*(5–6), 277–295. DOI: 10.1007/s12243-022-00914-x PMID: 37593439

Santos, A., Chinkes, E., Carvalho, M. A., Solórzano, C. M., & Marroni, L. S. (2023). The digital competence of academics in higher education: Is the glass half empty or half full? *International Journal of Educational Technology in Higher Education, 20*(1), 9.

Santos, C. L. (2021). Formação Complementar. In Leitão, P. (Ed.), *Transição entre o mundo académico e o mundo laboral:Um mundo de oportunidades* (pp. 59–74). Editora Rh.

Sarshartehrani, F., Mohammadrezaei, E., Behravan, M., & Gracanin, D. (2024, June). Enhancing e-learning experience through embodied AI tutors in immersive virtual environments: A multifaceted approach for personalized educational adaptation. In *International Conference on Human-Computer Interaction* (pp. 272–287). Cham: Springer Nature Switzerland. DOI: 10.1007/978-3-031-60609-0_20

Sato, S. N., Moreno, E. C., Rubio-Zarapuz, A., Dalamitros, A. A., Yañez-Sepulveda, R., Tornero-Aguilera, J. F., & Clemente-Suárez, V. J. (2023). Navigating the new normal: Adapting online and distance learning in the post-pandemic era. *Education Sciences, 14*(1), 19. DOI: 10.3390/educsci14010019

Sauvé, P. (2017). A First Look at Investment in the Final Act of the Uruguay Round. In *Globalization and International Investment* (pp. 241–252). Routledge.

Savov, S. A., Antonova, R., & Spassov, K. (2019, January). Multimedia applications in education. In Smart Technologies and Innovation for a Sustainable Future: Proceedings of the 1st American University in the Emirates International Research Conference—Dubai, UAE 2017 (pp. 263-271). Cham: Springer International Publishing.

Scherr, S. J., & McNeely, J. A. (2008). Biodiversity conservation and agricultural sustainability: Towards a new paradigm of 'ecoagriculture landscapes. *Philosophical Transactions of the Royal Society of London. Series B, Biological Sciences, 363*(1491), 477–494. DOI: 10.1098/rstb.2007.2165 PMID: 17652072

Schischke, E., Grevé, A., Ehrenstein, U., & Doetsch, C. (2024). Overview of Energy Storage Technologies Besides Batteries. In Passerini, S., Barelli, L., Baumann, M., Peters, J., & Weil, M. (Eds.), *Emerging Battery Technologies to Boost the Clean Energy Transition. The Materials Research Society Series.* Springer., DOI: 10.1007/978-3-031-48359-2_4

Schneider, C., Weinmann, M., & Vom Brocke, J. (2018). Digital nudging: Guiding online user choices through interface design Designers can create designs that nudge users toward the most desirable option. *Communications of the ACM, 61*(7), 67–73. DOI: 10.1145/3213765

Schroeder, R. (2016). Big data business models: Challenges and opportunities. *Cogent Social Sciences*, *2*(1), 1166924. DOI: 10.1080/23311886.2016.1166924

Schueller, M. S., & Histon, T. (2023, November 3). Digital tools for youth mental health. *Frontiers for Young Minds*, *11*, 1169684. DOI: 10.3389/frym.2023.1169684

Schumacker, R. E., & Lomax, R. G. (2016). *A beginner's guide to structural equation modeling* (4th ed.). Routledge., DOI: 10.4324/9781315749105

Schweiger, H.-G., Obeidi, O., Komesker, O., Raschke, A., Schiemann, M., Zehner, C., Gehnen, M., Keller, M., & Birke, P. (2010). Comparison of Several Methods for Determining the Internal Resistance of Lithium Ion Cells. *Sensors (Basel)*, *10*(6), 5604–5625. DOI: 10.3390/s100605604 PMID: 22219678

Secundo, G., Del Vecchio, P., Gianluca, E., Margherita, A., & Valentina, N. (2021). Threat or opportunity? A case study of digital-enabled redesign of entrepreneurship education in the COVID-19 emergency. *Technological Forecasting and Social Change*, *166*, 120565. DOI: 10.1016/j.techfore.2020.120565 PMID: 33518821

Sedlakova, J., & Trachsel, M. (2022). Conversational artificial intelligence in psychotherapy: A new therapeutic tool or agent? *The American Journal of Bioethics*, *23*(5), 4–13. DOI: 10.1080/15265161.2022.2048739 PMID: 35362368

Selwyn, N. (2016). Digital downsides: Exploring university students' negative engagements with digital technology. *Teaching in Higher Education*, *21*(8), 1006–1021. DOI: 10.1080/13562517.2016.1213229

Selwyn, N. (2016). *Education and technology: Key issues and debates*. Bloomsbury Academic., DOI: 10.5040/9781474235952

Selwyn, N. (2020). *Should robots replace teachers? AI and the future of education*. Polity Press.

Sergeev, S. (2010). Virtual Training Simulators: Problems of Theory and Methodology of Design. *Biotechnosfera*, *2*, 15–20.

Sharma, A., Lin, I. W., Miner, A. S., Atkins, D. C., & Althoff, T. (2023). Human–AI collaboration enables more empathic conversations in text-based peer-to-peer mental health support. *Nature Machine Intelligence*, *5*(1), 46–57. DOI: 10.1038/s42256-022-00593-2

Sharma, V. (2020). Digital platforms in post-COVID-19: Transform challenges into opportunities. *Teach. Learn. Dur., COVID-19*, 1–15.

Sharpe, W.F., Alexander, G.J. and Bailey, J.W. (2018). Investments. INFRA-M,1028

Shen, Z., Zhang, Y., Lu, J., Xu, J., & Xiao, G. (2020). A novel time series forecasting model with deep learning. *Neurocomputing*, *396*, 302–313. DOI: 10.1016/j.neucom.2018.12.084

Sheridan, K., & Kelly, M. A. (2010). The indicators of instructor presence that are important to students in online courses. *Journal of Online Learning and Teaching*, *6*(4), 767–779.

Shi, M., & Tan, C. Y. (2020). Beyond oral participation: A typology of student engagement in classroom discussions. *New Zealand Journal of Educational Studies*, *55*(1), 247–265. DOI: 10.1007/s40841-020-00166-0

Shiva, V. (2001). *Stolen harvest: The hijacking of the global food supply*. Zed Books.

Shiva, V., & Crompton, T. (1998). Monopoly and monoculture: Trends in Indian seed industry. *Economic and Political Weekly*, •••, A137–A151.

Simankov, V. S., Buchatskiy, P. Y., Teploukhov, S. V., Cherkasov, A. N., & Onishchenko, S. V. (2023). *"Designing an Intelligent Information and Analytical System for Evaluating Solutions in Renewable Energy Based on Digital Twins," 2023 Seminar on Information Systems Theory and Practice*. ISTP., DOI: 10.1109/ISTP60767.2023.10426987

Simon, A. R., & Shaffer, S. L. (2001). *Data warehousing and business intelligence for e-commerce*. Elsevier.

Simons, L., Martin, J., & Craven, M. (2015). Can simple digital tools promote wellbeing? *Proceedings of the King's Fund's digital health and care congress 2015*. Research Gate. DOI:DOI: 10.13140/RG.2.1.1707.1209

Simons, A., Weinmann, M., Tietz, M., & vom Brocke, J. (2017). Which Reward Should I Choose? Preliminary Evidence for the Middle-Option Bias in Reward-Based Crowdfunding. *Proceedings of the 50th Hawaii International Conference on System Sciences (2017), Hicss*. DOI: 10.24251/HICSS.2017.526

Singh, L. (2024). *Revolutionizing care: Developing an AI-human synergy mental health application for mental health worker's wellbeing* [Unpublished doctoral dissertation]. Capella University.

Singh, K. K. (2018). Bayer-Monsanto Merger and India's IP Approach to Agricultural Biotechnology: Navigating through a complex web of law and policy. *Asian Biotechnology and Development Review*, •••, 20.

Skinner, B. F. (1953). *Science and human behavior*. Free Press.

Skogberg, M., Mackiewicz, K., Mänd, K., Tuuling, L., Urdzina-Merca, I., Salanterä, S., & Pakarinen, A. (2022). Promoting the health and wellbeing of children: A feasibility study of a digital tool among professionals. *PLoS One*, *17*(3), e0265355. DOI: 10.1371/journal.pone.0265355 PMID: 35316279

Smalley, R. (2013). Plantations, contract farming and commercial farming areas in Africa: A comparative review. *Future Agricultures Working Paper, 55*, 73.

Snihir, I., Rey, W., Verbitskiy, E., Belfadhel-Ayeb, A., & Notten, P. H. (2006). Battery open-circuit voltage estimation by a method of statistical analysis. *Journal of Power Sources*, *159*(2), 1484–1487.

Solutions, L. (2023). *Student engagement strategies: Leveraging the power of AI.* https://www.lambdasolutions.net/en/blog/student-engagement-strategies-leveraging-the-power-of-ai

Song, Y. G., Cao, Q. L., & Zhang, C. (2018). Towards a new approach to predict business performance using machine learning. *Cognitive Systems Research*, *52*, 1004–1012. DOI: 10.1016/j.cogsys.2018.09.006

Song, Y., & He, Y. (2023). Toward an intelligent tourism recommendation system based on artificial intelligence and IoT using the Apriori algorithm. *Soft Computing*, *27*(24), 19159–19177. DOI: 10.1007/s00500-023-09330-2

Song, Z. X., Cheung, M. F., & Prud'Homme, S. (2017). *Theoretical Frameworks and Research Methods in the Study of MOOC/e-Learning Behaviors: A Theoretical and Empirical Review. New Ecology for Education — Communication X Learning*. Springer., DOI: 10.1007/978-981-10-4346-8_5

Song, Z., Tang, J., Liu, T. X., Zheng, W., Wu, L., Feng, W., & Zhang, J. (2021). XiaoMu: An AI-driven assistant for MOOCs. *Science China. Information Sciences*, *64*(6), 164101. DOI: 10.1007/s11432-019-9939-2

Spante, M., Hashemi, S. S., Lundin, M., & Algers, A. (2018). Digital competence and digital literacy in higher education research: Systematic review of concept use. *Cogent Education*, *5*(1), 1519143. DOI: 10.1080/2331186X.2018.1519143

Spiegel, S., Grant-Pillow, H., & Higgins, E. T. (2004). How regulatory fit enhances motivational strength during goal pursuit. *European Journal of Social Psychology*, *34*(1), 39–54. DOI: 10.1002/ejsp.180

Srivastava, U. (2006). Intellectual property rights and agro-biodiversity. *Indian Journal of Plant Genetic Resources*, *19*(01), 1–11.

Staffini, A. (2022). Stock price forecasting with a deep convolutional generative adversarial network. *Frontiers in Artificial Intelligence*, *5*, 837596. Advance online publication. DOI: 10.3389/frai.2022.837596

Stanphill, M. (2019, May 14). *Find your balance with new digital wellbeing tools.* Google. https://blog.google/outreach-initiatives/digital-wellbeing/find-your-balance-new-digital-wellbeing-tools/

Stein, H., Gurevich, I., & Gorev, D. (2020). Integration of technology by novice mathematics teachers—What facilitates such integration and what makes it difficult? *Education and Information Technologies*, *25*(1), 141–161. DOI: 10.1007/s10639-019-09950-y

Stephan Grabmeier. (n.d.). BANI versus VUCA: A new acronym to describe the world. https://stephangrabmeier.de/bani-versus-vuca/

Stewart, E., Milton, A., Yee, H., Song, M., Roberts, A., Davenport, T., & Hickie, I. (2022). eHealth tools that assess and track health and well-being in children and young people: Systematic review. *Journal of Medical Internet Research*, *24*(5), e26015. DOI: 10.2196/26015 PMID: 35550285

Stone, M. (1974). Cross-validatory choice and assessment of statistical predictions. *Journal of the Royal Statistical Society. Series B, Statistical Methodology*, *36*(2), 111–133. DOI: 10.1111/j.2517-6161.1974.tb00994.x

Stratton, E., Lampit, A., Choi, I., Calvo, R. A., Harvey, S. B., & Glozier, N. (2017). Effectiveness of eHealth interventions for reducing mental health conditions in employees: A systematic review and meta-analysis. *PLoS One*, *12*(12), e0189904. DOI: 10.1371/journal.pone.0189904 PMID: 29267334

Straub, D., Boudreau, M. C., & Gefen, D. (2004). Validation guidelines for IS positivist research. *Communications of the Association for Information Systems*, *13*(1), 380–427. DOI: 10.17705/1CAIS.01324

Streiner, D. L. (2003). Starting at the beginning: An introduction to coefficient alpha and internal consistency. *Journal of Personality Assessment*, *80*(1), 99–103. DOI: 10.1207/S15327752JPA8001_18

Strielkowski, W., Grebennikova, V., Lisovskiy, A., Rakhimova, G., & Vasileva, T. (2024). AI-driven adaptive learning for sustainable educational transformation. *Sustainable Development.*

Stringer, L. C., Fraser, E. D., Harris, D., Lyon, C., Pereira, L., Ward, C. F., & Simelton, E. (2020). Adaptation and development pathways for different types of farmers. *Environmental Science & Policy*, *104*, 174–189. DOI: 10.1016/j.envsci.2019.10.007

Sundaram, J. (2014). India's trade-related aspects of Intellectual Property Rights compliant pharmaceutical patent laws: What lessons for India and other developing countries? *Information & Communications Technology Law, 23*(1), 1–30. DOI: 10.1080/13600834.2014.891310

Sundberg, L., & Holmström, J. (2024). Fusing domain knowledge with machine learning: A public sector perspective. *The Journal of Strategic Information Systems, 33*(3), 101848. DOI: 10.1016/j.jsis.2024.101848

Sunstein, C. R. (2014). Nudging: A Very Short Guide. *Journal of Consumer Policy, 37*(4), 583–588. Advance online publication. DOI: 10.1007/s10603-014-9273-1

Sun, W., Ren, S., & Tang, G. (2025). In the era of responsible artificial intelligence and digitalization: Business group digitalization, operations and subsidiary performance. *Annals of Operations Research, 354*(1), 1–23. DOI: 10.1007/s10479-024-06453-z

Sweller, J. (1988). Cognitive load during problem solving: Effects on learning. *Cognitive Science, 12*(2), 257–285. DOI: 10.1207/s15516709cog1202_4

Székely, N., Weinmann, M., & Vom Brocke, J. (2016). Nudging people to pay CO2 offsets - The effect of anchors in flight booking processes. *24th European Conference on Information Systems, ECIS 2016, June.*

Tabachnick, B. G., & Fidell, L. S. (2019). *Using multivariate statistics* (7th ed.). Pearson.

Taber, K. S. (2018). The use of Cronbach's alpha when developing and reporting research instruments in science education. *Research in Science Education, 48*(6), 1273–1296. DOI: 10.1007/s11165-016-9602-2

Takona, J. (2024). AI in Education: Shaping the Future of Teaching and Learning. *International Journal of Current Educational Studies, 3*(2), 1–25. DOI: 10.46328/ijces.121

Tamim, B. (2023). Belgian woman blames ChatGPT-like chatbot ELIZA for her husband's suicide. *Interesting Engineering.* https://interestingengineering.com/culture/belgian-woman-blames-chatgpt-like-chatbot-eliza-for-her-husbands-suicide

Tavakol, M., & Dennick, R. (2011). Making sense of Cronbach's alpha. *International Journal of Medical Education, 2*, 53–55. DOI: 10.5116/ijme.4dfb.8dfd

Taylor, S., & Todd, P. A. (1995). Understanding information technology usage: A test of competing models. *Information Systems Research, 6*(2), 144–176. DOI: 10.1287/isre.6.2.144

Tena, R. R., Cejudo, M. C. L., Gutiérrez, M. P., & Sanchéz, R. B. (2021). The pandemic and changes in the self-perception of teacher digital competences of infant grade students: A cross sectional study. *International Journal of Environmental Research and Public Health, 18*(9), 4756. DOI: 10.3390/ijerph18094756 PMID: 33946972

Teploukhov, S. V., Lisov, D. N., Buchatskiy, P. Y., Bychkov, T. Y., & Onishchenko, S. V. (2024). "Development of the Module of Charging Stations Placement for Electric Transport Based on Genetic Algorithm," *2024 XXVII International Conference on Soft Computing and Measurements (SCM)*, Saint Petersburg, Russian Federation, 2024, pp. 123-126, DOI: 10.1109/SCM62608.2024.10554218

Thaler, R. H., & Sunstein, C. R. (2008). Nudge: Improving decisions about health, wealth, and happiness. In *Nudge: Improving Decisions about Health, Wealth, and Happiness*. DOI: 10.1016/S1477-3880(15)30073-6

Thaler, R. H., & Benartzi, S. (2004). Save more tomorrow: Using behavioral economics to increase employee saving. *Journal of Political Economy, 112*(S1), S164–S187. Advance online publication. DOI: 10.1086/380085

Thirupathi, L., Kaashipaka, V., Dhanaraju, M., & Katakam, V. (2025). AI and IoT in mental health care: From digital diagnostics to personalized, continuous support. In Joshi, H., Reddy, C. K. K., Ouaissa, M., Hanafiah, M. M., & Doss, S. (Eds.), *Intelligent systems and IoT applications in clinical health* (pp. 271–294). IGI Global., DOI: 10.4018/979-8-3693-8990-4.ch012

Thompson, B. (2004). *Exploratory and confirmatory factor analysis: Understanding concepts and applications*. American Psychological Association., DOI: 10.1037/10694-000

Thunström, L., Gilbert, B., & Ritten, C. J. (2018). Nudges that hurt those already hurting – distributional and unintended effects of salience nudges*. *Journal of Economic Behavior & Organization, 153*, 267–282. DOI: 10.1016/j.jebo.2018.07.005

Tietz, M., Simons, A., Weinmann, M., & VomBrocke, J. (2016). The decoy effect in reward-based crowdfunding: Preliminary results from an online experiment. *2016 International Conference on Information Systems, ICIS 2016*, 1–11.

Tondeur, J., van Braak, J., Ertmer, P. A., & Ottenbreit-Leftwich, A. (2017). Understanding the relationship between teachers' pedagogical beliefs and technology use in education: A systematic review of qualitative evidence. *Educational Technology Research and Development, 65*(3), 555–575. DOI: 10.1007/s11423-016-9481-2

Topol, E. J. (2019). *Deep medicine: How artificial intelligence can make healthcare human again*. Basic Books.

Torous, J., Bucci, S., Bell, I. H., Kessing, L. V., Faurholt-Jepsen, M., Whelan, P., Carvalho, A. F., Keshavan, M., Linardon, J., & Firth, J. (2021). The growing field of digital psychiatry: Current evidence and the future of apps, social media, chatbots, and virtual reality. *World Psychiatry; Official Journal of the World Psychiatric Association (WPA), 20*(3), 318–335. DOI: 10.1002/wps.20883 PMID: 34505369

Turnitin. (n.d.). How will AI impact skill development in education? Retrieved February 14, 2025, from https://www.turnitin.com/blog/how-will-ai-impact-skill-development-in-education

Tversky, A., & Kahneman, D. (1974). Judgment under uncertainty: Heuristics and biases. *Science, 185*(4157), 1124–1131. Advance online publication. DOI: 10.1126/science.185.4157.1124 PMID: 17835457

Twenge, J. M., & Campbell, W. K. (2018). Associations between screen time and lower psychological wellbeing among children and adolescents: Evidence from a population-based study. *Preventive Medicine Reports, 12*, 271–283. DOI: 10.1016/j.pmedr.2018.10.003 PMID: 30406005

Twenge, J. M., Haidt, J., Lozano, J., & Cummins, K. M. (2022). Specification curve analysis shows that social media use is linked to poor mental health, especially among girls. *Acta Psychologica, 224*, 103512. DOI: 10.1016/j.actpsy.2022.103512 PMID: 35101738

UCW. (2024, July 19). Advantages and disadvantages of AI in education. https://www.ucanwest.ca/blog/advantages-and-disadvantages-of-ai-in-education

Upadhyay, M., Neupane, S., & Dhaubadel, S. (n.d.). *Agreement on TRIPS: Legal Frameworkfor Plant Variety Protection*. https://www.fao.org/4/ae896e/ae896e08.pdf

Vaibhav, A., & Gupta, P. (2014, December). Gamification of MOOCs for increasing user engagement. In *2014 IEEE International Conference on MOOC, Innovation and Technology in Education (MITE)* (pp. 290–295). IEEE. DOI: 10.1109/MITE.2014.7020290

Vaideeswaran, V., Bhuvanesh, S., & Devasena, M. (2019). *"Battery Management Systems for Electric Vehicles using Lithium Ion Batteries," 2019 Innovations in Power and Advanced Computing Technologies (i-PACT)*. Vellore., DOI: 10.1109/i-PACT44901.2019.8959965

Valafar, F. (2002). Pattern recognition techniques in microarray data analysis: A survey. *Annals of the New York Academy of Sciences, 980*(1), 41–64. DOI: 10.1111/j.1749-6632.2002.tb04888.x PMID: 12594081 van Buuren, S. (2018). *Flexible imputation of missing data* (2nd ed.). Chapman and Hall/CRC., DOI: 10.1201/9780429492259

van Olmen, J. (2022). The promise of digital self-management: A reflection about the effects of patient-targeted e-health tools on self-management and wellbeing. *International Journal of Environmental Research and Public Health, 19*(3), 1360. DOI: 10.3390/ijerph19031360 PMID: 35162383

Vanden Abeele, M. M. (2021). Digital wellbeing as a dynamic construct. *Communication Theory, 31*(4), 932–955. DOI: 10.1093/ct/qtaa024

Vashishth, T. K., Sharma, V., Kumar, B., & Sharma, K. K. (2023). Cloud-Based data management for behavior analytics in business and finance sectors. In *Data-Driven Modelling and Predictive Analytics in Business and Finance* (pp. 133–155). Auerbach Publications.

Vatansever, Ç., & Ceylan, F. (2021). Gençlerin kariyer gelişiminde e-mentorluk sürecinin etkisi [Effect of the e-mentoring process on youth's career development]. *Istanbul Management Journal, 90*(90), 91–114. DOI: 10.26650/imj.2021.90.0005

Venkatesh, V., Morris, M. G., Davis, G. B., & Davis, F. D. (2003). User Acceptance of Information Technology: *Toward a Unified View. Management Information Systems Quarterly, 27*(3), 425–478. DOI: 10.2307/30036540

Venkatesh, V., Thong, J. Y. L., & Xu, X. (2012). Consumer Acceptance and Use of Information Technology: Extending the Unified Theory of Acceptance and Use of Technology. *Management Information Systems Quarterly, 36*(1), 157–178. DOI: 10.2307/41410412

Viana, C. M., Freire, D., Abrantes, P., Rocha, J., & Pereira, P. (2022). Agricultural land systems importance for supporting food security and sustainable development goals: A systematic review. *The Science of the Total Environment, 806*, 150718. DOI: 10.1016/j.scitotenv.2021.150718 PMID: 34606855

Vidanaralage, A. J., Dharmaratne, A. T., & Haque, S. (2022). AI-based multidisciplinary framework to assess the impact of gamified video-based learning through schema and emotion analysis. *Computers and Education: Artificial Intelligence, 3*, 100109. DOI: 10.1016/j.caeai.2022.100109

Vlachopoulos, D. (2020). COVID-19: Threat or opportunity for online education? *Higher Learning Research Communications, 10*, 16–19.

Voorhees, C. M., Brady, M. K., Calantone, R., & Ramirez, E. (2016). Discriminant validity testing in marketing: An analysis, causes for concern, and proposed remedies. *Journal of the Academy of Marketing Science*, *44*(1), 119–134. DOI: 10.1007/s11747-015-0455-4

Vrabie, C. (2023). Education 3.0 – AI and gamification tools for increasing student engagement and knowledge retention. *Lecture Notes in Business Information Processing*, *495*, 74–87. DOI: 10.1007/978-3-031-43590-4_5

Vygotsky, L. S. (1978). *Mind in society: The development of higher psychological processes*. Harvard University Press.

Waag, W., Fleischer, C., & Sauer, D. U. (2014). Critical review of the methods for monitoring of lithium-ion batteries in electric and hybrid vehicles. *Journal of Power Sources*, *258*, 321–339. DOI: 10.1016/j.jpowsour.2014.02.064

Wadhwa, M. (n.d.). *Digital mental health initiatives in India* [ICT India Working Paper #60]. [Columbia University] CSD Working Paper Series. Available at: https://csd.columbia.edu/sites/default/files/content/docs/ICT%20India/Papers/ICT_India_Working_Paper_60.pdf

Waliszewski, K., & Zięba-Szklarska, M. (2020). Robo-advisors as automated personal financial planners – SWOT analysis. *Journal of Finance and Financial Law*, *3*(27), 155–173. DOI: 10.18778/2391-6478.3.27.09

Walker, S. (2001). *The TRIPS agreement, sustainable development and the public interest: discussion paper* (No. 41). IUCN.

Wallis, J. A. M., Riddell, J. K., Smith, C., Silvertown, J., & Pepler, D. J. (2015). Investigating patterns of participation and conversation content in an online mentoring program for northern canadian youth. *Mentoring & Tutoring*, *23*(3), 228–247. DOI: 10.1080/13611267.2015.1072395

Wang, J., Ma, Y., Zhang, L., Gao, R. X., & Wu, D. (2018). Deep learning for smart manufacturing: Methods and applications. *Journal of Manufacturing Systems*, *48*, 144–156. DOI: 10.1016/j.jmsy.2018.01.003

Wang, L., Pertheban, T. R. A., Li, T., & Zhao, L. (2024). Application of business intelligence based on big data in E-commerce data evaluation. *Heliyon*, *10*(21), e38768. DOI: 10.1016/j.heliyon.2024.e38768 PMID: 39524730

Wang, S., Fan, Y., Stroe, D.-I., Fernandez, C., Yu, C., Cao, W., & Chen, Z. (2021). Battery system active control strategies. In *Battery System Modeling* (pp. 313–340). Elsevier., https://doi.org/10.1016/B978-0-323-90472-8.00006-8

Wang, X., Ma, L., Yan, S., Chen, X., & Growe, A. (2023). Trade for food security: The stability of global agricultural trade networks. *Foods, 12*(2), 271. DOI: 10.3390/foods12020271 PMID: 36673361

Wang, Y., Derakhshan, A., & Zhang, L. J. (2021). Researching and Practicing Positive Psychology in Second/Foreign Language Learning and Teaching: The Past, Current Status and Future Directions. *Frontiers in Psychology, 12*, 731721. DOI: 10.3389/fpsyg.2021.731721

Wang, Y., Gu, X., & Zhao, Y. (2024). The impact of AI usage on university students' willingness for autonomous learning: An expectation-confirmation perspective. *PLoS One, 19*(10), e0292756. DOI: 10.1371/journal.pone.0292756

Wang, Z., Crook, J., & Andreeva, G. (2020). Reducing estimation risk using a Bayesian posterior distribution approach: Application to stress testing mortgage loan default. *European Journal of Operational Research, 287*(2), 725–738. DOI: 10.1016/j.ejor.2020.04.020

Wang, Z., Feng, G., Zhen, D., Gu, F., & Ball, A. (2021). A review on online state of charge and state of health estimation for lithium-ion batteries in electric vehicles. *Energy Reports, 7*, 5141–5161. DOI: 10.1016/j.egyr.2021.08.113

Watal, J., Taubman, A., & Cottier, T. (2015). Revisiting the TRIPS Negotiations. In *WTO, The making of the TRIPS agreement: Personal insights from the Uruguay round negotiations. pp- 3-13.* WTO.

Wei, L. (2022). Intelligent analysis of e-government influence factors based on improved machine learning. *Neural Computing & Applications, 34*(15), 12241–12256. DOI: 10.1007/s00521-021-06657-5

Weinmann, M., Schneider, C., & vom Brocke, J. (2016). Digital Nudging. *Business & Information Systems Engineering, 58*(6), 433–436. DOI: 10.1007/s12599-016-0453-1

Weizenbaum, J. (1966). ELIZA—A computer program for the study of natural language communication between man and machine. *Communications of the ACM, 9*(1), 36–45. DOI: 10.1145/365153.365168

Wies, B., Landers, C., & Ienca, M. (2021). Digital mental health for young people: A scoping review of ethical promises and challenges. *Frontiers in Digital Health, 3*, 697072. DOI: 10.3389/fdgth.2021.697072 PMID: 34713173

Wigfield, A., & Eccles, J. S. (2000). Expectancy-value theory of achievement motivation. *Contemporary Educational Psychology, 25*(1), 68–81. DOI: 10.1006/ceps.1999.1015

Willbanks, J. E. (2014). E-mentoring: Examining the feasibility of electronic, online, or distance mentoring. *Journal of Higher Education Theory and Practice, 14*(5), 24–28.

Willems, C., Renz, J., Staubitz, T., & Meinel, C. (2014), Reflections on Enrollment Numbers and Success Rates at the openHPI MOOC Platform, *Proceeding of the Second MOOC European Stakeholders Summit (EMOOCs2014),* Lausanne, Switzerland, 101-106

Williams, B., Onsman, A., & Brown, T. (2010). Exploratory factor analysis: A five-step guide for novices. *Australasian Journal of Paramedicine, 8*(3), 1–13. Advance online publication. DOI: 10.33151/ajp.8.3.93

Williams, S. L., & Kim, J. (Jin-H. (2011). E-mentoring in online course projects: Description of an e-mentoring scheme. *International Journal of Evidence Based Coaching and Mentoring, 9*(2), 80–95.

Winterbottom, A. (2021). Becoming "traditional" a transnational history of neem and biopiracy discourse. *Osiris, 36*(1), 262–283. DOI: 10.1086/713897

Winter, L. (2010). Cultivating farmers' rights: Reconciling food security, indigenous agriculture, and TRIPS. *Vanderbilt Journal of Transnational Law, 43*, 223.

World Economic Forum. (2025). *The Future of Jobs Report 2025*. Geneva: World Economic Forum. Retrieved March 28, 2025 from https://reports.weforum.org/docs/WEF_Future_of_Jobs_Report_2025.pdf

Wright, B. D., & Pardey, P. G. (2006). The evolving rights to intellectual property protection in the agricultural biosciences. *International Journal of Technology and Globalisation, 2*(1-2), 12–29. DOI: 10.1504/IJTG.2006.009124

Wright, K., Levine, D. K., Salcido, M., Garringer, M., Almendra, T., Bazell, A., & Kaufman, M. R. (2024). From crisis to capacity: Lessons learned from youth e-mentoring during the COVID-19 pandemic. *Computers in Human Behavior Reports, 14*, 100400. DOI: 10.1016/j.chbr.2024.100400

WTO. (n.d.). TRIPS and the Conventions related to it. https://www.wto.org/english/tratop_e/trips_e/ta_docs_e/1_tripsandconventions_e.pdf

Wu, H. K., Lee, S. W. Y., Chang, H. Y., & Liang, J. C. (2013). Current status, opportunities and challenges of augmented reality in education. *Computers & Education, 62*, 41–49. DOI: 10.1016/j.compedu.2012.10.024

Wu, J., Liu, H., Yao, X., & Zhang, L. (2024). Unveiling consumer preferences: A two-stage deep learning approach to enhance accuracy in multi-channel retail sales forecasting. *Expert Systems with Applications*, *257*, 125066. DOI: 10.1016/j.eswa.2024.125066

Wykes, T., Lipshitz, J., & Schueller, S. M. (2019). Towards the design of ethical standards related to digital mental health and all its applications. *Current Treatment Options in Psychiatry*, *6*(3), 232–242. DOI: 10.1007/s40501-019-00180-0

Wynberg, R. (2023). Biopiracy: Crying wolf or a lever for equity and conservation? *Research Policy*, *52*(2), 104674. DOI: 10.1016/j.respol.2022.104674

Xerri, M. J., Radford, K., & Shacklock, K. (2018). Student engagement in academic activities: A social support perspective. *Higher Education*, *75*(4), 589–605. DOI: 10.1007/s10734-017-0162-9

Xu, S., Liu, P., Yang, Z., Cui, Z., & Yang, F. (2021). How does mentoring affect the creative performance of mentors: The role of personal learning and career stage. *Frontiers in Psychology*, *12*, 741139. DOI: 10.3389/fpsyg.2021.741139

Yaccob, N. S., Yunus, M. M., & John, D. S. (2023). Global education movement: English as a second language teachers' perceptions of integrating volatility, uncertainty, complexity, and ambiguity elements in lessons. *Frontiers in Psychology*, *14*, 1007970. DOI: 10.3389/fpsyg.2023.1007970

Yadegaridehkordi, E., Shuib, L., Nilashi, M., & Asadi, S. (2019). Decision to adopt online collaborative learning tools in higher education: A case of top Malaysian universities. *Education and Information Technologies*, *24*(1), 79–102. DOI: 10.1007/s10639-018-9761-z

Yazdanpanah, M., Tajeri Moghadam, M., Savari, M., Zobeidi, T., Sieber, S., & Löhr, K. (2021). The impact of livelihood assets on the food security of farmers in Southern Iran during the COVID-19 pandemic. *International Journal of Environmental Research and Public Health*, *18*(10), 5310. DOI: 10.3390/ijerph18105310 PMID: 34067638

Yin, H. (2018). What motivates Chinese undergraduates to engage in learning? Insights from a psychological approach to student engagement research. *Higher Education*, *76*(5), 827–847. DOI: 10.1007/s10734-018-0239-0

Yu, H., Miao, C., Leung, C., & White, T. J. (2017). Towards AI-powered personalization in MOOC learning. *npj Science of Learning, 2(1)*, 15. DOI: 10.1038/s41539-017-0016-3

Yu, D., Zhou, X., Noorian, A., & Hazratifard, M. (2025). An AI-driven social media recommender system leveraging smartphone and IoT data. *The Journal of Supercomputing*, *81*(1), 1–32. DOI: 10.1007/s11227-024-06722-5

Yüksek, G., & Alkaya, A. (2023). A novel state of health estimation approach based on polynomial model for lithium-ion batteries. *International Journal of Electrochemical Science*, *18*(5), 100111.

Zabolotska, O., Zhyliak, N., Hevchuk, N., Petrenko, N., & Alieko, O. (2021). Digital competencies of teachers in the transformation of the educational environment. *J. Optim. Ind. Eng.*, *14*, 25–32.

Zadorina, O., Burchak, L., Panas, O., Ardelian, O., & Apalat, H. (2023). Shaping the competencies of the future: The importance of developing soft skills in higher education. *Cadernos de Educação Tecnologia e Sociedade*, *16*(2), 361–371. DOI: 10.14571/brajets.v16.n2.361-371

Zakharov, Y. (2022). Comparative analysis of approaches to world concepts: SPOD Society, VUCA Society and BANI Society. *Society and Economy*, *64*(64), 149–158. DOI: 10.26565/2524-2547-2022-64-13

Zaky, Y. A. M. (2023). Chatbot Positive Design to Facilitate Referencing Skills and Improve Digital wellbeing. *International Journal of Interactive Mobile Technologies*, *17*(9), 106–126. Advance online publication. DOI: 10.3991/ijim.v17i09.38395

Zawacki-Richter, O., Marín, V. I., Bond, M., & Gouverneur, F. (2019). Systematic review of research on artificial intelligence applications in higher education - where are the educators? *International Journal of Educational Technology in Higher Education*, *16*(1), 39. DOI: 10.1186/s41239-019-0171-0

Zawati, M. H., & Lang, M. (2024). Does an App a Day Keep the Doctor Away? AI Symptom Checker Applications, Entrenched Bias, and Professional Responsibility. *Journal of Medical Internet Research*, *26*, e50344. DOI: 10.2196/50344 PMID: 38838309

Zeng, J., Sun, D., Looi, C. K., & Fan, A. C. W. (2024). Exploring the impact of gamification on students' academic performance: A comprehensive meta-analysis of studies from the year 2008 to 2023. British J ournal of Educational Technology.

Zepp, J., Dubé, C., & Melcher, D. (2021). A direct comparison of central tendency recall and temporal integration in the successive field iconic memory task. *Attention, Perception & Psychophysics*, *83*(3), 1337–1356. DOI: 10.3758/s13414-020-02187-9 PMID: 33389675

Zhang, X., & Jia, P. 2022, November. The Development Trend of Intelligent Logistics Based on Machine Vision. In *International conference on Advances in Materials and Manufacturing* (pp. 849-859). Singapore: Springer Nature Singapore.

Zhang, C., Bengio, S., Hardt, M., Recht, B., & Vinyals, O. (2021). Understanding deep learning (still) requires rethinking generalization. *Communications of the ACM*, *64*(3), 107–115. DOI: 10.1145/3446776

Zhang, Y., Llorente, A. M. P., & Sánchez-Gómez, M. C. (2021). Digital competence in higher education research: A systematic literature review. *Computers & Education*, *168*, 104212. DOI: 10.1016/j.compedu.2021.104212 PMID: 36568577

Zhang, Z. (2022). Toward the role of teacher empathy in students' engagement in English language classes. *Frontiers in Psychology*, *13*, 880935. DOI: 10.3389/fpsyg.2022.880935

Zhao, W. X., Zhou, K., Li, J., Tang, T., Wang, X., Hou, Y., Min, Y., ... "A survey of large language models." arXiv preprint arXiv:2303.18223 (2023).

Zhao, J., Zhu, Y., Zhang, B., Liu, M., Wang, J., Liu, C., & Hao, X. (2023). Review of State Estimation and Remaining Useful Life Prediction Methods for Lithium–Ion Batteries. *Sustainability (Basel)*, *15*(6), 5014. DOI: 10.3390/su15065014

Zhong, X., & Enke, D. (2019). Predicting the daily return direction of the stock market using hybrid machine learning algorithms. *Financial Innovation*, *5*(24), 24. Advance online publication. DOI: 10.1186/s40854-019-0138-0

Zhou, C., Wu, C., & Wang, Y. (2019). Dynamic portfolio allocation with time-varying jump risk. *Journal of Empirical Finance*, *50*, 113–124. DOI: 10.1016/j.jempfin.2019.01.003

Zimmerman, B. J. (2008). Investigating self-regulation and motivation: Historical background, methodological developments, and future prospects. *American Educational Research Journal*, *45*(1), 166–183. DOI: 10.3102/0002831207312909

Zohuri, B., & Moghaddam, M. (2020). From business intelligence to artificial intelligence. *Journal of Material Sciences & Manufacturing Research*, *1*(1), 1–10.

Zuboff, S. (2015). Big other: Surveillance capitalism and the prospects of an information civilization. *Journal of Information Technology*, *30*(1), 75–89. DOI: 10.1057/jit.2015.5

About the Contributors

Prakasha G S is an Associate Professor at the School of Education, Christ University, Bangalore, India. He has completed his MEd from the Regional Institute of Education, India. PhD from Bangalore University. He also completed an MSc (Chemistry) and MPhil in Education. He cleared the National Eligibility Test (NET) and was awarded a junior research fellowship (JRF). He holds IB Educator certificates in MYP and DP from DePaul University, Chicago, USA. He is a visiting research fellow, International University, Malaysia. He visited Germany, France, Switzerland, Canada, & USA. He led EdTech Project with Switzerland's Teacher-Ed University (hep, Vaud). He visited Miami University, University of Cincinnati-Ohio, University of Mount Union, Johns Hopkins University-Washington DC, Syracuse University, Binghamton University, Campbellsville University- Kentucky, New York University, & the Catholic University of America. His research areas are EdTech, TeacherEdu, & HigherEdu. He has published 50+ research articles indexed in SCOPUS and WOS databases. He wrote several book chapters, articles, & conference papers. He has presented papers worldwide, to name a few: Bangkok, Switzerland, Russia, University of Oxford-UK, & Binghamton University, USA. He is an active editorial board member for SAGE, SPRINGER NATURE, and Taylor & Francis. He received a research grant from Sophia University, Tokyo, JAPAN. At the Centre for Advanced Research and Training (CART), he has conducted many national-level research data analysis workshops at Christ University. He has trained PhD scholars, students, and faculty members in quantitative data analysis using SPSS. He has delivered lectures nationwide, including ICSSR- New Delhi, XLRI, Mount Carmel, Jyothi Nivas, Christ Nagar, Thiagarajar College, & St. Joseph's University. He conducts research activities that have national and international emphasis. He completed a Major Research PROJECT on Teacher & Teacher Educator Competencies. His PROJECT with NGO-GRAMA contributed a sustainable & replicable "Informal-Education-Model" to uplift rural education in India. He authored 6 books 1) Teaching Internship Models Around the World-

Routledge, 2) Educational Perspectives on Digital Technologies in Modeling & Management- IGI, 3) Digital Technologies- IGI: Insights in Education and Industry, 4) Diagnostic Testing, 5) Attention-deficit-behaviour, 6) Teacher-effectiveness by Lap Lambert, Germany.

Juan J. Leiva-Olivencia is Full Professor in the Department of Didactics and Educational Organisation at the University of Málaga (Spain). His professional career is linked to research projects on educational innovation, ICT, inclusive education, interculturality and teacher training. He combines his teaching and research work with the role of evaluator for various international agencies for the evaluation of quality in higher education.

***

Abdulgalimov Ramazan Medzhidovich graduated from the Faculty of Physics and Mathematics of the Dagestan State Pedagogical Institute in 1983, received a PhD in pedagogical Sciences in 2000, received a doctorate in Pedagogical Sciences in 2021, currently works as the head of the Department of Biophysics, Computer Science and Medical Equipment of the Dagestan State Medical University (Makhachkala). He is a member of the Academic Council of Dagestan State Medical University, a member of the Doctoral Dissertation Council of Azerbaijan State Pedagogical University, Chairman of the State Research Committee of Dagestan State Pedagogical University. He has published more than 40 articles in foreign publications, journals refereed by the Higher Attestation Commission and SCOPUS, is the author of the monograph "Professional Worldview of a modern doctor", 6 collective monographs. He has commendations and certificates of honor, the winner in the nomination of the radiant department in 2018, 2020, 2022, 2024.

Abdulgalimova Guriyat Nurakhmedovna graduated from the Biological Faculty of the Dagestan State Pedagogical Institute in 1985, received her PhD in pedagogical Sciences in 2010, and currently works as an associate professor at the Department of Medical Biology at the Dagestan State Medical University (Makhachkala). He has commendations and certificates of honor of national and Russian significance. He is a veteran of labor.

Sweeta Agrawal is a research scholar pursuing a PhD at Rama Devi Women's University. She is also a lecturer at a college affiliated with Sambalpur University. Her research interests include AI in education, technology adoption, cloud-based accounting

Beril Ceylan Nebioğlu has been working at the department of Computer Education and Instructional Technology on Education Faculty at Ege University since 2005. She had her Ph.D. degree in 2015. She is interested in technology integration, educational material design, and instructional design. She has many book chapters, articles, and presentations.

Merve Çolakoğlu has a master's degree in Computer Education and Instructional Technology at Balıkesir University. Her research interests include STEM education, student engagement, artificial intelligence and learning analytics in education.

Indrajit Ajit Doddanavar is a full-time Assistant Professor of Business Management at Jain College of Engineering and Research, Belagavi, Karnataka, India, affiliated to Visvesvaraya Technological University, Karnataka. In 2019, he earned a Ph.D. in 'Branding' from Jain Deemed-to-be University. His research and interest areas include topics in marketing, human resources, and general management that are relevant to industry and consumers. He has fourteen years of extensive expertise across many sectors of Industry. He wants to work as a researcher to investigate relevant and current research premises. He can be contacted at email: doddanavar.ia@gmail.com.

Kasimov Arif Kamalutdinovich graduated with honors from the Dagestan State Pedagogical Institute in 1992 with a degree in physics with an additional specialty in Mathematics, having received the qualification of a teacher of physics and mathematics. Since 1992, she has been working as a teacher at the Dagestan State Pedagogical University. In 2009, she defended her dissertation for the degree of Candidate of Pedagogical Sciences. Since 2015, he has been working as an associate professor at the Department of Physics and Teaching Methods of Dagestan State Pedagogical University. Since 2017, he has also been working as an associate professor at the Department of Biophysics, Computer Science and Medical Technology of Dagestan State Medical University. The scientific field is related to higher school pedagogy. The subject of the research is the physical training of students of the faculties of natural science education and medical universities. Scientific interest boils down to the formation of physical fitness as an integrative basis for the professional training of students of natural science education and medical students.; effective use of physical knowledge and advanced technologies in the professional training of medical students.

Natalya Kononova In 1999, she graduated from Stavropol State University with a degree in mathematics and computer science, majoring in Mathematics. From 1999 to 2001, she worked as a teacher of mathematics and computer science at the MOE Secondary School No. 6. From 2002 to 2004, he was a teacher of mathematics, computer science, and information technology at the Stavropol Cooperative College of Economics, Commerce, and Law. From 2004 to 2006, he was an assistant at the Department of Mathematics and Natural Sciences, from 2006 to 2008, he was a senior lecturer at the Department of Medical Sciences and Deputy Dean of the Faculty of Economics, Management and Service, from 2008 to 2012, he was an associate professor at the Department of Medical Sciences and head of the Department of Pre-university Education at the Stavropol Institute of Economics and Management named after O.V. Kaznacheyev. In 2008, she defended her thesis for the degree of Candidate of Physico-mathematical Sciences, in 2011, she was awarded the academic title of Associate Professor at the Department of Applied Mathematics and Computer Technology.

Kormakova Valentina Nikolaevna received her higher education in the specialty "teacher of history and social science" in 1987 at the Belgorod State Pedagogical Institute. She received her Doctor of Sciences degree in pedagogy from Belgorod State University in 2013. In 2002, she was accepted as a senior lecturer at the University of Belgorod and is currently a professor at the Department of Pedagogy at Belgorod State National Research University. She is the author of over 500 scientific publications on issues of professional and life self-determination of an individual, theory and methods of continuous education, educational management; application of digital technologies in education. She gives reports at international and all-Russian scientific conferences. She heads the Master's program "Management in Education". She was awarded the Badge "Excellent Worker of Public Education" (1995); Honorary certificates of the education departments of the Belgorod city administration and the Belgorod region (2012, 2014). Received a Letter of Gratitude from the Chairman of the Belgorod Regional Duma (2023).

Paulo Leitão is a master in organizational psychology and holds a postgraduate degree in digital enterprise management. He began his professional career in 2014, focusing primarily on organizational consulting, particularly in the areas of talent acquisition and development. As an independent researcher, Paulo is interested in the application of complexity and chaos theories to organizational realities, and in the utility of skills and technology as support for decision-making in recruitment and selection contexts.

Ooi Kok Loang, LLM, CMIIA, CIA, is with the Department of Finance at the University of Malaya, renowned for his expertise in behavioural finance and capital markets. His professional journey includes significant roles as an auditor and business consultant at KPMG (Big 4 Audit Firms) and Deputy Dean at two prestigious private universities. Dr. Ooi holds a Master in Finance and a PhD in Behavioral Finance from Universiti Sains Malaysia (QS137), a Master of Laws (Business Law) from the International Islamic University Malaysia (QS Top 300 by subject), and is a US Certified Internal Auditor. With a strong portfolio of publications and active contributions as a reviewer for top international SCI and Scopus-indexed journals, Dr. Ooi is keen to mentor passionate PhD students, offering a vibrant environment for research and academic excellence.

M. Iswarya is an Assistant Professor in the Department of Management Sciences at Hindusthan College of Engineering and Technology, Coimbatore, India. A gold medalist during her undergraduate studies, she possesses expertise in General Management, Human Resources, and Systems. Dr. Iswarya has contributed to the academic community through her publications in both national and international journals. She holds a doctoral degree focused on the area of Positive Organisational Behaviour. Her current research interests lie at the intersection of emerging technologies and business, specifically exploring gamification, the metaverse, artificial intelligence, and business analytics. Beyond her academic role, she is also an active trainer.

Nesin S M holds a PhD in Neurophysiology and interested in studying cognition and consciousness. Her publications include neuromarketing and neuroscience

Achyuta Nanda Mishra is a Professor at School of Law, Christ University, Bangalore. He has 25 years of teaching experience. He has his research interests in the areas of Constitutional law, Legal theory, Democracy and Globalization. He has supervised a Postdoctoral research scholar and is currently guiding Ph.D. scholars under his supervision.

Semiral Öncü, PhD, is currently a faculty member at the Department of Educational Sciences, Necatibey Faculty of Education, Balikesir University, Türkiye. His research interests involve student engagement, the design and implementation of e-learning/blended learning environments along with effective and interactive use of ICT in educational settings.

Onishchenko Stefan Vladimirovich, Research Engineer of the Renewable Energy Sources Laboratory, Assistant of the Department of Automated Information Processing and Control Systems of the Faculty of Engineering and Physics of the Federal State Budgetary Educational Institution of Higher Education "Adygeya State University". In 2022 graduated from the Institute of Computer Technologies and Information Security of the Institute of Engineering and Technological Academy of the Southern Federal University in the direction of training 09.04.01 Informatics and Computer Science, with the award of a master's degree. The sphere of scientific interests of Onishchenko S.V. includes problems of building systems with renewable energy, the use of artificial intelligence in energy systems. He has a number of publications in highly rated publications peer-reviewed in scientific bases Web of Science, Scopus, VAK, RINC and RSSI journals. Participates with reports at all-Russian and international conferences on the subject of the research.

Ananya Pandey is a Research Scholar in School of Law at Christ University, Bangalore. She has her research interests in the field of Constitutional Law, Environment Law, Human Rights, Food Security, Intellectual Property Law and Jurisprudence.

Tatiana Alexandrovna Peleshenko graduated with honors from the North Caucasus Federal University in 2014 with a degree in Integrated Information Security of Automated Systems with the qualification of Information Security Specialist. In 2019, she defended her dissertation for the degree of Candidate of Technical Sciences. From 2014 to the present, he has been working as an associate professor at the Department of Computational Mathematics and Cybernetics of the North Caucasus Federal University. Since 2014, he has been engaged in research work in the field of information security and IT technologies, according to the results of which, over the past 3 years, he has 17 articles, including scientific articles in leading peer-reviewed scientific journals and publications included in the List of the Higher Attestation Commission of the Russian Federation - 2. Since 2022, he has been the head of the professional retraining program "Information security. Technical protection of confidential information".

Shubhangi Rote is a MBA in HR specialization. Planning to pursue a Ph.D. in Management. She research and interest areas include topics in human resources, and general management that are relevant to industry and consumers. She wants to work as a researcher to investigate relevant and current research premises. She can be contacted at email: s.doddanavar@gmail.com.

R.V.Naveenan, is working with Christ University, Bengaluru, India as Associate Professor. He has two postgraduation to his credit, which includes MBA in Finance and MA in Economics. He has obtained his Ph.D. in Management (Finance) from Manonmaniam Sundaranar University, Tamil Nadu, India. He has thirteen years of experience in academics and research. He is a passionate teacher and enthusiastic researcher. He has presented his research ideas in various national and international conferences. He has published around 15 research papers in high quality indexed national and international journals publications. His area of research is Green Banking, Financial Inclusion and Fintech.

Aashi Saraiya is an undergraduate psychology student at Mithibai College of Arts in India. At the same time, she is also working as a psychology intern at Infiheal, which is a Healthtech company.

Srishti Srivastava is an entrepreneur and health technologist who holds a B.Tech and M.Tech degree in Chemical Engineering from IIT Bombay (2014) and an Executive MBA from Cornell University (Cornell Maha60 2023–2024). Her journey into psychology began after personal struggles with anxiety following her appearance on TV Quiz Show India's Child Genius, fueling her dedication to mental health. In the public sector, Ms. Srivastava served as a Chief Minister's Fellow in Maharashtra, launching state wide government scheme "MahaTribes" to support tribal artisans. Her private sector experience includes roles as a financial risk consultant at KPMG and leading clinical data analysis at Krixi Corporation, a US-based health tech startup. In 2022, she founded Infiheal to provide accessible, stigma-free mental healthcare, which was applauded for its AI-based scalable approach by Prime Minister Narendra Modi in his 108th national Address. Ms. Srivastava has since contributed to public policy, speaking at the World Bank Roundtable in 2023, presenting her research on digital mental health to 10 US Senate offices in 2024, and advising the Government of Maharashtra on entrepreneurship and skill development. Ms. Srivastava's expertise and commitment to mental health and policy reform continue to drive meaningful, transformative impact.

Amit Subramanyam, an adept Faculty in Commerce & Management and a diligent Research Scholar, is associated with the D.M.S Mandal Group of Education in Belagavi, Karnataka, India. He brings a rich academic background including an M.B.A, M.Com., and is currently pursuing his Ph.D. Teaching and Professional Experience Mr. Amit Subramanyam has accumulated over four years of teaching experience across various institutions. His expertise spans General Management, Human Resources, Organizational Behavior, Statistics, and Research Methodology. Notably, his teaching journey includes roles at DMS Mandal's Bhaurao Kakatkar College, SSS Samiti's College of Business Administration, Yashwantrao Chavan Maharashtra Open University, Nasik, and Jain College of MBA and MCA. Before embarking on his academic career, Mr. Amit Subramanyam honed his skills in the corporate sector, working as a Supervisor, Office Associate, and Team Leader across different organizations from 2008 to 2019. Scholarly Contributions and Workshops He has actively participated in various Faculty Development Programs (FDPs) and workshops focused on research methodology, SPSS application, open-source tools, and innovative writing strategies. His commitment to academic excellence is further highlighted by his contributions to conference proceedings and journal publications on themes like Green HRM, organizational sustainability, and urbanization challenges. Research and Community Engagement Mr. Amit Subramanyam's research interests have led to noteworthy publications in peer-reviewed journals and presentations at international conferences, where he has been recognized with a Best Paper Award. Additionally, he has drafted chapters for books focusing on educational reforms and sustainable higher education systems. Beyond academia, Mr. Subramanyam has played a pivotal role in community service and extension work, including career counselling in rural schools and serving as an evaluator for Rotary Club initiatives. In sum, Mr. Amit Subramanyam exemplifies a blend of academic excellence, research prowess, and community service, contributing significantly to the fields of Commerce and Management.

Sruthi Suresh is a doctoral research scholar at CHRIST (Deemed to be University) and works as a Psychology Educator and TOK Coordinator at Knowledgeum Academy, Bengaluru. As a doctoral scholar, she has engaged with the department as a Research Assistant, Teaching Assistant, Impact Assessment Research Team Lead, and Department Magazine Editor. Sruthi was also one of Project FUEL Fellows in December 2023. She writes and presents widely on child and adolescent mental health issues, teacher education, and social-emotional learning.

Jarin T is currently working as Professor in Department of CSE at Jyothi Engineering college under APJ Abdul Kalam Technological University. He has completed his BE, ME and Ph.D. in faculty of Electrical Engineering and Information and Communication Engineering from Anna University. His areas of interest include Switching Strategies, Deep Brain Stimulation, Beetle Search based Optimization, electromagnetic Radiation and its Hazards, and Electric Vehicles etc. He has 15 years of experience in teaching and research and published more than ninety-five papers in leading peer reviewed International Journals. He has four patents to his credit.

Kashvi Tuli is a Dance Movement Therapy Facilitator, registered under Creative Movement Therapy Association of India, which is a member of American Dance Therapy Association. She has received her BA Psychology Honors from Indraprastha College for Women, University of Delhi, New Delhi in 2024. Currently, she is also pursuing a Post Graduate Diploma in Guidance and Counseling from Jamia Milia Islamia University, New Delhi.

Anushka Varma received her MSc in Counselling Studies from the University of Edinburgh in 2022, after earning her B.A. (Honours) in Psychology from the University of Delhi in 2021. She worked in crisis intervention at the Edinburgh Crisis Centre for two years before transitioning to her current role as a psychologist at Infiheal, a mental health startup based in Mumbai, India. Her primary research interests include AI and mental health, crisis intervention, and chronic mental health conditions.

Index

A

B

C

D

E

F

G

H

I

U

V

www.ingramcontent.com/pod-product-compliance
Lightning Source LLC
LaVergne TN
LVHW062330300926
849161LV00007B/1637